The Species Problem

The general notion of species is one of the most fundamental in biology. But the idea of species is also one of the most persistent unresolved obsessions of biologists and philosophers. This new book investigates the multifaceted species problem as a "conceptual envelope" of that notion. Contemporary conceptualism and evolutionary epistemology allow for a fresh look by analyzing the framework of history viewed as changes of this problem ordered by changing philosophical-scientific contexts. In this analysis, the species problem is characterized in a nontrivial pluralistic manner, in contrast to an "accepted" monistic attitude.

Key Features

- Provides new insights into the persistent species "problem".
- Focuses on the conceptual history and identifies pivotal landmarks in the history of the concept of species.
- Argues for a scientific consistency of species pluralism.
- Discusses the "evolving specieshood" in the context of new essentialism.

The Species Problem
A Conceptual History

Igor Ya. Pavlinov

CRC Press
Taylor & Francis Group
Boca Raton New York London

CRC Press is an imprint of the
Taylor & Francis Group, an **informa** business

First edition published 2023
by CRC Press
6000 Broken Sound Parkway NW, Suite 300, Boca Raton, FL 33487-2742

and by CRC Press
4 Park Square, Milton Park, Abingdon, Oxon, OX14 4RN

CRC Press is an imprint of Taylor & Francis Group, LLC

ISBN: 9781032386812 (hbk)
ISBN: 9781032386836 (pbk)
ISBN: 9781003346296 (ebk)

DOI: 10.1201/9781003346296

Typeset in Times
by KnowledgeWorks Global Ltd.

If we do not realize what species is, nothing will secure us from delusion.

Boëthius

A species is exactly what a given definition claims it is.

Steven Bartlett

Contents

Preface

The general notion of species is one of the most fundamental in biology, like those of organism, gene, evolution, ecosystems, etc. So the species problem, caused by multiplicity of the treatments of this notion, is one of hotspots vividly discussed by biological philosophers and theoreticians. As a matter of fact, several dozen books have been published on this subject during the last decades, and some of them contained historical accounts of species.

Accordingly, a question arises quite reasonably, why this another one? The answer in a general form is as simple as it may be: just because the themata related to the species notion cannot be exhausted to the extent that it is among the most fundamental ones, so each turn of its conceptual history opens new perspectives in its comprehending.

A peculiar feature of this notion is intrinsic multiplicity of its meaning that makes it "nomadic" (Alessandro Minelli), by adopting different guises in different substantive contexts with preserving the same name. With this, the latter's persistence may cause an impression the notion has a certain uniform content preserved in all its particular treatments—but is it really so? Might it be that "eliminative pluralism" (Mark Ereshefsky) is correct in suggesting that the current general notion, inherited by biology from medieval scholasticism, is just a confound of more meaningful particular conceptions, so should not it be abandoned in favor of the latter, that do not bear such historical load and are more adequate to the currently dominating worldview?

For these and related questions to be answered properly, it is necessary, from time to time, to reconsider this entire subject with a fresh open-minded look and to try once more to understand why there are diverse hyposteses of this "nomadic" notion and if there is something essential in common to them that would justify their denoting by the same term "species".

To meet this challenge, this new book suggests an essentially different look at the whole of the species issues, considered in a historical vein. It investigates not the species notion as such, but the multifaceted species problem as its "conceptual envelope", with the main emphasis on the historical dynamics of both multiplicity of the meanings of this notion and its causes. Among the latter, of prime importance are the changes in scientific-philosophical contexts, underlying different views of the structure of Nature and the ways of its cognition and conceptual representation.

A new look at the historical development of the species problem I suggest here is based on a combination of the contemporary conceptualism and evolutionary epistemology standpoints. According to their provisions, conceptual history of the species problem is viewed as part of the historical changes in the three-partitioned cognitive situation, which is basically ordered by the changes in the just-mentioned contexts. Such a dual standpoint allowed me to uncover certain important aspects of the problem in question that have been dominating at different stages of its conceptual history but remained not highlighted previously, such as genus-species uncertainty, partonomic-taxonomic dualism of species, its natural-classificatory dualism, etc. All these findings evidently show that the conceptual history of the species problem is not so much simple and straightforward as it is usually acknowledged by a "received opinion".

Before getting on with reading this book, it would be perhaps sound to think about one of the attractive features of the conceptual history of any scientific discipline, which is globally endless and largely unpredictable. Indeed, physicists were sure a few decades ago that they had comprehended the universe in total, so it remained for them just to specify numerical values of the world constants and to uncover a dozen new stars and a couple of new particles. And—what an unpleasant surprise!—they realized that what they were sure to have comprehended nearly completely constituted but about 15% of the total universe, and they called the remainder, which they have not even suspected to exist, "dark matter" and "dark energy".

Why have I told this story? In the 18th century, Carl Linnaeus was sure at the beginning of his scientific career that all the species had been created by God; at its finishing, he believed that God's creations had been prototypes of the orders, while the species appeared afterward due to the former's mutual crossing; however, now most scientists believe in the origin of species by gradual evolution. In the 1940s, E. Mayr believed that "the importance of the species as such is reduced", while in the 1960s and afterward he declared systematics to be "the science of species". In the 1960s and 1970s, pheneticists declared that they have resolved all the problems with species and said an "ultimate word" in the science of taxonomy—and who remembers their declaration now, besides several historians of systematics? Currently, phylogeneticians pretend for the same—and who knows where will they appear with their next "ultimate" version in a decade or so?

Indeed, Matthew Barker considers the species problem "Revving Up" rather than "We Are Done"; David Hull sees the most recent success in its resolving but in a clearer comprehension of the points of disagreement, while Scott Atran is inclined to count it one of the basic set of scientific puzzles and measure by its figuring out success of theoretical biology. Thus, when reflecting on the conceptual history of this problem, why not imagine each of the latter's successive stages, including the current one, as just a transition between its previous and future developments?

So, I suggest that you read my book with this attitude in mind and, when finished begin to wait for another one that would continue this endless fascinating discussion of the whole of the species problem.

Igor Ya. Pavlinov
Zoological Museum at Lomonosov Moscow State University
Moscow, Russia

About the Author

Igor Ya. Pavlinov was, until his retirement in 2018, a leading researcher and chief of the Mammal Division at the Zoological Museum at the Lomonosov Moscow State University. He is currently the curator of mammals at this museum. His principal research interests are in theoretical taxonomy and phylogenetics, systematics of mammals, and morphometrics. He is the author of several dozen books on natural history, including: *Cladistic Analysis: Methodological Aspects* (1990), *Evolution of Life* (2001), *Systematics of Mammals of the World* (2003), *Foundation of the Contemporary Phylogenetics* (2005), *Nomenclature in Systematics* (2015) and its successor *Taxonomic Nomenclature—What's in a Name: Theory and History* (2022), *Foundation of Biological Systematics* (2018) and its successor *Biological Systematics: History and Theory* (2021), *Mammals of Russia: A Reference and Guide* (2019), several books on particular orders and families of mammals, etc.

About the Author

[illegible]

Introduction

The *species notion*, according to a predominating viewpoint, is among the most fundamental in biology, along with the notions of gene, cell, ecosystem, evolution, etc. The biological species, a.k.a. *biospecies*, understood in its most general sense, is most usually thought of as a peculiar natural phenomenon inhered in the living [Zavadsky 1961, 1968; Volkova and Filyukov 1966; Mayr 1982, 1988; de Sousa 1989].[1] According to an extreme natural-philosophical standpoint, it is "a way of implementation of life" [Komarov 1940; Zavadsky 1961, 1968; Shcherbakov 2013], therefore, "without the species notion, all [biological] science turns into nothing" [Engels 1964: 189]. Apprehension of this idea constitutes the basis of the so-called *species thinking* [Richards 2010], which can be regarded as an important part of the *biological way of thought* [Beckner 1959].

However, according to an opposite point of view, the species notion is redundant in biology. It is inherited from a "prebiological era" of the historical development of natural philosophy, so it is morally outdated and should be excluded from the basic thesaurus of biological science [Burma 1954; Ehrlich 1961; Ereshefsky 2001; Kober 2008], just as the phlogiston notion appeared to be excluded from the thesaurus of modern chemistry [Davidson 1954].

A gap between these two extremes is occupied by a point of view, according to which there are different particular *species conceptions*, and this conceptual diversity reflects a complex multiaspected nature of biological diversity, a.k.a. *biodiversity* [Mishler and Donoghue 1982; Dupré 1993, 1999; Sluys and Hazevoet 1999; Pavlinov 2009a, 2013a, 2017, 2021a, 2022a].[2] This array of conceptions, known as *species pluralism*, generates a kind of *species uncertainty* that shapes the *species problem*, in its most general understanding, which is in itself a multiaspected conceptual phenomenon [Mayr 1957a; Hey 2001a; Stamos 2003; Pavlinov 2007, 2009a, 2013a, 2017, 2022a; Saikia et al. 2008; Amitani 2010; Richards 2010; Chambers 2012; Bartlett 2015; Ereshefsky 2017; Wilkins et al. 2022]. This problem is as fundamental as the species notion itself because it concerns a comprehension of such essential natural phenomena studied by biology as the structure of biodiversity, the process of biological evolution, ecosystem structure and functions, etc. [Stamos 2003; Pavlinov 2018; Barraclough 2019]. This word combination appears at the beginning of the 20th century [Robson 1928; Hawkins, 1935], whereas an awareness of the very problem is usually dated by the second half of the 19th century and is tied to the rise of the Darwinian evolutionary theory [Hey 2001a; Pigliucci 2003; Wilkins 2009a]. However, if we suppose that it is tied not with this particular theory but rather with the subsistence of different understandings of species phenomenon, then we have to admit that it has persisted since antiquity, when the cognitive foundations of

[1] This term does not refer to a particular conception that appears in the first half of the 20th century and which in fact is *reproductive* one (see Section 5.4).

[2] Hereafter, the *notion* refers to an understanding of species in general, the *concept* refers to a broad, and the *conception* to a narrow formalized generalization about species.

DOI: 10.1201/9781003346296-1

both natural philosophy and science began to be laid [Amitani 2010; Pavlinov 2018, 2022a]. Adherents of the species notion regard species pluralism to be an evidence of its vitality [Ellis 2011], while its opponents believe that an inability to elaborate a single unified definition for this notion testifies against it [Stanford 1995].

One of the key reasons for such considerable diversity of opinions about biospecies lies in the very fundamental nature of this notion. Indeed, one of the paradoxes of science is that the more general and fundamental a certain notion or concept is, the less rigorously and unambiguously can it be understood and defined; this inverse relationship is reflected by the *principle of incompleteness of knowledge* [Antipenko 1986]. A fundamental nature of the species notion implies that its discussion is addressed to the very foundations of the structure of living matter (ontology) and the ways of its cognition (epistemology). Therefore, the species problem is in many ways not so much biological as scientific-philosophical, and it affects such basic issues as the reality or nominality of the hierarchical structure of biota and of species as its element, the latter's ontological status, etc. [Volkova and Filyukov 1966; Ghiselin 1974, 1997; Morgun 2002; Stamos 2003; Richards 2010; Slater 2013; Pavlinov 2018, 2022a]. Therefore, as said in Preface, the discussion of the species problem is doomed to be endless, and "ongoing inquiry into the 'species problem' surely counts as one of the basic set of scientific puzzles by which to measure success" of theoretical biology [Atran 1987: 270].[3] According to the philosopher of biology David Lee Hull (1935–2010), who published a lot on the species problem, there is but a small success in its resolving reached so far, which consists in that "the points of disagreement are much clearer now than they have been previously" [Hull 2002: 119]. So the contemporary biology, perhaps, finds itself only at the beginning of the way toward a real understanding of what the causes and possible ways to solve the species problem are or might be.

This circumstance determines and explains a great attention that modern philosophically minded biologists and biologically oriented philosophers pay to the discussion of both the notion and the problem of species. Since the mid-20th century, many hundreds of articles and dozens of books have been published on this general topic; among the latter, there are monographs [Volkova and Filyukov 1966; Ghiselin 1997; Hey 2001a; Morgun 2002; Stamos 2003, 2007; Wilkins 2009a; Richards 2010; Slater 2013; Zachos 2016; Sigwart 2018; Mishler 2021], thematic collections [Mayr 1957b; Iwatsuki et al. 1986; Parmasto 1986a; Roger and Fischer 1987; Otte and Endler, 1989; Ereshefsky 1992a; Claridge et al. 1997; Wilson 1999a; Wheeler and Meier 2000; Alimov and Stepanyanz 2009; Pavlinov 2013b; Wilkins et al. 2022], and dissertations [Sloan 1970; Wilkins 2003a; Reydon 2005a; Kober 2008; Amitani 2010; Talpsepp 2013]. Many books on the philosophy of biology include sections on the species issues [Grene 1974a; Mayr 1982, 1988, 1997; Ruse 1989; Atran 1990; Wolters and Lennox 1995; Mahner and Bunge 1997; Sterelny and Griffiths 1999; Sober 2000, 2006; Ellis 2001; Ereshefsky 2001; Hull 2001; Grene and Depew 2004; Pigliucci and Kaplan 2006; Ruse 2008; Ayala and Arp 2010].

---

[3] However, an empirical mind views this otherwise: the debates over plurality of species conceptions are "vainly beating the air; [...] neither basic nor applied science would seem to have been slowed appreciably because the species-concept debate remains unresolved" [Winker et al. 2007: 30].

A theoretical framework of the species notion is proposed to subsume as *doctrine of species* or *theory of species* [Komarov 1940; Zavadsky 1961; Volkova and Filyukov 1966; Sucker 1978], and acknowledgement of its fundamental importance is reflected in the proposal to delineate a special biological discipline dealing with all species issues to be called *eidology* [Zavadsky 1968; Parmasto 1986b; Morgun 2002] or *eidonomy* [Dubois 2011] (after Greek ἑίδοσ, *eidos* = Latin *species*). According to the botanist Kirill Mikhaylovich Zavadsky (1910–1977), "an establishment of a distinct discipline for studying species is a question that has been overdue for two or three decades ago, [and] the doctrine of species should become a special branch of biology" [Zavadsky 1961: 4]. This "new science of species should investigate all aspects of the species problem" [*op.cit.*: 5], and "the question of the universality of species as a phenomenon of nature" should be considered the most important [*loc.cit.*]. This discipline has not been recognized so far, but it will be occasionally mentioned in our text. A kind of apotheosis of a special attention to biospecies is the foundation of the *International Institute for Species Exploration* in 2017 (https://www.esf.edu/species/mission.htm).

* * *

The first necessary step toward solving a problem is its understanding, which defines the choice of the path on which the first and subsequent steps are to be taken in a "due" direction. However, in the case of the species problem, there is even no consistency of the opinions regarding this important precondition: the main content of the multiaspected species problem is understood in different ways by different authors [Ereshefsky 1992b, 2001; Lidén 1992; Ruse 1995; Simonetta 1995; Boyd 1999; Sloan 2002; Stamos 2003; de Queiroz 2005a; Reydon 2005a; Pavlinov 2007, 2009a, 2013a, 2017, 2021a; Wilkins 2009a, 2018; Richards 2010; Ellis 2011; Chambers 2012; Kunz 2012; Bartlett 2015; Zachos 2016; Nathan 2017; Maxwell et al. 2020; Mishler 2021]. Some pay attention to the metaphysical issues involving species ontology (real or nominal) and origin (creation or evolution). Others are interested in the biological content of the problem, viz., the general structure of biodiversity (ranked or rankless), the essential characteristics of species (evolutionary, ecological, genetic, etc.), its nonequivalence in different groups of organisms, etc. Still others pay special attention to the contradictions that arise when developing species classifications based on different species conceptions. Finally, according to a quite radical viewpoint, opposing to the just listed, this problem is thought of as a result of theorists' psychological intention to look for a problem where there is none, so it is rather a "pseudo-problem".

In this book, the main attention is paid to the historical development of the species problem, considered in its full capacity resulting from its centuries-old history and in a broad scientific-philosophical context. As far as the problem in question is a conceptual construct regarded at a theoretical and partly philosophical levels, its historical representation can rightly be referred to as a *conceptual* (also *intellectual* or *philosophical*) *history* [Mandelbaum 1965; Wilkins 2003a; Pavlinov 2018, 2022a].

The reason of the great attention to the latter is that "the history and philosophy of biology and natural history has often been told by tracing the long thread of 'species concepts'" [McOuat 1996: 473]. The significance of such view on the species problem is that it obliges to pay much attention to philosophical doctrines that affect its formation. There are several important books among those mentioned above that are especially devoted to the historical issue [Stamos 2003; Wilkins 2009a; Richards 2010].

In our book, the historical development of the species problem is investigated in a more comprehensive fashion and in a rather specific context as compared to those just mentioned. In general, the standpoint adopted herewith is framed by *nonclassical philosophy of science* [Rozov 2002; Stepin 2005], the essence of which, with respect to the principal aims of our enterprise, can be very briefly summarized as follows: (a) Nature is a tremendously complex phenomenon and does not allow for simple descriptions in the form of a single theory or concept, (b) the cognition of Nature is intentional and selective, and (c) it is carried out within the framework of a cognitive situation in a certain scientific-philosophical context with the help of cognitive models (d) developed by cognizing subjects as the active participants of knowing. Upon a closer examination, this general standpoint is shaped by three scaffolding constructs. One of them is *contemporary conceptualism*, which presumes a specific look at the studied object [Quine 1969; Wartofsky 1979; Swoyer 2006]: it asserts that any natural phenomenon can be comprehended by means of a certain conceptual apparatus, which makes results of comprehension context relative. The second construct is *evolutionary epistemology*, which likens the process of conceptual history to biological evolution [Toulmin 1972; Popper 1974, 1984; Hull 1988, 2001; Abachiev 2004; Bradie and Harms 2016]. A combination of these two scaffolding constructs provides understanding of the conceptual history of the species problem as part of the historical development of the respective cognitive situation, which is regulated by the historical changes in the general scientific-philosophical context. The third construct is being developed by *cognitive science*, which places a special emphasis on the role of a cognizing subject in the cognitive activity, including the formation of a particular cognitive situation [Toulmin 1972; Haack 1979; Mikeshina 2005, 2007; Velichkovsky 2006].

A basis for selection of the path, which is suggested here to follow in our exploration of conceptual history of the species problem, also includes an understanding that the *history of science is part of science*. As a matter of fact, all theories and concepts, as well as particular problems ("scientific puzzles"), are historical entities subject to development over time, directed by a certain historical dynamics of scientific-philosophical contexts [Kuhn 1962; Lakatos 1978; Kuznetsova 1998; Maienschein 2000; Rozov 2002; Stepin 2005]. An insight of their historicity grows from regarding any cognitive situation as a developing nonequilibrium system, with one of its important properties being its endowment with a special kind of "historical memory", which manifests itself in a *continuity* of its development [Prigogine and Stengers 1984]. In our case, the latter means that in the conceptual history of the species problem, the content of the species notion at a certain stage of its development depends to more or less degree on its previous understandings. In other words, the notion in question is "historical" in the sense that it is historically conditioned and therefore historically transient [Volkova and Filyukov 1966; Zavadsky 1968; Mayr 1982]; more specifically, it is contextually conditioned, and its scientific-philosophical contexts change over time forcing it to develop accordingly. So, without a detailed historical analysis of the varying contexts of this notion, it is hardly possible to comprehend the species problem as a "historical entity" in all its capacity, including correct recognition of the main landmarks in its history and the causes of its contemporary content.

While studying relevant literature and delving into conceptual history of this problem, it became apparent to me that its standard "textbook" presentations, reduced to the key steps personally associated with Aristotle, Linnaeus, and Darwin (e.g., [Zavadsky 1968; Mayr 1982; Leroux 1993; Richards 2010; Zachos 2016]), is essentially deficient. Its closer consideration revealed some important trends and events and their interconnections, which escape attention in the case of its usual rough-grained presentation. For instance, an important historical fact is omitted by the contemporary "received view" that the development of the general idea of species in systematics of the 16th–17th centuries was based on the genus-species classification scheme inherited from scholasticism, which had been elaborating it during the 4th to the 12th centuries, based on the onto-epistemology of Neoplatonism [Pavlinov 2018, 2021b]. Thus, in order to comprehend how and why the species notion was acquired and then developed by early systematics, it is requested to analyze how and why it was developed by scholasticism and what was the latter's relation to Neoplatonic ideas.

* * *

The general structure of the book is as follows.

Chapter 1 is introductory: it outlines a brief idea of cognitive situation and its structure, which expresses in a concentrated form a conceptualistic view of Nature and the general ways of its cognition. The main emphasis is placed on the structure of cognitive situation under our consideration, and the principal notions shaping it are characterized: species, species (meta)concept, and species problem. On this basis, the main historical trends and stages in a conceptual history of the species problem are outlined.

Chapter 2 discusses the development of the species notion within antique onto-epistemology. It is shown how its particular treatments are formed at its classical stage and how they are refined, formalized, and detailed at its postclassical stage. Partonomic-taxonomic dualism is shown to be a characteristic feature of the antique understanding of species.

Chapter 3 describes the main issues of the genus-species problem developed by medieval scholasticism. The versions of species, generally understood as universal, are grouped into three basic cognitive standpoints, viz., realism, conceptualism, and nominalism. An emphasis is placed on the development of the species notion in the context of genus-species cognitive scheme.

Chapter 4 considers the development of the species issues influenced by initial formation of natural science during the Renaissance and the early Modern Times. The first part of the chapter continues illumination of philosophical issues discussed in Chapter 3. Its second part characterizes the development of genus-species classification approach and the beginning of its transformation into a biologically meaningful species problem within the incipient systematics as a biological discipline. The Herbal epoch begins this development, while scholastic and early postscholastic systematics continue it in their own ways.

Chapter 5 first reviews the formation of early conceptual prerequisites for the development of an idea of species as a biological unit. It begins with the objectification of this idea in the context of biblical world picture and maturation of generative idea that explains species integrity. Then, the main manifestations of the general trend of biologization of the species problem during the 19th–20th centuries are characterized. This includes the rise of evolutionary and ecological ideas, emergence of the *species question* within the framework of biosystematics, recognition of multiplicity of particular species conceptions, and formation of the *species problem* in its contemporary account.

Chapter 6 briefly discusses three issues that are among most important in the contemporary species problem: (a) how to elaborate a kind of summative definition of species in its general understanding, allowing to unite all its particular interpretations; (b) how to understand ontology of species, taking into account the diversity of its modes in different groups of organisms, and (c) how natural species can be interpreted in the context of "new essentialism". Appearance of this account here is justified by understanding the historical development as an endless process, of which the contemporary stage is but a transition between former and future considerations.

1 The Species Problem

In Broad Strokes on a Conceptual Historical Canvas

According to an initial assumption based on a monistic idea of the universe, the realized history of Nature, human society, science, etc., is unique; however, it can be interpreted and expounded in different ways. These interpretations and expositions turn "history in general" into particular narratives depending on how the content of science and its history is understood [Rozov 2002].

Since our book deals with the *conceptual* history (as well as with the *pre*history) of the species problem in biology, the focus of our attention is on the historical changes in the theoretical constructs and contexts that shape this problem. For understanding a general course of these changes, one of the basic provisions of evolutionary epistemology is important, according to which the historical development of science, including its theories and concepts, can be likened to biological evolution (see Introduction). From this standpoint, if speciation is the key act in biological evolution, then *conceptuation* corresponds to it in the conceptual history. Each scientific concept appears "in its own time and place" according to its relevance to the developing general scientific-philosophical context, and when this concept loses its significance because of change in the context, it "dies out", i.e., expelled into the background of the developing cognitive situation.

This chapter begins with a brief presentation of a cognitive situation and its structure, which in a concentrated form expresses a conceptualistic view of Nature and the general way of its cognition. The main emphasis is quite naturally placed on the cognitive situation, which is built around the species notion and problem. Further, the basic notions comprising the conceptual framework of this situation are briefly characterized: what is species, what is species concept, and what is species problem in their general understanding. On this basis, the main content of the conceptual history of the species problem is briefly outlined and the most important historical trends and stages in it are characterized.

1.1 DEVELOPING COGNITIVE SITUATION

According to a basic presumption of the classical philosophy of science, natural science is aimed at studying Nature as such, as the entire objective reality that exists outside and besides a cognizing subject. However, in a real research activity, this global cognitive supergoal is fundamentally unfeasible. The reason is that Nature as a whole, in its "absolute" comprehensive account, is inaccessible to cognition in a full overwhelming capacity due to the limited possibilities of human cognitive tools [Faust 1987; Russell 1992]. Therefore, human knowledge is always *selective* and therefore *intentional*: it is addressed not to global Nature as a whole, but to its certain local

DOI: 10.1201/9781003346296-2

manifestations (aspects, fragments, etc.), which are isolated as local cognizable objects based on some considerations by a cognizing subject [Knyazeva 2006; Kull 2009; Amitani 2010; Gontier and Bradie 2017]. This means an inevitably reductionist nature of human knowledge, which makes it basically partial and incomplete with respect to the cognized objects; this is summarized by the principle of incompleteness of knowledge [Antipenko 1986].

This general understanding is formalized by the concept of *cognitive situation*, within which all cognizing activity is carried out; it is fundamental for conceptualist philosophy of science [Yudin 1997]. It is constructed around some specific cognitive task that a researcher formulates with respect to a selected manifestation of Nature set as a cognized object. In our case, this task is generally formulated as *cognizing of the structure of species diversity of living matter.* Due to selective, intentional, and reductionist character of knowledge, this task cannot be formulated in a single exhaustive manner, so it is not so much a task as a *problem* with a multiplicity of prospects of solvability. Thus, a cognitive situation is always a *problematic situation*, and its development involves the formulation and attempts to solve particular scientific problems. In our case, we are talking about the *species problem*, which is built around the general *species notion* as a focal point of the respective cognitive situation.

Of primary importance is the understanding that a cognitive situation is shaped and developed not by itself but in a certain scientific-philosophical context [Miller 1996; Stepin 2005; McCray 2006]. It is the latter that prescribes which particular problems (tasks) in natural science are regarded significant, how they can be formulated and solved in a scientifically meaningful fashion, etc. The changes in this context largely determine transitions from one problem (task) to another, and in the most pronounced form, they have the character of *scientific revolutions* [Kuhn 1962]. For example, in the conceptual history of the species problem, postscholastic revolution in systematics at the turn of the 18th–19th centuries becomes of paramount importance to serve as a powerful stimulus to the biologization of this problem [Pavlinov 2018, 2021b].

It follows from the above that a comprehension of the conceptual history of the species problem can be reached if it is based on the study of historical dynamics of the whole cognitive situation embracing it, which is in turn explored in the context of the historical development of its scientific-philosophical context. Accordingly, this book begins with summarizing, in a quite concise form, a general idea of what a cognitive situation is, how it is structured, and how it functions with regard to the problem of our concern [Pavlinov 2011, 2017, 2018, 2021b].

1.1.1 Structure of Cognitive Situation

The basic structure of any cognitive situation is three-partitioned: it includes ontic, epistemic, and subjective components that interact with each other in a complex way.

The **ontic** (Greek όντος, being), or the *object* component refers to *what* precisely is to be investigated, i.e., to the object itself, by indicating its general properties and characteristics that are significant from the point of view of a particular research task. The scale of this component varies greatly from the general world picture (e.g., creationist or evolutionary) to the specific ideas about biodiversity. It is essential to emphasize that the basis of this component is shaped not by a natural object itself in its suchness (e.g., biodiversity or natural species), but by its *conceptual representation* serving as its *cognitive model* [Wartofsky 1979; Miller 1996]. This representation configures a *conceptual reality*, so it is the latter that is identified, under certain assumptions, with the ontic component and which a cognizing subject actually deals with. The natural object corresponds to Popper's "first world", while the conceptual reality corresponds to his "third world" [Popper 1974]. According to the principle of *ontological*, or *context relativity*, which is fundamental for contemporary conceptualism [Quine 1969; Dupré 1993; Miller 1996; Swoyer 2006], a particular cognizable object (say, biodiversity) is not given in a certain "absolute" form, but framed by the respective conceptual assumptions, which are set within a certain scientific-philosophical context by cognizing subject. For example, under a realistic (or conceptualistic) position, biodiversity can be thought of as a hierarchy of monophyletic groups or of local ecosystems, while under nominalistic provisions, it is reduced to a "nondimensional" set of organisms.

Thus, in our case, the main element of the ontic component is not natural biospecies itself (no matter how it is understood), but its cognitive model in the form of species notion, which content may vary from a commonsense intuitive image to a scientifically sound more or less formalized concept. It is this notion, or a rather respective concept, on which basis a cognizing subject decides what can be identified as *the* species in the overall structure of biodiversity [Pavlinov 2013a, 2017, 2022a; Bartlett 2015]. In order to make this concept an exploratory tool, it should embed those characteristics of species that are considered essential from a certain standpoint (phylogenetic, ecological, pragmatic, etc.). According to the *principle of constructiveness*, if some characteristic is not incorporated in a cognitive model of the studied object, it "falls out" of the cognitive situation and cannot be constructively (scientifically) investigated [Kuraev and Lazarev 1988; Margolis and Laurence 2012]. Since it is basically impossible to include all characteristics of a natural object (in our case, biospecies) in its cognitive model, it is obvious that the latter is simpler than the object itself: this means that a model is unavoidably *reductionist* [Wartofsky 1979]. As a result, an *ontological gap* arises between an object and its representation [Williamson 2000], which is often not taken into account when the species conceptions are discussed both from theoretical and practical perspectives. It is evident that the less complete is a set of characteristics fixed by a cognitive model of natural species, the more reductive it is and thus the less adequately it reflects natural species as a cognizable object. From this viewpoint, for example, phenetic conception turns out to be essentially incomplete for the exploration of species as an evolutionary unit, and phylogenetic conception is insufficient for its exploration as an ecological unit (on these and other conceptions, see Section 5.4).

The **epistemic** (Greek ἐπιστήμη, knowledge) component determines *how* a conceptualized object should be investigated in one way or another. The sphere of epistemology involves the criteria of truthness and scientificity of knowledge about species, including evaluation of scientific consistency of particular species conceptions, definitions, etc. An essential part of epistemology, which puts it closer to the practical needs of research activity, is methodology: on the basis of criteria proposed by it, the consistency of methods for studying species diversity is assessed.

A very significant contribution of this component to the structure of cognitive situation is the justification of an inevitably reductive procedure of individuating the studied object; in our case, this is biospecies as a unit of the overall structure of biodiversity. The entire operation of reduction begins at the highest levels of the structure of Nature and proceeds stepwisely down to the lowest levels of the organismal diversity. As a result of this sequential reduction, a deductive (descending) *reduction cascade* arises, which is an irremovable part of cognitive situation. At each step of this cascade, a particular conception is elaborated for a particular phenomenon being distinguished at this step (living or inert matter, process of evolution or biodiversity structure, species or biomorph, etc.) by indication of (a) its place in the process or structure fixed at a given step and (b) its own essential characteristics. Elaboration of each conception is substantiated by certain theoretical considerations, which makes the total reduction cascade conceptualized from the very beginning to the end. In our case, when the species notion is introduced at a certain step of conceptual reduction, it is to be justified prior to it what is, say, phylogeny or ecological community and how it is structured, and then, at the next step, how species is to be treated as an element of the respective structure, be it phyletic lineage or an ecological "cell". It is apparent that any such consideration is valid in the context of a biologically substantive theory (higher-order conception) elaborated at a previous step of the reduction, which in our example is either evolutionary or synecological theory.

The reduction cascade is characterized by two important properties.

First, since a cognitive model is simpler than a studied object, and each step in the cascade adds something of its own to an overall reduction, the more steps are involved in the reduction operation leading from a highest-order concept to a particular conception (say, from biota to ecospecies), the less a cognitive model defined by the latter is "natural" (refers to something actually existing in nature) and the more it is "artificial" (refers to something nonexisting out of a particular cognitive situation). Consequently, the longer such cascade is, the broader becomes an ontological gap between a natural phenomenon itself and its cognitive model in the form of a conception. Thus, a species notion, even quite general, is much simpler in comparison with a presumed multifold natural unit as an element of Nature. This is all the more true with regard to its particular interpretations, which reduce this general notion to its particular manifestations (phylogenetic, ecological, genetic, etc.).

Second, since each complexly organized natural object can be reduced in different ways and represented by different finite cognitive models [Wartofsky 1979], conceptions of higher levels are fragmented at each step of their reduction. Such sequential detailing of a general theory/concept, from which the all reduction procedure begins, leads to an increase in the total number of conceptions as we move from top to bottom along the reduction cascade. This multiplication of species conceptions yields a kind of *conceptual pyramid* as a figurative representation of the ontic component of a cognitive situation (see Section 6.1).

An irremovable occurrence of the epistemic component in the cognitive situation inevitably leads to a specific *epistemic pitfall*, which is caused by ignoring or downplaying the above-considered ontological gap [Williamson 2000]. Due to this, accordingly to one of Zen Buddhism aphorisms, "a person is inclined to mistake his finger pointing at the moon for the moon itself". In our case, when biologists analyze the species diversity, they hardly take into account that it is a particular species conception (a "finger") that makes them seeing in Nature a certain phenomenon they call "species" (the "moon") [Bartlett 2015; Pavlinov 2018, 2022a].

The **subjective**, or *subject* component includes a cognizing subject—the one *who* conducts research. When analyzing this component, the main issues are the causes and character of a subject's impact on the structure of cognitive situation; these issues are studied by cognitive science [Wartofsky 1979; Velichkovsky 2006; Mikeshina 2007]. The role of a subject (in a broad sense) in a cognitive situation is rarely taken into account, but it is very significant, diverse, and unavoidable [Toulmin 1972; Fleck 1979; Haack 1979; Kholodnaya 2004]. This is because it is the cognizing subject that individualizes both the very object and its particular manifestation to be studied, guided by certain personal intentions (Popper's "second world") [Quine 1969; Mikeshina 2007; Slater 2013]. In our case, it is the subject that (a) "extracts" biodiversity from the surrounding Nature, (b) fixes certain aspects of its consideration (evolutionary or ecological or structural or else), (c) individualizes biospecies according to a particularly fixed aspect, and (d) recognizes specific characteristics of species that reflect most closely its "place" in a fixed aspect of biodiversity.

At an individual level, the effect of the subjective component is manifested primarily in the *thinking styles*, in which personal cognitive preferences and intentions are accumulated. In our case, we should mention *species thinking* [Richards 2010], *population thinking* [Mayr 1942, 1959a, 1982, 1988], *individual thinking* [Ereshefsky 2010a; Haber 2016], etc. At a higher level, an effect of the subjective component manifests itself in the social regulation of preferences in choosing research priorities for the reasons unrelated to science itself [Rozov 2002].

These three components of cognitive situation complexly interact with each other, and by this, they influence indirectly each other. It is to be emphasized that the interaction between the ontic and epistemic components is mediated by the subjective component, which fixes a cognizable object and construes its cognitive model guided by certain epistemic ideas. The latter encompasses realistic or nominalistic cognitive intentions, which are formed on a subjective basis and dictate the ways of perceiving the structure of Nature. A reductive influence of the epistemic component on the ontic one is evident in this "dictate". The reverse influence is manifested in that, for instance, the choice of methodology (phylogenetic or phenetic) for studying species diversity depends on the acknowledgement or denial of evolution as a cause of this diversity.

As a result of such interrelations, the ontic and epistemic components are united by an *onto-epistemic correspondence* to form a complexly structured *onto-epistemic basis* of cognitive situation. This basis may be thought of as a kind of *conceptual space* (*conceptual framework*), in which the cognitive model of a studied object is defined and functioning [Botha 1989; Gärdenfors 2004; Efremov 2009]. An array of the concepts involved in the shaping of conceptual space defines its *content* and *dimensionality*: for instance, the understanding of biospecies can be phenetic, essentialist, or evolutionary ecological.

The interdependence of the basic components of a three-partitioned cognitive situation allows us to represent the latter metaphorically in the form of a *cognitive triangle*: its vertices correspond to these components, and the edges reflect their interrelationships. This figure is constructed based on ideas about the fundamental triadic structure of cognitive activity [Barantsev 1983]. A certain notion/theory/concept is placed in the center of this triangle, and around it, the entire cognitive situation is built; in our case, this is the general species notion with its own specific "envelope" in the form of the species problem.

* * *

The above ideas about the structure of cognitive situation may seem too "philosophical", but they closely concern to certain serious issues of the species problem, which are no less philosophical in their content.

One of them is a dilemma of "reality *vs.* nominality" of species, which is actively discussed since Antiquity and constitutes the main content of the species problem up to the end of the 19th century [Volkova and Filyukov 1966; Wilkins 2003a, 2009a]. Currently, this issue is more often clothed in a dilemma of "objectivity *vs.* subjectivity" of species (e.g., [Mayr 1957a, 1969, 1982; Ghiselin 1987; Ruse 1988; Daston and Galison 2007]). Both dichotomies are usually treated according to the classical tradition in an "absolute" fashion: species is either a real (objective) natural phenomenon or a nominal (subjective) mental artifact.[1] However, this classical dichotomy is hardly correct from the conceptualistic standpoint. The ontological relativity makes no sense in discussing this dichotomy out of a particular cognitive situation: what "exists" in one conceptual reality may be "absent" in a different one. This issue is aptly illustrated by various treatments of species in different taxonomic theories, making their input in a long-lasting disagreement between realists and nominalists [Pavlinov 2018, 2021b].

Close to the preceding is a fundamental issue of "objectivity" of knowledge about the structure of species diversity. It is obvious that irremovable occurrence of a subject in a cognitive situation means that an "objective knowledge" in its classical sense is fundamentally impossible. If we talk about "truth" of that knowledge, we face a serious problem of multiplicity of the ways of defining the truth and its criteria, which are specific to different scientific-philosophical doctrines [Levin 2011; Glanzberg 2013]. Looking at this issue with the above-discussed ideas in mind, it might be reasonable to think of the knowledge about species diversity as the latter's specific *information model*, so the "truth" of this knowledge turns out to be an adequacy (isomorphism) of the model to the diversity in question [Wartofsky 1979; Fraassen 2008; Frigg and Nguyen 2016]. However, even in this case, the problem of ontological relativity in an analysis of both diversity itself and constructing its cognitive models still remains most relevant.

1 Somewhat different is a dichotomy "non-arbitrary *vs.* arbitrary" modes of species delineation [Simpson 1961], but this approach is not so much "philosophical" as "practical".

1.1.2 Species, Species Concept, Species Problem

A core cognitive *unit idea* (in the sense of Lovejoy (1959) and Knight (2012)) of our interest is an *idea of species*, as it is generally shaped by an array of onto-epistemological systems historically developed by the human cognitive activity. Considered in such fashion, it is a kind of nonformalized *primitive* (in the sense of Wierzbicka (1996)), and it can be treated as a specific *disciplinary matrix* (in the sense of Kuhn (1962) and Boyd (1999)) integrating those systems and providing for their mutual continuity regardless of their differences.

Accordingly, the core question around which the entire species problem is built and the searches for answers to which direct its conceptual history is addressed basically to ontology: *what is species*? [Cronquist 1978; Mayr 1982, 1996; Eldredge 1993; Mahner 1993; Grene and Depew 2004; Wilkins 2003a, 2009a]. This question has been formulated by antique philosophers and was being actively discussed since then, following the precept of the antique thinker Boëthius: "If we do not realize what species is, nothing would secure us from delusion" [Boëthius 1906]. As it becomes evident from the previous section, the possible answers to it depend on the contents of respective cognitive situations and are ultimately shaped by the scientific-philosophical contexts, in which these situations are immersed. All this means that the sought answers (a) depend on those particular contexts, (b) are not only scientific but also philosophical, and (c) therefore cannot be simple and unequivocal.

In its most general understanding, neutral with respect to any particular onto-epistemology, species is thought of as a collection of similar things and/or their properties, constituting a kind of "diffusive unit", which is distinguished in the overall structure of the observed diversity of Nature. This understanding originates on an inductive basis in indigenous human communities [Atran 1990] and initially remains intuitive and thus a nonformalized image. In a common language of science and philosophy, it figures as a generalized *species notion*, which rationalization transforms this image into a more or less formalized general *species concept*. Both original (intuitive) species image and transformed (rationalized) species concept are context-relative: the former depends more on pragmatic contexts and the latter is shaped by conceptual contexts. Bearing in mind what is said in the previous section, it should be emphasized that it is the species notion/concept (a mental unit), and not species itself (a natural unit), that constitutes the basic element of the respective cognitive situation.[2]

> The terms "notion" and "concept" are rather close in their general meaning, and the relationship between them can be summarized as follows [Voyshvillo 1989; Murphy 2002; Margolis and Laurence 2011]. The notion (a signifier) is addressed directly to a certain manifestation of the cognized reality (a significate) itself, whereas the concept is an interpretation of the notion, and it functions as its rationally organized "shell".

[2] M. Hatch believes that the species concept is a product of the 18th-century rationalism [Hatch 1941]. J. Hey counts "the modern era of species concepts" from the 1942 book by E. Mayr [Hey 2006]. It is entered in the active circulation in the early 20th century (see Section 5.3).

The difficulties begin when we detail the above basic question and begin to ask more particular queries that have specific onto-epistemic contents; the main points of these queries are as follows. Does species exist really (in Nature) or mentally (only in a cognizing mind)—i.e., does it belong to ontology or epistemology? If species is real (objective), is it a universal phenomenon of Nature (of the entire material world) or is it specific to its different sections ("chemical species", "biological species", etc.)? If it is real, what are the causes for its existence as a specific unit of the structure of Nature? Again, if species is real, how does it relate to other units of the diversity of Nature, which in a sense are comparable to species (e.g., biomorphs), but essentially are "nonspecies"?

From a certain realistic point of view, "what species really are, biologically, depends on what causes their integration, their continuity, and their distinction from each other" [Van Valen 1988: 61]; although L. Van Valen writes about the particular species, this declaration is possible to consider true regarding species in its general meaning. However, this most general answer to the basic question is not as simple as it might seem, as it involves reflections on the causalities of natural phenomena. It becomes far more complicated if we take into account that the possible particular answers depend on specific scientific-philosophical contexts. And since there are many of the latter, there may also be many answers: adherents of different world pictures, reflecting on Nature and its structure, respond to that question in significantly different ways. Simplifying our cognitive situation, the possible answers to this question can be reduced to three general options that correspond to three basic natural-philosophical doctrines—nominalism, realism, and conceptualism.

According to a nominalistic account, only individual things (particulars) exist in Nature, while any groupings of them (universals) and the endowment of these groupings with some special qualities (essences) are conjectured based on the similarity of things, so they exist only in a cognizing mind as some mental images. These groupings (including species) are distinguished as classification universals: they allow describing the observed diversity in a unified way, "imposing" on it a structure that actually does not exist. This account of species, according to which it performs a purely cognitive (classificatory) function, is acknowledged in almost all logical systems and currently is supported by positivist philosophy.

Realists believe that species really (objectively) exists as a natural unit generated by certain natural causes, and it occupies a certain place in the overall structure of Nature. In its extreme expression (naive naturalism), a realistic standpoint is manifested in a belief that "we do not define species, we *see* them" ([Wilkins 2018: 169], italics in original); similarly, it is assumed that "there are *observable* lineages out there in the natural world" ([Mishler and Brandon 1989: 77], italics added).[3] Obviously, this is not true: actually, a species as such and even more so a lineage, though under an assumption of its reality, cannot be perceived (seen, observed) in the same way as a certain concrete thing (physical body) is perceived [Assis 2011]. This means that we do see or observe not species or lineages as such, but just concrete things, from observations of which and reflections upon which we infer metaphysical ideas about their groupings supposed to be natural.

[3] If this is so, why not to establish the *Society of Lineage Watchers*?

Conceptualism considers the structure of Nature based on the aforementioned idea of ontological relativity. In this case, it is assumed that Nature is complexly structured, and its various manifestations are individualized as cognizable objects through certain conceptions that partition a general image of Nature into particular conceptual realities. The latter can be regarded as objective to the extent that they can be associated with certain processes structuring Nature and generating manifestations of its overall structure, which are fixed mentally according to particular cognitive tasks [Quine 1969; Putnam 1991; Mamchur 2004]. Each conception appears as a part of the general cognitive model that describes some basic properties of the cognized manifestation (aspect, fragment, etc.) of Nature. Regarding a conceptual reality corresponding to the biodiversity, one of its basic elements is the *species concept*, viz., a cognitive model of a lower order, individuating in this reality a particular unit, which corresponds to a supposed natural species. This concept (or conception) serves as a kind of heuristic, which, when referring to a cognized reality, allows distinguishing the particular species in it and making assumptions about some of their properties [Brigandt 2003a; Pavlinov 2013a, 2017, 2022a; Bzovy 2016; Wells et al. 2021]. Accordingly, it is a substantive scientific theory or concept that turns any observation into a relevant scientific fact [Popper 1974; Faust 1987], which holds true for the abovementioned "seeing" of species or lineages.

Regardless of particular onto-epistemologies, a general understanding of species as a certain individualized unit implies an assumption of its certain integrity. Accordingly, a core of the species concept includes reference to supposed principal factors responsible for this integrity. It is thought of as a causal agent to be invoked as an organizing principle [Mishler and Brandon 1987], which is hereafter designated as *species integrator*. Usually, certain relations among conspecific organisms (similarity, kinship, reproductive, behavioral, etc.), as well as their specific relations to their environment (ecological role, etc.), are considered in this capacity. The integrator's action ensures an internal *cohesion* of species and provides a kind of "glue" necessary for its stability and discreteness as an individualized object [Caplan 1980; Brooks and Wiley 1986; Barker and Wilson 2010; Kunz 2012; Barker 2019a,b; Kendig 2022]. Thus, the notion of a species integrator is introduced in order to explore an array of species conceptions on a single causal basis by treating them as particular interpretations of the same general concept that differ in particular meanings of that integrator.

An important part of contemporary conceptualizations of the species notion involves an explicit demarcation of two aspects of its consideration. One of them is *species category* signifying a certain level of generality in the structure of biodiversity, and another is *species unit* allocated to this level; in taxonomic classifications, they correspond to the notions of rank and taxon, respectively. It is suggested to distinguish in the first half of the 20th century by the philosopher of biology, Joseph Henry Woodger (1894–1981) [Woodger 1937], and it quickly becomes acknowledged by biologists. This distinction introduces a *category-unit dualism* in the considered cognitive situation, and its analysis concerns the general issue of the reality of species as a natural phenomenon (see Chapter 6).

Thus, we can conclude all the above by acknowledging that the answers to the above basic question about species, both in its most general sense and about species of living organisms in particular (hereafter, we will be talking only about it), can be purposefully searched only in a certain cognitive situation with the help of a certain species concept/conception that includes a reference to the species-specific integrators. Figuratively speaking, each such concept/conception serves as those "spectacles", with which researchers examine Nature and try to discern species in its structure: so, with which particular concepts/conceptions Nature is explored, the respective particular units, commonly called species, are individuated in it [Pavlinov 2013a, 2021a, 2022a; Bartlett 2015].

* * *

It follows from the said above that a starting point in the search for possible answers to the basic question "what is species?", after the formation of a general conceptual framework, is an elaboration of a certain species concept. Accordingly, this basic question central to the species problem now should be reformulated as follows: "*what is the species concept*?"; it focuses more on epistemology.

In a certain general sense, a concept (Latin *conceptio*, comprehending) is a certain mental image of some natural phenomenon, representing it in a more or less formalized form in a cognitive situation [Margolis and Laurence 2012; Solonchak and Pesina 2015]. Such a vision of the latter's structure constitutes the so-called *conceptual thinking* [Körner 1959]. As said above, in the historical development of any cognitive situation, a concept is preceded by a less formalized general image (understanding) of the phenomenon in question.

Until the 18th century, the core of the species problem is shaped by the species concept treated as a predominantly classification universal, which was inherited from scholasticism. In the 18th century, its distinct biologically meaningful conceptualization begins to develop based on a generative idea (see Section 5.1.2); simultaneously with this, the species rank get fixed in the hierarchy of the System of Nature. As a result, the 19th century witnesses the rise of a comprehension of the fundamental status of biospecies as a specific natural phenomenon, differing from all kinds of "species" of inanimate nature, as well as from the units of other levels of generality. Thus, the natural philosopher-materialist Friedrich Engels writes in his *Dialectic of Nature* in the second half of the 19th century that "without the notion of species, all [biological] science turns into nothing" [Engels 1964: 189]. In the middle of the 20th century, the Russian botanists designate species as a special "way of realization of life" [Komarov 1940: 10; Zavadsky 1961: 3], reflecting a clearer awareness of natural causes of species as a fundamental element of the structure of the evolving biota. The very term "species concept" in its biological meaning becomes acknowledged in the first half of the 20th century (see Section 5.3).

The principal way to conceptualize the species notion is its *theoretical definition*, which is *intensional* by referring to the essential (primary) characteristics of species as a unit of the structure of Nature. This yields a *theoretical concept*, in which the primary characteristics shape a core, while its periphery is shaped by a set of secondary characteristics of species. A definition that includes the total set of possible primary characteristics provides a multidimensional *general species concept*, repeatedly mentioned above, or a *meta-concept*. The latter gets split into several *particular theoretical conceptions* based on particular substantive interpretations of primary characteristics according to the particular onto-epistemic contexts—classificatory, evolutionary, ecological, etc. Speaking more formally, this yields a splitting of the whole conceptual space defined by species meta-concept into local subspaces with respective particular species conceptions. *Operational conceptions* are elaborated based on the secondary characteristics to produce criteria suitable for the identification of the particular species in the empirical research.

The species notion/concept performs a specific function of the so-called *boundary object* [McOuat 2009; Star 2010]; in other terms, it may be regarded as a kind of *proxi* or a *prototype* [Amitani 2010, 2015, 2017]. It allows the people, operating within the same cognitive situation, to communicate intelligibly with each other by using the same notion, which is presumably accounted more or less similarly. Indeed, when phylogeneticians and pheneticians discuss the species problem, they refer to the same general species notion, though they interpret it in a quite different manner.[4] With respect to the meta-concept, its particular interpretations, including operational conceptions, may be treated as performing the function of some *correspondence rules* (in the sense of Rudolph Carnap) that serve as a means of solving specific research tasks in different departments of biology, such as systematics, ecology, phylogenetics, etc., by placing them within the same conceptual framework roughly outlined by general species meta-concept [Richards 2008, 2010].

Thus, it seems from the point of view of conceptualism that the question "what is species concept?" is more constructive and therefore more meaningful than the preceding question "what is species?" This important property of a concept is determined by that it can be subjected to an estimation of its scientific efficiency by certain criteria developed within the epistemic component of the cognitive situation. For example, D. Hull specifies three such criteria of general kind, calling them "philosophical": *theoretical significance* (can be built into a meaningful scientific theory), *universality* (applicability to a wide subject area), and *applicability* (the possibility of operationalization) [Hull 1969, 1997, 1999].

The former criterion is of special importance, as it presumes reformulation of the most important question of the whole species problem, viz., about the existence of species. Indeed, in a classical form, it is addressed directly to ontology and looks like that: is there really a specific phenomenon inherent in living nature, which is designated as biospecies? In its conceptualistic account, it refers to the complex onto-epistemic basis of the cognitive situation and looks like that: to what extent the species concept is needed for building some substantive theory (evolutionary, ecological, etc.) that would be of fundamental importance in the whole of biology? [Collier 1985; Cracraft 1987, 1989a, 2000; Ruse 1988; Hull 1997, 1999]. In a more "scholastic" form, this query looks like this: "if true statements of biological sciences are committed to species terms, then species are real" [Collier 1985: 17].

---

[4] An "elephant" of a well-known ancient Indian parable about blind men, who touch its different parts and imagine it differently, which is mentioned sometimes with respect to the species problem [Brooks and McLennan 1999; Winston 1999], may be taken as a vague illustration of such boundary object.

It is evident from the preceding that this "conceptualistic" question makes any sense only if it is set and answered with respect to a certain substantive theory. Regarded in its quite general biologically meaningful sense, it could be addressed as a theory that treats biota as a self-developing nonequilibrium system; it will be discussed in more detail elsewhere (see Section 6.1). In a more particular sense, it presumes that these are certain particular conceptions that are subject to evaluation, and as far as they function differently in different substantive theories, the particular criteria for their evaluation should also be different [Richards 2010]. Accordingly, there will be at least as many contextually specified questions about the significance of species conception and possible answers to them as there are such theories. For instance, one of them is such taxonomic theory, in which species category is fundamental [Blackwelder 1967; Mayr 1969, 1988; Stuessy 2008]. Another is evolutionary theory, in which species is considered a final stage of the speciation process [Hull 1977; Mayr 1982, 1988; Ghiselin 1997; Shaw 2012]. The third is synecological theory, in which species figures as an element of community structure, occupying a certain ecological niche in it [Odum 1953; Schwartz 1980; Chase and Leibold 2003]. Besides, a specific context for evaluating species conceptions is set by the pragmatics of biodiversity protection issues [Sarkar 2005; Barraclough 2019].

While realists' and conceptualists' answers to these questions are generally affirmative, then nominalists' answers are as generally negative. For instance, proponents of the positivist philosophy of science propose to develop a kind of "biology without species" [Burma 1954; Kober 2008]. In systematics, this philosophical position is adopted by phenetic theory [Sokal and Sneath 1963; Sneath and Sokal 1973], while cladistics rejects species concept from another standpoint [de Queiroz and Donoghue 1988, 1990; Mishler 2010, 2021; Zachos 2016]. When studying the functional structure of ecosystems, according to one standpoint, it is biomorph and not the species that is considered its basic unit [Levushkin 1976; Chernov 1991; Kirpotin 2005]. In the projects dealing with biodiversity, a proposal to abandon species concept is justified by that it has no universal rigid definition, so it is to be replaced with some operational conceptions [Stanford 1995; Pleijel and Rouse, 2000; Sarkar and Margules 2001].

* * *

Generally speaking, the fragmentation of conceptual space of the species problem generated by the diversity of particular conceptions is caused by the need to solve different research tasks related to the different manifestations of the structure of biota [Mishler and Donoghue 1982; Kitcher 1984a; Ereshefsky 1992b; Bachmann 1998; Amitani 2010; Pavlinov 2013a, 2022a; Zachos 2016]. With regard to each of these manifestations, species appears not as a general notion or concept, but as a narrow conception, which basis is shaped by a particular interpretation of the species integrator presupposed by this manifestation. From a certain point of view, the multiplicity of conceptions and their historical dynamics can be considered an indicator of the vitality of the species notion, which allows for different elucidations stipulated by a more profound understanding of the multifold structure of biota [Ellis 2011].

The development of many particular conceptions, each concretizing the general species concept in a specific fashion, leads to the displacement of the prototypical understanding of "species in general" by its various conceptualizations and associated formalizations [McOuat 2001; Pavlinov 2009a, 2018, 2021a, 2022a; Amitani 2013, 2015]. For this reason, the general species notion loses its presumed (though a vague) concreteness: its intuitively grasped content splits into different (more rigid) interpretations, which leads to a *conceptual species uncertainty* [Hey et al. 2003; Bartlett 2015; Pavlinov 2022a]. Because of this, the entire cognitive situation takes a guise of a patchwork or mosaic [Novick and Doolittle 2021].

The fixation of particular manifestations of the overall structure of biodiversity, which is responsible for the conceptual species uncertainty, is produced on different grounds. It is important for us to identify two of them, probably most general—*aspect-based* and *object-based* [Pavlinov 2011, 2018, 2021b]: these terms equally refer both to the aspects themselves and the species conceptions corresponding to them (for other categorizations of species conceptions, see Section 5.3). In the first case, we fix different aspects of the diversity of the same pool of organisms, which results in the recognition of particular aspect-based manifestations of their overall diversity. In the second case, the fragmentation is resulted from separating different pools of organisms of various levels of generality, in which particular aspect-based manifestations of their diversity are considered.

Among the ways of aspect-based fragmentation of the overall diversity of life, the most fundamental is the fixation of its *partonomic* (meronomic) and *taxonomic* aspects [Meyen and Shreyder 1976; Meyen 1977; Tversky and Hemenway 1984; Tversky 1989, 1990; Lansing 1995; Lyubarsky 1996; Pavlinov 2018, 2021b]. Their descriptions are formalized by respective logical systems, *intensional* and *extensional*. Partonomic aspect is studied by *partonomy* or *mereology* (meronomy) [Mahner and Bunge 1997; Simons 2000], and it involves analyses of certain properties of organisms and their aggregates, which are designated as *partons* (Latin *pars*, part) or *merons* (Greek μέρος, part). Here belongs, for instance, distinguishing between homologies and analogies, identification of essential features of organisms and their aggregates (including species), formation of traits for their description and comparison, etc. Taxonomic aspect is studied by *taxonomy*, which deals with the aggregates of organisms of different levels of generality, i.e., *taxa* (in the general sense).[5] In the antique and medieval onto-epistemologies, these two aspects are closely interconnected by an idea of the essential multilevel organization of Nature. Therefore, the units recognized in each of these aspects (partons and taxa) of a certain level of generality are designated by the same term, *έίδοσ* (Greek) or *species* (Latin) (see Sections 2.1 and 3.1). On this basis, a *partonomic-taxonomic dualism* of a total understanding of "species" emerges, which is preserved to a certain extent in the early (scholastic) systematics.

In the realm of systematics, this dualism appears at a macro level due to reductionist *vs.* holisitic worldviews. In the first case, as just said, any groupings of organisms

[5] In another meaning, taxonomy refers to the theoretical section of biological systematics [Simpson 1961; Pavlinov 2021b].

are treated as aggregates described in terms of set theory. In the second case, Nature is regarded as an integrated superorganism, and separation of the groups of organisms is interpreted as a division of a whole into parts, which are described in the terms of mereology. This view of Nature is inherent in Platonic eidology and in some versions of medieval natural philosophy based on it (Augustine, Boëthius, etc.); in Modern Times, it is most characteristic of organismic natural philosophy (Schelling, Oken, etc.). An evident instance of partonomic division at a macro level is geographic zoning, which is the division of the earth's surface into fragments of different levels of generality [Meyen and Shreyder 1976; Meyen 1977]. From this point of view, a cartographic representation of classifications of organisms [de Jussieu 1789; Giseke 1792; Strickland 1841] makes them not taxonomic, but partonomic [Chebanov 2007]. Hierarchical structure of biodiversity, if interpreted holistically in terms of phylogenetic pattern, may also be treated as partonomic [Eldredge and Cracraft 1980; Pavlinov 2015].

Fixation of species as a natural unit seems to be object-based, whereas all its principal conceptions are aspect-based [Pavlinov 2011]. Phylogenetic and ecological aspects of biodiversity are usually indicated as the most general ones [Eldredge and Salthe 1984; Schulze and Mooney 1994; Faith 2003]; they can be further split, and referring to them results into conceptual *species pluralism*, including recognition of different *kinds of species* (see Section 5.4). With some reservations, those conceptions can also be included here that differently account for the ontology of species, for example, as a natural kind or a quasi-individual (see Section 6.2).

The discrimination of different aspect-based species conceptions is not arbitrary to the degree that their totality reflects a fundamental property of the living nature, namely, a complex multidimensional nature of the biotic structure. The latter's multiaspectedness arises for two main reasons. One of them is "external": biota is structured differently by different physical and other causes that do not coincide in their mechanisms and effects, with each factor generating a specific aspect of the biota structure, and thus different aspects are not reducible to each other or to some more general and unified aspect. Another reason is "internal": when elaborating species conceptions, the emphasis is placed on different species integrators. In any case, if a certain conception reflects some biologically substantive aspect of the diversity of biota and is biologically meaningful in this sense, it can be regarded as realistic and having the right to exist [Mishler and Donoghue 1982; Kitcher 1984a,b; Dupré 1993, 1999; Sluys and Hazevoet 1999; Vasilyeva 2002; Wilkins 2003b; Pavlinov 2009a, 2013a, 2017, 2018, 2021a, 2022a].

The species conceptions based on the analysis of similarity as such are partly ontically and partly epistemically motivated: such qualification of them proceeds from a recognition that similarity relation is not an attribute of the objective reality [Tversky 1977; Sober 1984; Rieppel and Kearney 2002; Pavlinov 2018]. Of such "mixed" onto-epistemic character is also the aspect-based concurrent consideration of species in natural and classificatory aspects, which yields a *natural-classificatory dualism* of species.

The object-based are not so much the abovementioned species conceptions themselves as the spheres of their application. This means that in different groups of organisms with significantly different biological organizations, including the structure of their diversity, most relevant are different aspect-based conceptions (see Section 6.3). On the other hand, considered in such milieu may be the relations involved in the hierarchical structure of biodiversity: (a) between species and genus and (b) between species and infraspecific categories. The first option is relevant to medieval scholasticism and early scholastic systematics, which operate with a rankless sliding hierarchy and generates *genus-species uncertainty* (see Sections 2.1.3 and 3.6). The second option is relevant to the studies of species diversity in the first half of the 20th century, and it generates *ranking species uncertainty* within the hierarchy of "around-species" units [Bessey 1908; Greene 1910] (see Section 5.2.3).

The pragmatic species conceptions stand apart by being generated at the "cross" of subjective and epistemic components of the cognitive situation. This standpoint minimizes substantiation of the conceptions' content by reference to the structure of biodiversity as such and therefore, strictly speaking, they cannot be treated as either aspect- or object-based. Nevertheless, they also contribute to the shaping of species pluralism and, thus, to the general species problem.

* * *

The preceding consideration of how the general species notion can be thought of and how it is split into particular interpretations brings us close to the understanding of the main content of the *species problem.* The latter summarizes in a specific way an array of possible answers to the basic questions "what is species?" and "what is species concept?". Different researchers highlight in them different particular issues according to their cognitive interests. Some of them discuss the essential properties and ontology of species as a classification universal or a natural phenomenon; others consider mechanisms of its generation and sustainable existence; some are interested in the ways of individuating species as a structural unit of biodiversity; and there are those who deal with the concept-based or concept-free methods of distinguishing and classifying the particular species.

Thus, notwithstanding that there are different understandings of the species problem with respect to its content (see Section 5.4), it is reasonable to consider its origin due to a variety of particular treatments of the general species notion, which yields the abovementioned species uncertainty and is most usually designated as the abovementioned species pluralism. From a conceptualistic standpoint, particular conceptions generate their specific cognitive situations, and in each of the latter, there is no species problem for an obvious reason: any particular situation thus framed is, that is to say, "monoconceptual". So the problem in question is an attribute of a conceptual construct of a higher level of generality, viz., a cognitive *meta-situation* that somehow encompasses all the diversity of particular situations. With this, since the general reference system for particular conceptions is the general notion of species, which is closely related to the notion of species category, one can speculate that the species problem deals not so much with the individual conceptions themselves as with the species category that functions as a kind of "unifying agent" of the latter and a basis of their comparison and summation.

The species uncertainty in its accepted "one-dimensional" understanding boils down to acknowledging the existence of different species conceptions. But actually, this is only a part of the problem, and there are different reasons for its emergence and persistence. As a matter of fact, included here are: the worldview issues about the origin of natural species (creation or evolution); discussions of the ontology of species (real or nominal, natural kind or quasi-individual); biological issues concerning the structure (ranked or rankless) of species diversity or the nature of species integrators or the application of particular conceptions in different groups of organisms; finally, practical issues of interrelations of species classifications based on different conceptions [Mayr 1982; Ruse 1995; Sloan 2002; Stamos 2003; Wilkins 2003a, 2009a; Reydon 2005a,b; Richards 2010; Ellis 2011; Chambers 2012; Bartlett 2015; Pavlinov 2018]. Thus, the species problem is multidimensional in itself, and it can be split in various hypostases [de Queiroz 2005b; Ellis 2011; Ereshefsky 2017; Pavlinov 2018; Wilkins et al. 2022]. Therefore, it is permissible to talk not about one, but about several problems that rise from different understandings of both the general species notion and the problematic issues related to it, which collectively structure the cognitive meta-situation. Taking all this into account, the latter's core attribute is not a species problem in its narrow sense but a multidimensional and multilevel *species problematics*; so the latter term will be used throughout this book. At any rate, from a philosophical perspective, the recognition of species pluralism in its modern account (see Section 5.4) is significant by its obvious anti-positivist emphasis: it is equivalent to recognizing that the structure of biota is complex and multidimensional and does not allow its description by means of a single biologically meaningful "superconception".

Based on the above ideas about the three-component structure of the cognitive meta-situation, three main aspects can be distinguished in the species problematics.

When the latter is considered in an ontic vein, the principal issues refer to the basic question "what is species?", including an issue of its ontology: they shape the principal content of the entire species problematics throughout its conceptual history [Stamos 2003; Grant 2011; Kunz 2012; Bartlett 2015; Ereshefsky 2017; Pavlinov 2018]. Another issue, set by a contemporary conceptualistic view of the species problem, includes finding out whether there really exist particular aspects of the structure of biota that are described by the relevant species conceptions. In an epistemic guise, the basic question of the species problematics is "what is species concept?": how is it developed, how is its scientific validity evaluated, how are theoretical conceptions translated into operational ones, etc.

An interesting context for discussing the epistemic components of the species problematics is provided by a notion of *cognitive themata* suggested by the philosopher of science Gerald Holton [Holton 1973, 1996; Faust 1987]. According to it, this problematics can be structured in a specific manner by recognizing most fundamental themes ("themata") in it, which "serve as organizing constructs that aid in translating complex phenomena into a familiar and relatively simple" ideas [Faust 1987: 23]. In the case of the problematics in question, the most evident themata are provided by conjoint dyads or triads of antinomies in the treatment of species, i.e., its reality or nominality, its integration by essence or by generation chains, etc. Most of such themata become conscious at the very beginning of the conceptual history, and they persist with its development, just changing their philosophical milieu and priorities.

In a subjective aspect, the focus is on how subjective world views, cognitive intentions, etc., affect both the multiplication of particular treatments of species and the preferences given to them. Due to this, a "psychologism" of the species problem comes to the fore [Ghiselin 1966, 1987; Holsinger 1987; Hey 2001a,b; Amitani 2010; Richards 2010; Ellis 2011; Pavlinov 2022a]. On the one hand, theoreticians tend to see problems where there might be none, and they turn their discussions into a fascinating (for them) "mind game" [Ellis 2011]. On the other hand, pragmatic practitioners are not much different from theoreticians in this respect, as they make their contribution to the species problematics by suggesting their own "rules of game" in it [Seifert 2014; Stijn 2018]. An important part of this "psychologism" is constituted by a contradiction between personal inclinations to species monism or pluralism, i.e., between an aspiration to develop some unified understanding (definition) of species and an inability to achieve this end [Pavlinov 2017, 2021a, 2022a].

1.2 HISTORY ON A CONCEPTUALISTIC ACCOUNT

In the conceptual history of the species problem (or problematics), the main events are changes in the conceptual framework of the studies on species diversity. They concern the structure of the cognitive situation shaped by the species issues of various contents, levels of generality, scientific or pragmatic significance, etc. In our case, the general route of the history in question is shaped by changes of the ideas about the origin and basic features of biota, including how and why does it emerge and become structured, whether species (in whatever meaning) is a part of the biotic structure, and if it is, what are specific mechanisms of its integrity, etc. [Gross 1988; Hey 2001a,b; Pigliucci 2003; Stamos 2003; Pavlinov 2009a, 2013a, 2017, 2022a; Wilkins 2009a; Richards 2010, 2013; Grant 2011].

As emphasized in the previous section, the content of the species problem is determined not by itself but by the entire scientific-philosophical context of the cognitive situation. Accordingly, its conceptual history is governed by the historical dynamics of that context due to changes in the specific accents that are placed on certain cognitive tasks by the scientific-philosophical community. Therefore, in the reconstruction of the conceptual history of the multifaceted species problematics, its historical changes can be correctly comprehended only in the context of the historical dynamics of its scientific-philosophical milieu. The latter's "governance" is most evident through its impact on each of the three basic components of the cognitive situation and their interdependence.

The historical dynamics of the ontic component is caused basically by the changes in the world pictures. These are most clear-cut in transitions from antique natural philosophy to biblical one and from the latter to the natural science world picture. This involves changes in the understanding of what Nature is, what are its structure and the causes for this structure, and in our case, whether there is anything in Nature that we call "species". The dynamics of the epistemic component includes changes in the ways of describing the structure of Nature, which are framed as cognitive and research programs. In their development, *rationalization* of the cognitive activity is certainly its primary factor, which takes different forms at different stages of the conceptual history. This rationalization includes, before all, *conceptualization* of the general species notion, which involves the development of conceptions as a specific means for describing various manifestations of the structure of Nature. It emerges in the Antiquity, and in its postantique development, the first most important is the appearance of *classificatory program* based on the genus-species logical scheme, and the next becomes its subordination to the biologically sound tasks, which means *biologization* of the species problematics. In the development of the subjective component, the main events involve the changes in motivations of the relation of humans to Nature affecting both other components. The first is the transition from consuming to cognitive intentions; the next are the changes in the cognitive dominants from a comprehension of the laws of Nature to a "positive" description of facts, and afterwards again to the laws of Nature.

Regarding conceptual history of the species problem in the context of evolutionary epistemology mentioned in the Introduction, two main trends are quite naturally identified in it, *integration* (*unification*) and *diversification* [Pavlinov 2018, 2021b]. The first is set by the persistence of the disciplinary matrix shaped by an idea of species as a certain unit of the structure of Nature or of its mental representation. The second is set by diverse detailing of this idea, which eventually yields the development of particular species conceptions. The whole process can be assumed ordered to the extent that the changes in the general world picture are ordered, proceeding from the mythological through the natural-philosophical to the natural-scientific ideas.

An integration trend can be likened to a directed aspect of biological evolution usually called *anagenesis*: it involves consequential, more or less orderly changes in the general understanding of species as a specific cognizable phenomenon. It involves the latter's conceptualization, and as far as biological issues are concerned, also its biologization. A diversification trend can be likened to *cladogenesis*, and it involves the development of different species conceptions that yields gradual patterning and complication of the species problematics. The different particular lineages of the historical development of the diversified species notion "cross" within the framework of the general integration trend, which may be likened to *reticular evolution*. Various superimpositions of these trends and their impacts on each other provide a complex integral picture of the conceptual historical development of the problematics in question.

One of the important features of the integration trend is an *inertia* of the conceptual history, which is generally inherent in the long-term dynamics of any complex nonequilibrium systems, including cognitive ones [Rozov 2002; Rozova 2014]. This inertia manifests itself in a *historical continuity*, which means that the same or essentially similar fundamental themata of the species problem (such as reality *vs*. nominality) continue to be discussed at different stages of its conceptual history. Some of their treatments grow directly out of the preceding ones, either supporting or rejecting them. Others arise as if *de novo* (i.e., without explicit references to their predecessors), but, in fact, they may represent manifestations of the *iterative evolution*. This means that they obey some general "logic" of the conceptual history of the species problematics, which necessarily invokes similar ideas at different phases of its development. Such iterative pattern is exemplified by the generative idea, which repeatedly appears in various guises in the course of the considered conceptual history.

* * *

Representation of the conceptual history of the species problem as a narration is more or less subject-motivated. The latter's effect is most conspicuous in the placement of accents in this narration depending on the commitment of historiographers to one of two basic cognitive standpoints, *presentism* and *antiquarism* [Foucault 1970; Rozov 2002; Kuznetsova 2009; Pavlinov 2018]. In the former, a certain concept that arose in the past is discussed in the context of the current state of the respective cognitive situation, i.e., through a prism of the problems and tasks existing in it today. In the latter, the concept is regarded in a scientific-philosophical context that existed at the time of its origin and, generally speaking, gave rise to it.

For example, presentism obliges contemporary biologists, devoted to an evolutionary idea, to evaluate the early history of the species problematics (from Antiquity to the Middle Ages) in terms of their contribution to the initial development of that idea.[6] However, such presentism bears a hidden sign of a teleonomic interpretation of history, which is allegedly directed by a certain final cause [Gorman 1992], which in our case is nowadays dominating understanding of species. Accordingly, only those conceptual constructs that correspond to the "how it should be" from an evolutionary standpoint are considered worthy of attention, while the rest is discarded as worthy of oblivion. From the point of view of antiquarism, such standpoint is hardly correct: in fact, the philosophers of that time, for the most part, were not concerned with any evolutionary idea in its contemporary sense and elaborated the respective cognitive situation based on quite different ontic premises. Therefore, what they denoted as species, including understanding of its content and causes of its origin and persistence, though formally coinciding with the modern one, have little in common semantically with the latter [Pavlinov 2018; Brower 2021; Minelli 2022].

It is obvious that presentism inevitably occurs to a certain extent in every historical reconstruction of a concept that is at the center of a narrator's attention. Indeed, this book examines the history of ideas about biospecies, according to which the ideas of the antique and scholastic thinkers are interesting to its author to the extent that they relate to that notion. However, this does not mean that these ideas should be necessarily evaluated as "progressive" *vs.* "regressive" or "obsolete" with respect to what contemporary biologists think of species as a natural phenomenon (and they think of it so differently!). In fact, the significance of the antiquarism position, from a conceptualistic viewpoint, is that it allows one to judge with minimal (as far as possible) bias how the philosophers of the former times, followed by naturalists, understood the causes and structure of the world they intended to comprehend. This means a kind of "immersion

[6] One of the curious examples of this kind is a reproach by E. Mayr to antique and medieval thinkers that they were essentialists ("typologists") instead of being evolutionists [Greene 1992; Powers 2013].

in the past", which serves as an indispensable precondition for an understanding of the contextual causes of arise and development of what was then and is now commonly called "species".[7] And it is the careful "antiquarist" tracing of the history in question that makes it possible to understand how and why the ways of conceptualizing the general species notion were paved and changed, and the corresponding cognitive situation transformed accordingly at particular stages of its conceptual history.

With this regard, one of the primary preconditions of proper reconstructing the conceptual history of the species problem is to follow Aurelius Augustinus in his appeal to "believe in order to understand" (*credo ut intelligam*): in our case, to believe in what the early creators of the ideas about species believed in letting the contemporary concepts eventually grew from them. And this is not an easy task, because, when exploring conceptual history, contemporary natural scientists most usually analyze not primary sources in their original versions but their subsequent translations and comments. However, the latter's authors and their nowadays readers, especially if they are guided by presentism, "see" in the texts what they anticipate to behold corresponding to their understanding of the respective topic [Connell 2001]. And if we take into account what were historical intricacies that the works of antique and early medieval thinkers went through to have become available to the contemporary researchers [Maurer 1982; Koyré 1973; Marenbon 1988; Copleston 1993a,b; Gilson 1991, 2010], it remains only to wonder that we pretend to think we understand the true thought movements of great teachers of the past. Indeed, a deficiency of this understanding is evidenced by modern literature that clarifies or even rethinks previous interpretations of the antique sources marking the beginning of conceptual history explored by us [Boylan 1983; Gotthelf and Lennox 1987; Kullmann and Föllinger 1997; Lennox 2001; Berti 2016].

1.3 A HISTORICAL NARRATIVE OF THE SPECIES PROBLEM

There is an opinion that the species problem arises in the second half of the 19th century stimulated by the development of an evolutionary idea (e.g., [Hey 2001a; Pigliucci 2003; Wilkins 2009a]). But if we imagine the structure of this problem as an array of the abovementioned Holton's themata fragmenting it in a special way, then it becomes clear that it has a very long history rooted in Antiquity. This makes the question of its *persistence* an important part of the entire species problematics: why does it continue throughout the centuries? [Reydon 2004, 2005a; Amitani 2010]. So, in general, its conceptual history looks like a persistence added with a certain dynamics due to changes in the emphases on different themata.

In antique natural philosophy and partly in medieval scholasticism, the main content of the species problematics is determined by partonomic-taxonomic dualism, according to which species is accounted context—dependently as either a parton (a property of things) or a taxon (a set of things). However, due to acknowledging their

7 That is why the verbs using in a narration of the conceptual history of the species problem are applied in this book in present tense: this creates an effect of "presence here and now" in the epochs being described.

ontic unity by the then philosophers, this dualism is not perceived problematic; it turns out to be so in the eyes of the subsequent researchers who disjoint the intensional and extensional aspects of considering any natural phenomenon [Cassirer 1950; Carnap 1969; Pellegrin 1990; Shuman 2001].

In the Middle Ages, a contradiction between realistic, nominalistic, and conceptualistic accounts of species turns out to be the most significant: species (in whatever guise) either exists objectively, or is a product of human mental activity, or it is individualized against a totality of various manifestations of the manifold diversity of things by means of certain conceptions. In addition, an idea of a cognitive chain consisting of particular forms-species makes a certain contribution to the vagueness of species treatments by medieval scholastics (see introduction to Chapter 3). A realistic-nominalistic opposition gets embodied in natural-classificatory dualism, which becomes especially noticeable in postscholastic systematics since the mid-18th century.

For the biologists of the 19th century, the main part of the problem becomes a question of the origin of biological species, interpreted now strictly taxonomically: it is either eternal as an embodiment of Divine archetype or arises "locally" as a result of gradual evolution. Acceptance of the latter in a Darwinian version yields an important question of the hierarchical structure of natural species, including interrelations of population units of various levels of generality.

Finally, in the 20th century, the multiplicity of particular species conceptions becomes the principal part of the problem, as they differently explain biological mechanisms responsible for both species stability and variability. It is in this account that the species problem becomes realized and designated in the first half of the 20th century [Robson 1928; Hawkins 1935; Faegri 1937; Turrill 1938] and then actively discussed in biological and biological-philosophical literature [Mayr 1957b, 1982; Ruse 1995; Wilson 1995; Ghiselin 1997; Wilkins 2009a; Richards 2010; Sloan 2013; Ereshefsky 2017].

As noted above, three main stages are traditionally distinguished in the conceptual history under consideration, which are marked by the key figures of Aristotle, Linnaeus, and Darwin. The philosopher of biology John Wilkins, actively studying the species issues, designates these stages as *premodern*, *early modern,* and *modern* and "coarsens" them eventually to two, *prebiological* and *postbiological* [Wilkins 2003a, 2009b]; I would rather call the latter *biological.* The importance of this two-phase division is emphasizing a nonbiological meaning of the term "species" to designate scholastic *universalia,* which is typical for the studies up to the 16th century, and in some philosophical treatises even later [Wilkins 2009a; Richards 2010; Pavlinov 2018].

From a scientific-philosophical standpoint adopted in our book, three other macrostages can be distinguished in the conceptual history of the species problematics that largely coincide with the history of basic cognitive programs shaping comprehension of the structure of diversity of the living: *prerational*, *prebiological rational*, and *biological rational* [Pavlinov and Lyubarsky 2011; Pavlinov 2018, 2021b]. In a rationally arranged history, transitions from one stage to another are marked by the local scientific revolutions directing its anagenetic trend, which are Cesalpinean, Tournefortian–Linnaean, Buffonian, Darwinian, etc. At each of these stages, a diversification trend is evident in the appearance of particular interpretations of species. In Antiquity, it is caused by the divergence of Platonic and Aristotelian worldviews; in the Middle Ages, it is shaped by realistic, conceptualistic, and nominalistic accounts of universals, and recently, this trend is associated primarily with the emphasis on different species integrators.

* * *

Human cognition of Nature is rooted in *folk systematics* that belongs to the early prehistory of the comprehension of the diversity of life. It involves the distinction of certain units in this diversity based on a kind of intuitive "classification instinct" [Foucault 1970; Atran 1990]. One of these units, distinguished at a lower level of this diversity, is identified with species by modern researchers, who treat it mostly taxonomically. However, this very initial phase of the entire "species story" is by no means conceptualistic, as the indigenous people produce their folk classifications on an intuitive background. Therefore, it is not considered in our book; its reviews can be found elsewhere (e.g., [Atran 1990, 1998; Berlin 1992; Pavlinov 2013c, 2018, 2021b]).

The subsequent development of the ideas about species is associated with the rationalization of cognitive activity by antique thinkers [Asmus 1976; Russell 1992]. In their onto-epistemology, the conjoint notions of *genos* and *eidos* are given the status of universals and become of prime importance. The basic issues include clarification of their existence outside and before (Platonic eidology) or in the things (Aristotelian ousiology). Their partonomic (form, essence of things) and taxonomic (grouping of things) accounts are intimately interrelated, with the latter conceived primary with respect to the former, which yields a partonomic-taxonomic dualism. The late antique Neoplatonists (Porphyry, etc.) and the earliest medieval scholastics (Augustine, Boëthius, etc.) rationalize this onto-epistemology in a way resulting in a specific combination of Platonic and Aristotelian views on the genus-species interrelations.

In the development of cognitive situation by medieval scholasticism, a contradiction between realistic and nominalistic accounts of genus and species as universals takes a focal place. Along with this, elaboration of the *genus-species scheme* of division of notions with sliding hierarchy of these basic universals (Albert the Great, Pierre Abelard, Thomas Aquinas, etc.) becomes of paramount importance. Because of genus-species interchangeability inhered to it, the *genus-species problem* emerges with a *genus-species uncertainty* as its significant ingredient. An interesting conception of *species multiplication* grows as a specific "dead-end" offshoot of scholastic epistemology (Aquinas).

The premises of postscholastic epoch are associated with the formation of a classificatory cognitive program, according to which classification is treated as the principal form of representation of the qualitative structure of Nature [Rozova 1980, 1986]. On this basis, renaissance herbalism (Gesner, Bauhin, etc.) first develops in the 15th–16th centuries, followed by early scholastic systematics (Cesalpino, Jungius, etc.). In them, the notions of genus and species still retain their predominantly logical meaning of conjoint *classification* units, and the diversity of all three "kingdoms of Nature", viz., animals, plants, and minerals, is described based on the same unified genus-species scheme (Linnaeus). At the same time, these units become understood principally as taxa (aggregates of things), which results in the elimination of genus-species uncertainty and *taxonomization* of the genus-species problematics. Because of this, partonomic-taxonomic dualism declines and *natural-classificatory dualism* emerges (G. Leibniz) as a specific reflection of the realism-nominalism dilemma.

In a more advanced scholastic systematics, an important shift toward contemporary understanding of species becomes noticeable, which involves substantiation of its real (objective) status by reference to the biblical world picture (Ray, Linnaeus). This serves a prerequisite for (a) reinforcement of the understanding of species as a predominantly *natural* unit and (b) "decomposition" of genus-species relativity by introducing fixed taxonomic ranks, one of which is taken by species (Tournefort, Linnaeus). All of this results in an *autonomization* of species as both natural and cognizable objects, so the previous genus-species problematics becomes a more particular *species* one. A predominating one-trait (essentialist) account of species is replaced with multitrait one, which gives born to a *phenetic idea*. With G. Leibniz's and G.-L. Buffon's explicit accentuating on the generative unity as one of the substantial (along with similarity) criteria of species, the basis for maturation of the latter's understanding as a *biological* unit is laid [Zavadsky 1961, 1968; Mayr 1969; Wilkins 2003a, 2009a; Zachos 2016].[8]

Notwithstanding this rather new content, a conceptualist core of the species problem remains largely the same in being reduced basically to the dilemma of its reality *vs.* nominality. These traditional debates are supplemented with the dilemma of discreteness *vs.* continuity of the structure of Nature, which is characteristic of the natural philosophy of the early Modern Times (Locke, Leibniz, etc.) [Koyré 1985; Gaydenko 2000]. Accordingly, one of the main issues involves a possibility of drawing clear-cut natural boundaries between the particular species. This dilemma becomes reinforced in the 18th century due to controversy between the hierarchical *System of Nature* and linear *Ladder of Nature* [Lovejoy 1959; Wilkins 2003a, 2009a; Pavlinov 2018]. Their confrontation is subsequently overcome by the loss of interest in "Ladder" natural philosophy due to wide acknowledgement of the general idea of Nature as a complexly structured and hierarchically arranged system. An important role in this conceptual shift is played by the organismic natural philosophy of the first half of the 19th century (Schelling, Oken, etc.).

* * *

An ultimate awareness of a fundamental difference between living and inanimate nature at the turn of the 18th–19th centuries transforms both the species notion and the entire problem from basically logical (classification) to biological [Wilkins 2003a, 2009a; Pavlinov 2018, 2021b].[9] In it, the generative idea is fixed as the basic one, in which species unity and discreteness are justified by a continuity of the chains of generations (Buffon, etc.). This means a significant advance towards biologization of the general notion of biospecies as a distinct natural unit inherent only in living matter, and it marks a significant move from scholastic to postscholastic systematics.

---

[8] These authors indicate J. Ray as a founder of the "generative" understanding of species; the rationale for our treatment is set out in Section 5.1.2.

[9] As a matter of historical fact, this awareness is traced back to antique and biblical natural philosophies (see Section 5.1).

In the second half of the 19th century, a fundamentally new conceptualization of the species problem in a biologically meaningful fashion yields a significant change in the structure and dimensionality of the respective cognitive situation. One of its principal novelties is brought by the microevolutionary model of Darwin, in which species is treated as an *evolutionary* unit, i.e., as both a participant and a result of the continuous divergent evolutionary process. At the beginning of the 20th century, an ecological idea is added, which treats species as an *ecological* unit being a structural element of the natural communities [Turesson 1922].

An idea of the gradual nature of evolution with intraspecific variation as its prerequisite yields a comprehension of species as but a final stage of differentiation of intraspecific varieties (races, subspecies, etc.). A new taxonomic theory, *classification Darwinism*, rises based on this idea, in which "Linnaean" species loses its special salient status. However, this taxonomic school still remains bounded by the tradition of "discrete" description of the diversity of organisms, so it faces a peculiar issue of distinguishing and ranking of the species and the intraspecific forms. Its consideration gives rise to a new wave of species nominalism and to the so-called *species question* [Johnson 1908; Greene 1910], with a *ranking species uncertainty* being its focal point.

The situation changes closer to the mid-20th century due to the development of the *synthetic theory of evolution* (STE): in it, speciation, i.e., formation of new species, is interpreted as a nodal event of the evolutionary process. This leads to a revival of the salient status of species treated realistically as a natural unit taking a fixed position in the hierarchy of biota. In this evolutionary theory, the main emphasis is made on the interpretation of species as a *syngameon* [Lotsy 1931], and *reproductive* (originally called *biological* in a narrow sense) conception is developed on its basis [Dobzhansky 1935; Mayr 1942] to become the first explicitly recognized species conception in the history of the species problematics.

With the development of the cognitive situation in such way, it becomes clear that since the particular groups of organisms possess different reproduction systems, this conception has limited relevance. As a result, an idea of recognition of different *kinds of species* (narrowly treated biospecies, ecospecies, genospecies, quasispecies, etc.) emerges, with fixing each kind by a corresponding conception [Du Rietz 1930; Dobzhansky 1935; Huxley 1942; Mayr 1942]. This "multiplication of essences" yields *species pluralism* as a significant part of the species uncertainty and awareness of impossibility of the latter to be generalized by some unified concept gives rise to the contemporary *species problem* [Robson 1928; Hawkins 1935; Faegri 1937].

In the mid-20th century, an infiltration of positivist philosophy of science in biology leads to a strengthening of nominalistic account of species. This notion is declared a "relic of the past" [Burma 1954; Sokal and Sneath 1963], and the respective cognitive situation is suggested to reshape based on its predominately pragmatic account. Accordingly, various empirical conceptions are proposed subsequently to solve purely practical classification tasks (e.g., [Sneath and Sokal 1973; Pleijel and Rouse 2000; Seifert 2014]).

Towards the end of the 20th century, another attack on the species notion is undertaken by a cladistic systematics based on an idea of the rankless phylogenetic pattern generated by a continuous process of phylogeny [Eldredge and Cracraft 1980; de Queiroz and Gauthier 1992]. As a result, species category is once again deprived of its distinguished (salient) status, and with it indirectly species unit as well [Zachos 2016; Mishler 2021].

At the end of the 20th century, an active discussion of the plurality of particular species conceptions, increased in their number, makes a contradiction between species monism and pluralism an essential part of the species problem. Attempts to streamline some way this plurality give rise to an idea of their hierarchical arrangement formalized as a *hierarchical pluralism* [Mayden 1997; Richards 2010; Pavlinov 2013a, 2022a; Sloan 2013]. With this, a kind of apotheosis of the general trend of biologization in the comprehension of species in the conext of "new essentialism" becomes an idea of its inherent quiddity called *specieshood*, which is part of species natural history, evolves with it and is specific to different groups of organisms [Pavlinov 1992, 2009a, 2013a, 2022a; Griffiths 1999; LaPorte 2017].

2 Antique Natural Philosophy

Species as Eidos

The important initial elements of rationality, forming a conceptual framework of cognitive activity in modern science, are developed during Antiquity. This rationality involves the elaboration of the principles of operations with generalizing notions to denote not sensually perceived objects (humans, animals, plants, etc.), but certain unobservable intelligible essences (humanity, animality, vegetability, etc.). Striving for this becomes an integral part of the development of general ideas about the causes and structure of Nature and the ways to comprehend and describe it in the most general fashion. Regarding living nature, one of the manifestations of this structure is species diversity; accordingly, the contemporary ideas about species as one of the basic elements of biotic structure are rooted in the antique onto-epistemology.

The latter's conceptual history is traditionally divided into three main periods—early (the seventh to the sixth centuries BC), classical (the fifth century BC to the second century AD), and postclassical (the third to sixth century AD). The first one does not especially concern us here. The second stage (mainly Plato and Aristotle) involves the formation of basic concepts related to the species problematics. During the third stage, these concepts are formalized and detailed by Neoplatonism.

2.1 CLASSICAL PERIOD: THE BIRTH OF *GENOS* AND *EIDOS*

To begin with, it is worthwhile at least mentioning, in brief, the early period, which is very rich in philosophers (Thales, Pythagoras, Parmenides, Democritus, Empedocles, Anaxagoras, etc.) and initial ideas about the structure and causes of the universe and the ways of its comprehension [Asmus,1976; Losev and Takho-Godi 1993; French 1994a]. In particular, the notion of έίδοσ is developed by them to denote a certain exterior appearance (habitus, look) of things: it is this notion that will become that of *species*, as it will be understood by later philosophers and naturalists.

The classical Antiquity is most important to us for its natural-philosophical world pictures developed in the works of two great philosophers, Plato and Aristotle. They are based on a very rich understanding of Nature as *Physis* (φύσις), which means both an active root cause of everything that exist and the existing everything themselves in their integrity [Akhutin 1988; Losev and Takho-Godi 1993; French 1994a; Gaydenko 2003; Chiaradonna and Galluzzo 2013]. Its fundamental attributes are designated by very polysemic notions, which particular meanings are set in the antique treatises contextually. An extensive literature is devoted to their discussion, and the points of view on this subject are very diverse, as well as their translations by the authors committed to different natural-philosophical views [Balme 1962; Woods 1993; Lennox 2001, 2006; Orlov 2011]. Therefore, to avoid their undesirable

DOI: 10.1201/9781003346296-3

incorrect identifications with their widely adopted modern derivative notions, those most important to us will be given below not in their traditional translations, but in simple transliterations: γένοσ = *genos* (plural *genê*) usually translated as *genus* or *kind*, έίδοσ = *eidos* (plural *eidê*) usually translated as *idea* or *forma* or *species*, and οὐσία = *ousia* (plural *ousiai*) usually meaning *essence* or *substance*; besides, more two will be given in the traditional terminology: ὕλη = *matter* and μορφή = *form* (usually in a metaphysical sense).

2.1.1 Plato's Eidology

The core of Plato's (429–347 BC) world picture, most consistently presented in his *Sophist*, is a general idea of *The One*, or *Theos*, as the beginning and the root cause of everything, which contains the whole plenum of being. The transcendent *One* incarnates into the universe and transforms it into an ordered Cosmos through the successive emanation of the eternal and unchanging *eidê* contained in it [Losev and Takho-Godi 1993; Orlov 2011]. An initial emanation of the *One* generates *eidê* of the highest orders (ideas), which are *megista genê* (greatest genera); they emanate lower-order *eidê* (forms), which generate far more particular ones and thus become *genê* (genera) in relation to them. The last steps of emanation are *ousiai* (essences), which interact with an unstructured *matter*, and concrete material (sensible) things appear due to this, with *ousiai* incarnated in them to make them what they are by their nature.[1] Thus, Plato's world picture is basically *eidology*, as it is accentuated on *eidê*, or *genology*, as it is accentuated on *genê*; it is also called a "theory of ideas" or a "theory of forms" [Losev and Takho-Godi 1993; Gaydenko 2008; Chiaradonna and Galluzzo 2013]. In contemporary literature, the identification of Platonic *eidê* with *ideas* serves as a basis for designating his ontology as idealistic, and their identification with *essences* gives a reason to consider it the beginning of *essentialism*.

It is of interest to note the following point related to the species problem. According to Plato, not all sensible things can be regarded as embodiments of transcendent *eidê*: such, for example, are human artifacts (*techne*), which *ousiai* are rooted in people's intentions [Chiaradonna and Galluzzo 2013]. The domestic breeds of plants and animals brought to existence due to human activity should be allocated to this category. From this, C. Linnaeus' position will quite naturally follow, who will ontologically separate species from varieties in his *Philosophy of Botany*: the former are "work of Nature", the latter are "work of Art" [Linnaeus 1751] (see Section 4.3.3).

It is important to emphasize that the hierarchy of *eidê* of different levels of generality originated through the emanation of The One is most properly treated as a division of the whole into parts, in which the whole itself does not lose its integrity. Thus (in the modern terminology, see Section 1.1.2), this division is inherently partonomic,

---

1 This world picture is very similar to the fundamental ontological doctrine of Taoism, expressed by the aphorism "Tao gives birth to One, The One gives birth to Two, The Two gives birth to Three, The Three gives birth to all universal things" [Reninger 2017]. It is noteworthy that both Platonic and Taoist world pictures are formulated independently in the same epoch.

which makes Platonic eidology a *mereology*, i.e., the doctrine of the relation of parts (*meros*) to the whole and between themselves (see also on Aristotle below).

In Plato's cosmology, transcendent *eidê* and *ousiai* exist before and outside material things: their eternal and unchanging being belongs to *Physis* as an ideal world existing prior to the things, just as, say, an idea of a jug in a master's head exists prior to a ready jug and outside it. This analogy indicates that each eternal *eidos* is more fundamental and more "real" in a sense, as an element of *Physis*, as compared to the things, which are just its particular temporary embodiments. With this, the further a certain *eidos* "falls away" from The One and gets closer to its ultimate incarnation, the less completely and perfectly it embodies the One. Accordingly, human cognition should be directed not to particular things or lower *eidê* (*ousiai*), but to *Physis* as it encounters the eternal primary *eidê*, relying not on perceptions of things but rather on rational reasoning about the *eidê*. With this, the human cognition thus directed is to follow a deductive reasoning: it begins with the reflections about the *Physis* and its root causes and ends with the particular *ousiai*. For example, there is an eternal and unchangeable higher *eidos* "animality", which emanates lower *eidos* "mammality", and the latter's emanation produces *ousia* "horseness", which embodiments are the particular horses. Thus, for the naturalists to understand what are particular horses "by their nature", it is necessary to comprehend first what "animality" is as the cause of "horseness" [Kitts 1987].

Plato's eidology is developed by Middle Platonists, whose treaties provide an important groundwork for Neoplatonic philosophical tradition. The first of them is Plato's disciple Speusippus: he attempts to apply basic elements of Platonic "theory of ideas" to the elaboration of something like an initial "theory of classification" [Tarán 1981; Dillon 1996]. There are two treatises by Speusippus that are devoted to this theme, *Divisions and Hypotheses relating to the Resemblances* and *Essays on the Genera and Species of Examples* [Lang 1964; Tarán 1981]. Of them, only the first one is partially preserved, while the second is known by its mentioning in the book of the late Antique historian Diogenes Laërtius, which contains descriptions of lives and works of about 80 most prominent antique philosophers [Laërtius 1853]. Speusippus asserts it that a complete understanding of the causes of the universe can be achieved through a comprehension of the hierarchical structure of the diversity of *eidê* based on the reflections on mutual similarities and differences of things. On this account, Speusippus recognizes particular *genê* of plants and animals of different levels of generality, following a deductive dichotomous scheme and indicating their distinguishing features. Aristotle questions the eligibility of this scheme, but it is the latter that will be developed by scholastics and adopted by early systematics. Also noteworthy is Speusippus' successor Xenocrate, among whose works is a lost treatise with remarkable title *Of Genus and Species* [Dillon 1996]. All this indicates a great attention paid by Platonists to these two notions of our interest.

2.1.2 Aristotle's Ousiology

An outstanding written legacy of Aristotle (384–322 BC) includes several dozen volumes that become known after his death and will be repeatedly reprinted and commented subsequently, until now. In his world picture, *genos*, *eidos*, and *ousia* are given a fundamentally different existential (ontic) content in comparison with that of Plato's and his direct followers. In Aristotle's ontology, all these categories do not exist outside the things, but are contained in them, immanent to them, and in this sense may be thought of as the primary reified beings [Losev and Takho-Godi 1993; Politis 2004; Lennox 2006; Athanasopoulos 2010; Tipton 2014]. Of them, the most fundamental is *primary ousia*, which is treated as a *substantial being* [Lennox 2006; Tipton 2014]. These are the *primary ousiai* that are at the very beginning of *Physis*: they are causes of both themselves and everything else, so Plato's *genê* and *eidê* are stem from their being and secondary with respect to them. Accordingly, these Platonic categories are called the *secondary* and higher-order *ousiai* [Aristotle 1962]; therefore, Aristotle's ontology is *ousiology* [Marx 1977; Shatalkin 1996]. As Aristotle states in his *Categories*, "secondary substances [are] those within which, being species, the primary or first are included, and those within which, being genera, the species themselves are contained" [Aristotle 1962: 19]; according to our terminology, we should read in this quote *ousiai*, *eidê*, and *genê*.

Thus, an organizing principle of Aristotelian *Physis* is an ascending cascade of causalities, in which the lower-order essences are causes of the more general ones: in the above example, the first is "horseness", then goes "mammality", and then "animality". Nevertheless, there is a certain supreme creative principle of the universe in Aristotle's ontology, which is a transcendent *Prime Mover*, which place in the structure of the universe can be likened to that of Plato's The One [Merlan 1946].

Late antique and medieval commentators of Aristotle focus on his philosophical treatises (*Categories*, *Metaphysics*, *Topics*, etc.), while his zoological works (*On the Parts of Animals*, *On the History of Animals*, etc.) are mentioned cursorily as an application of his general philosophical ideas. The contemporary naturalists, beginning from the early 19th century, get most interested in his classificatory reflections [Zavadsky 1961, 1968; Mayr, 1969; Hey 2001a; Fürst von Lieven and Humar 2008; Pietsch 2012; Voultsiadou et al. 2017], creating thereby an impression that he and his closest disciples (like Theophrastus, see below) deal primarily with taxonomic classifications of living beings. However, the studies in Aristotle's biology during the last decades result in a quite different understanding of Aristotle's attention to the diversity of living nature: he accentuates basically on its partonomic aspect, and his *genê* and *eidê* refer primarily to *ousiai* (essences) of animals [Grene 1974b; Boylan 1983; Gilson 1984; Pellegrin 1986; Gotthelf and Lennox 1987; Deverreux and Pellegrin 1990; Kullmann and Föllinger 1997; Gotthelf 1999; Lennox 2001, 2017; Lennox and Bolton 2010; Tipton 2014; Falcon and Lefebvre 2018]. With this, his interests in the "applied" tasks of dividing animals into "genera" and "species", as they are predominantly understood today, are of the second or even third order for him relative to the analysis of their essences.

Aristotle follows Plato in adopting a realistic account of intelligible *genos* and *eidos*: for him, they are as real in the eternal ideal world as the particular things are real in the transitory material (sensible) world [Lloyd 1962; Sucker 1978]. Their reality is in that they exist outside (before and besides) the human cognitive activity: this constitutes their radical difference from general notions and concepts that arise and exist in a cognizing mind as the mental images of *ousiai*. In accordance with the ontic *principle of perfection* [Lovejoy 1959], everything existing in the ideal world is arranged into an overall hierarchy of *genê* and *eidê* and is exhausted by it; in other words, there can be nothing in the ideal world that would be out of a certain *genos* or *eidos*. As Aristotle emphasizes in his *Metaphysics*, "the essence of being will not be in anything else that is not an *eidos* of a *genos*, but only in them alone" (cited after [Shatalkin 2012: 186]). This ontic tenet will be embodied subsequently in the scholastic logical scheme of the division of notions, which will be later on inherited by classical systematics.

It is of importance to emphasize that the *ousia* (essence), as a reified being of *eidos*, is understood by Aristotle functionally and teleologically, i.e., as a goal or a purpose of a material thing in which it is embodied, so that the thing exists in *Physis* in order to commit a goal or purpose presumed by the respective *ousia* [Balme 1962, 1980; Lennox 1980; Gaydenko and Petrov 2002; Politis 2004; Gotthelf 2012; Tipton 2014]. In the living nature, the thing is an organism or its certain part (organ) or a feature (physiological, behavioral, etc.), which is designed for the implementation of its vital functions. The harmonious union of different parts and features as embodied *primary ousiai* makes an organism a single integrated whole. To explain this organismal integrity providing for its vitality, Aristotle, following his earlier antique predecessors, endows an organism with a *soul* (ψυχή, *psychē*) [Gaydenko and Petrov 2002; Polansky 2007]: he believes "the soul is a cause and a governing principle of the living body. [...] the soul is the cause as that for the sake of which" [Aristotle 2010: 55]. According to this, Aristotle metaphorically calls an organism the *greatest genos* to emphasize its integrity as a single whole of interrelated *primary ousiai* reified in its particular functional parts and features [Lennox 2009; Gotthelf 2012].[2] From this point of view, Aristotelian ousiology, as well as Platonic eidology, can be considered a mereology, i.e., the doctrine of the whole and its parts [Pellegrin 1990].

According to Aristotle, everything potentially and really existing in the universe is a result of the interactions between *primary matter* and *eidê*, with the letter being treated as *form* [Gotthelf and Lennox 1987; Carr 1997; Gaydenko and Petrov 2002; Tipton 2014; Berti 2016]. Aristotle is sure that "if neither matter nor form is eternal, then there will be nothing at all" (cited after [Madigan 1994: 86]). The *eidê* belong to the world of Platonic ideas and are active principles, while the primary matter is unstructured being with an infinite amount of potentialities, and with respect to the *eidê*, it acts as a single universal *genos*: this is because the primary matter is thought of as a latent "generative being" which interacts with the multiple active *eidê* [Carr 1997; Heider 2014]. Thus, Aristotelian ontology of *genos* and *eidos* differs fundamentally from Plato's: according to the latter, they are interchangeable and differ just in their relative position in the overall sliding hierarchy of "ideas", whereas according to Aristotle, they are identified with significantly different attributes of *Physis*, viz., with the passive primary matter and the active form, respectively [Rudebusch

[2] This explains the great importance that early systematicians of the 16th–17th centuries (Cesalpino, etc.) will ascribe to the identification of the "main essences" of organisms, as the embodiments of their souls, in order to arrange them in the Natural System [Pavlinov 2018; see Section 4.3).

and Muniz 2020]. With this, the matter-*genos*, being infinite and potentially all-encompassing, is more general in relation to particular finite *eidê*; this understanding will later become characteristic of Neoplatonists and many scholastics. Aristotle's *genos* and *eidê*, as the ontic root causes of actual existence of the things, are fundamentally independent of each other, in the sense that no one of them is a cause of another. However, they become inextricably tied due to their interactions: none of them has an independent meaning as a cause of the material being, but only their conjugation leads to the appearance of things, each endowed with its own particular quiddity, its *whatness* (τί ἐστι). To formalize this general idea, Aristotle develops the concept of *hylemorphism*: everything that actually exists, including living organisms, appears as a result of joining *eidos* as an active principle with *genos* as a passive principle.

* * *

In his epistemic system outlined in *Categories*, Aristotle provides *genos* and *eidos* with quite different meanings as compared to their ontic treatment, though there is a certain correlation between them [Aristotle 1962]. Epistemically treated, they become *predicates*, i.e., they predicate (tell) about something, with the latter becoming *subject of predication*, i.e., of being what is predicated about [Balme 1962; Marx 1977; Pellegrin 1986, 1990; Gotthelf and Lennox 1987; Evangeliou 1996; Falcon 1997; Orlov 2011]. In such account, *genos* is still regarded as something more general in relation to *eidos*: the *genos* is what is predicated in essence about its *eidê*. This means that judgments about *genos* are primary relative to those about *eidos*, so in any cognitive scheme, it is necessary to cognize first the qualitative certainty of a *genos*, which is the high-order *secondary ousia*, and then to consider how it differentiates quantitatively into lower-order *secondary ousiai*. The last step of this scheme corresponds to the lowest predicates, which are not further divided and therefore manifests themselves as *indivisible* (ἄτομοη) *eidê* identified with *primary ousia*. This scheme is applicable to any intelligible objects: for example, Aristotle in his treatise *On the Soul* examines in this way a question of what is admissible to consider the greatest *genos* for the soul and whether the latter's specific realizations differ by *genos* or by *eidos* [Aristotle 2010].

Thus, the Aristotle's deductive cognitive scheme represents a fixed conceptual system of notions, including *genos* and *eidos* as predicates tied by clearly fixed logical relations. In substantiating this scheme, Aristotle proceeds from an idea that rational cognition is focused on the general (*eidê*, secondary *ousiai*), and the perceptions of particular things embodying the first *ousiai* serve as a kind of "row data" for comprehending that general (universal) that causes them. As a result, in Aristotle's onto-epistemology, two mutually corresponding "counter" hierarchies coexist, reflecting causal relations between its basic elements: the ontic (natural) hierarchy is ascending and corresponds to the hierarchy of *ousiai* and *eidê* as such, while epistemic (logical) hierarchy is descending and corresponds to the hierarchy of *genê* and *eidê* as predicates being instruments of cognition [Lyubarsky 2018].

Discussing in the *Topics* subordination of *genê* and *eidê* in this scheme, Aristotle establishes an important difference between them: the former are characterized qualitatively and denote general notions, while the latter are characterized quantitatively and refer to particular substances [Sloan 1970; Sucker 1978; Pellegrin 1986; Falcon 1997; Orlov 2011; Shatalkin 2012; Witt 2021]. According to this, they are different cognitive tools: in the analysis of the diversity of organisms, consideration of the *genos* focuses on qualitative *communality* (συναγωγή), while consideration of the *eidos* highlights quantitative *difference* (διαίρεσις) [Balme 1962; Grene 1974b; Shatalkin 2012]. A totality of the latter is implicitly present in the whole, which means understanding a *genos* as a unity of pluralities of *eidê*. Accordingly, to the extent that the latter represent some quantitative variations of a qualitatively characterized *genos*, they do not have an independent meaning; they always exist not by themselves but as "*eidê* of a certain *genos*" [Balme 1980; Gotthelf and Lennox 1987].

This understanding of interrelations between *genos* and *eidos* as predicates entails several important corollaries. First, the whole deductive scheme of argumentation, based on the Aristotelian categories, turns out to be (in the terminology adopted here) *genus-species* scheme: "genus" and "species" are regarded in it not separately, but conjugately. Second, this conjugacy makes the problem of our interest not species per se, but rather an integrated *genus-species problem*. Lastly, this scheme is a sequential division of the same generic common, determined at the very beginning of the cognitive process: thus, this scheme contains a germ of an idea of the *single basis of division*. All these basics of the Aristotelian *Organon* will be developed by scholasticism and inherited by early systematics.

In the future, the notions of *genos* and *eidos*, treated onto-epistemically, will be generalized by the fundamental notion of *universal* as, according to one of the most popular accounts, an intelligible essence, or form. It will become one of the most actively discussed in the whole of European philosophy in general [Gilson 1955; Grene 1974a; Gaydenko and Smirnov 1989; Motroshilova 2000] and in connection with the foundations of the species problem in particular [Sloan 1970; Wilkins 2003a, 2009a,b; Richards 2010; Shatalkin 2012].

2.1.3 Aristotelian Polysemy: The Problem Arises

The Aristotelian general understanding of *genos* and *eidos* encompasses several significantly different meanings, which can be designated as an *Aristotelian pluralism* [Bzovy 2016]. It is permissible to consider it an origin of the just-mentioned genus-species problem, which will develop throughout its centuries-old conceptual history, changing its accents, but generally preserving the general polysemic content that is laid by the philosophers of classical Antiquity.

From the point of view of the main subject of this book, this pluralism is of importance to the degree it involves species, so its principal appearance looks as a multifold genus-species uncertainty in its wide sense. It presumes two ways of paired considerations of the categories in question: in one of them, their partonomic and taxonomic accounts are combined, and in another, their natural and classificatory accounts are involved. This becomes the main content of the genus-species problematics in the antique onto-epistemology that will pass from it to scholastic philosophy. As noted above, contemporary researchers interpret *genos* ("genus") and *eidos* ("species") very narrowly, mostly in a taxonomic natural fashion, thus omitting a fundamental fuzzy multidimensionality of their understanding by Aristotle. Because of this, an "accepted view" of the initial development of ideas about these categories turns out to be largely distorted [Pratt 1982, 1984; Pellegrin 1990; Stamos 2003; Grene and Depew 2004; Wilkins 2009b; Pavlinov 2018].

Aristotelian partonomic account of *genos* and *eidos* means that they are treated intensionally, i.e., not as aggregates of things (in particular, organisms), but as their essences (forms in a philosophical sense, qualities, properties, etc.) [Lennox 1980; Pellegrin 1985; Gotthelf and Lennox 1987; Lansing 1995]. In a more "material" account, this consideration is focused on particular parts or properties (partons = merons) of things in which these essences are embodied. This emphasis makes it seemingly understandable why one of the main zoological works by Aristotle is entitled *On the Parts of Animals*, and not just *On Animals* [Aristotle 2001]: it deals primarily with *ousiai* that are embodied in the parts (properties in general) of the animated beings [Lennox 1980, 2001, 2006; Orlov 2011; Richards 2010]. Accordingly, the Aristotelian hierarchy of essences of different levels of generality, like Plato's ideas, is described in terms of partonomy or mereology [Pellegrin 1990].

> This thesis is of key importance for our version of the conceptual history of the species problem, and it should be emphasized because there is another point of view. Thus, the philosopher Ernst Cassirer suggests that Aristotle, when describing the diversity of animals, operates with the extensional logic of classes [Cassirer, 1950], and Jody Hey perhaps agrees with him, when writes that Plato and Aristotle "refer to categories, or kinds, of things" [Hey 2001a: 6].

Understanding *genê* and *eidê* as aggregates of things characterized by the corresponding essences means their taxonomic (extensional) consideration, which provides a taxonomic classification of their diversity. Accordingly, at the level of epistemology, these categories act as classificatory universals that allow presenting the general structure of the diversity of Nature in a hierarchical form, dividing the things (including organisms) into groups of different levels of generality [Lloyd 1961, 1981; Pellegrin 1985; Richards 2010; Tipton 2014; Berti 2016].

Such dual understanding of *genos* and *eidos*, which appear in the cognitive situation as both partons and taxa, can be designated as a *partonomic-taxonomic dualism*. For antique philosophers, there is no special problem in this dualism at all, since both these hypostases are closely interrelated causally. Indeed, they are but two of numerous different manifestations of an overall interdependence of everything with everything in a single complexly structured *Physis*. Their particular account, either partonomic (intensional) or taxonomic (extensional), is context-dependent, so one has to be aware of the respective context to comprehend an implied meaning of these categories properly. In such an obscure account, this dualism will play a significant role in the medieval philosophy, in which partonomically accounted "species" (as a universal) is regarded as a form. Subsequently, it will penetrate early (scholastic) systematics, while in the latter's postscholastic version, it will be undermined by an adoption of the taxonomic treatment of species.

However, for those philosophical analysts who strive for an unambiguous semantic fixation of the notions, this specific dualism will become an important part of the genus-species problematics and will sometimes yield remarkable conceptual suppositions. For instance, from a perspective of the contemporary exclusively taxonomical treatment of species, it is permissible to assume that Aristotle builds a kind of "zoology without species" [Pellegrin 1985], though certainly not in the sense of the contemporary advocates of a "speciesless biology" (see Section 5.4). An attempt to resolve this dualism in a mixed antique-modern fashion provides a *reductionist* species conception: essentially defined taxa as supra-organismal aggregates are possible due to an intraorganismal existence of essences as partons [Sloan 1970].

Within the framework of an analytical cognitive intention implying a separation of these two considerations of *eidos*, it is possible to specify a fairly clear criterion for their distinguishing in the works of Aristotle and his followers. A partonomic account implies that *eidê* are *contained in* the things as their essences (substances). A taxonomic account presumes that the things *belong to eidê* according to the commonality of their essences (substances) contained in them. In addition to this, the philosopher David Mowbray Balme (1912–1989), keeping in mind the Aristotelian teleonomy, points out the following important difference: Aristotelian *eidos* appears as an essence (parton) if it implies a purposeful interpretation, and as a "species" (taxon) if there is no such implication [Balme 1980].

The second of the abovementioned pairs of oppositions is evident in Aristotle' distinguishing and fixing terminologically natural and logical accounts of the categories in question [Sloan 1970; Grene 1974b; Pratt 1984; French 1994a; Wilkins 2003a, 2009a]. On the one hand, treating them as natural units, Aristotle means that the organisms belonging to them beget "like from like" by breeding, and this results in the continuous chains of generations treated as *geneis* and *phyleis* (γενέίσ, φιλέίσ). In the treatise *On the Generation of Animals*, these categories, persisted in *Physis* due to continuous chains of generations, are designated by Aristotle as *homogenesis* and *homophylesis* (ὁμογενέσίν, ὁμόφυλέσίν), respectively [Aristotle 1943]. On the other hand, the same categories are regarded as classification units, which encompass the alike things endowed with the same essences and thus are *homoousios* or *homoeidos* (ὁμοούσιος, ὁμοειδῆ); these units are denoted by standard terms *genos* and *eidos*. Outlining in the context of the genus-species problematics, a distinction between these two accounts signifies an important *natural-classificatory dualism*, according to which *genos* and *eidos* are comprehended simultaneously in two modes of existence.[3] Ontically, they are specific natural units, *homogenesis* and *homophylesis*, which integrity and persistence is provided by a specific natural mechanism (begetting). Epistemically, they are specific classification universals being *homoousios* and *homoeidos* due to certain essential similarities of their specimens.

[3] This distinction will be explicitly made by G. Leibniz in the 17th century (see Section 4.2).

It is noteworthy that it is the second pair of terms in their subsequent Latin translation (genus and species) that will be fixed in systematics, while the first one, more biologically meaningful, will be forgotten, due to terminological legacy of scholasticism [Pavlinov 2021b]. This is quite clear illustration of a peculiar character of the "historical memory": it preserves such notions in the conceptual space that, by their original semantics, contradict ontological foundations of contemporary biology.

At the level of ontology shaped by natural-classificatory dualism, the first way of positing the categories of *genos* and *eidos* corresponds to the dynamic aspect of being, and the second way corresponds to its stationary aspect. Therefore, with an expanded metaphysical account of the natural and classificatory distinction, one may fantasize about its rooting in Plato's general contemplation of a global dual structure of the universe. Intelligible *genê* and *eidê* constitute the world of unchangeable essences, whereas sensually perceived things (including living organisms) constitute the world of becoming [Fine 1993]. Developing this idea, it is possible to speculate that in Aristotelian world picture, the dualism in question appears as a manifestation of a general idea of *Physis* as a Great Chain of Being [Lovejoy 1936], which lowest-level finite "links" (mortal organisms) beget consequently, while both the infinite Chain and *ousiai*, passing from one "link" to another in it, remain self-identical and, in this sense, stationary [Akhutin 1988; Gaydenko 2000].

Such deliberation provides an ontological justification for an infinite persistence of *ousiai* despite a very limited duration of the organisms embodying them [Lennox 1985; Stamos 2003; Smith 2009]. By this, it seems to resolve one of the most serious controversies of Aristotelian ousiology and thus directly relates to an understanding of the antique genus-species problematics. Indeed, the potentially endless continuations of unchanging essences, provided for by the unbroken continuity of the chains of generations of reproducing organisms, also ensure a potentially endless persistence of "genera" and "species" as aggregates of these organisms [Lones 1912; Sloan 1970]. Such mechanism of persistence, which is expressed by the previously mentioned formula "like begets like", according to Aristotle, is a special natural law to which all the living beings obey [Sloan 1970, 1972; Berti 2016]. It underlies a *generative idea* that will be formalized as a generative species conception, with its popular contemporary phylogenetic version [Wilkins 2003a, 2009a, 2010] (see Section 5.1.2).

* * *

A hallmark of Aristotle's classificatory method is that in each individual cognitive act is associated with the identification of a *genos* and its division into *eidê*; these categories retain their subordination, but when describing particular groups of organisms, they do not have any fixed position in the overall hierarchy [Lloyd 1961, 1981; Balme 1962; Sloan 1970; Grene 1974b; Sucker 1978; Pellegrin 1982, 1986; Gotthelf and Lennox 1987; Tipton 2014]. This means that they are endowed not with an "absolute", but with a *relativistic* meaning: an *eidos* individuated within a *genos* at a certain level of generality, in turn, becomes a *genos* if a property implied by it is further divided into more detailed characteristics and the next lower-level *eidê* are distinguished accordingly. Such a relative sliding character of subordination of interchangeable *genê* and *eidê* deserves being designated as *genus-species uncertainty* in its narrow sense: none of these categories has a fixed position in the genus-species hierarchy, so it is determined contextually and fixed terminologically in each case study. Thus, in *The Parts of animals* Aristotle identifies the greatest *genos* of animals with blood, it is divided into *eidê* of viviparous and oviparous animals, and within the latter, the lower *eidê* of crocodiles and snakes are recognized; at the same time, in the *History of Animals*, he distinguishes greatest *genê* of oviparous and viviparous fish, lizards, frogs, etc. [Aristotle 1883]. In addition, in *Metaphysics*, he often uses the term *genos* in an ordinary sense to designate breeds and races of domestic animals. This fundamental property of the genus-species hierarchy will later become an important part of the classificatory method of scholastics (see Section 3.5).

In *The Parts of Animals*, Aristotle specifically focuses on how classification categories should be distinguished [Sloan 1970; Gotthelf 1991, 2012; Charles 1997]. First, he definitely links their recognition with the essences, which become their *differentiae*. Thus, with regard to *eidê*, he writes that "if the *eidê* are indivisible, and the differentiae are indivisible, and if no differentiation is common to several groups, the number of differentiae must be equal to the number of *eidê*" (cited after [Wilkins 2009b: 15–16], who uses "species" for *eidê*). This rule will become of paramount importance in the genus-species classificatory scheme to be elaborated by scholastics and adopted in scholastic systematics (see Section 4.3.2). Second, he writes that if we follow a dichotomous division recommended by Plato and the Middle Platonists, then implying different essences yields different divisions of organisms into *genê* and *eidê*, i.e., their different classifications. However, according to monistic natural philosophy, the *Physis* is unified in its organizational basis; therefore, it is divided in the only way to provide what will later be called natural groups; as Plato figuratively states, in order to distinguish them, *Physis* should be "carved at its joints". This contradiction, sometimes referred to as an "Aristotle's problem" [Sloan 1972], is responsible for a potential multiplicity of classifications, and it is permissible to regard as the first conspicuous discrepancy between taxonomic pluralism and monism [Henry 2011; Pavlinov 2018].

To resolve this problem, Aristotle considers it necessary to distinguish *genê* and *eidê* within a single hierarchy not dichotomically by the sole property, but simultaneously by several ones. This approach makes it possible to identify those complexly organized integral "souls" that are the causes of organisms and determine their existence in the habitat to which they fit [Sloan 1970, 1972]. This idea of Aristotle seems to be an attempt to overcome a natural-classificatory dualism by making classification units substantially as much close as possible to the natural ones.

In one of the studies of Aristotelian metaphysics, for an essentialist and a more formal classificatory considerations of *genos* and *eidos* be distinguished terminologically, they are proposed in the latter account to term *quasigenus* and *quasispecies* ([Lloyd 1962], D. Barnes' comments in [Porphyry 2003]). Analyzing in this way metaphysical meaning of Aristotle's differentiation of male and female, Stella Sandford treats them as quasispecific subcategories of the quasigeneric category of biological sex, and summarizes this proposal as a concept of *quasi-taxonomy* [Sandford 2019].

2.1.4 Theophrastus' *Enquiry into Plants*

A remarkable example of the application of Aristotelian method to describe the diversity of living beings are the works of Aristotle's disciple, Theophrastus (ca. 370–ca. 288). In his philosophical treatise *On the First Principles* (also known as *Metaphysics*), the terms έίδοσ and γένοσ are used throughout, which in Latin and English translations appear as *species* and *genus*; he also writes there about nature (φύσισ) as an essential property, i.e., as the *nature of things* [Theophrastus 2010]. In his botanical encyclopedia *Enquiry into Plants*, these notions are also widely implicated [Theophrastus 1916].

In this encyclopedia, the categories of *eidos* and *genos* are applied mostly in the taxonomic meaning; therefore, in its English editions, the respective Greek terms are freely translated as *kind* and *class*, respectively [*op.cit.*]. Their position in the overall rankless hierarchy is not fixed, and they sometimes are treated interchangeably. When dividing plants into principal groups of trees, shrubs, semishrubs, and grasses, Theophrastus calls them *eidê* and ranks the latter "greatest", while designates each particular group (kind) of aquatic and land plants as *genos* [*op.cit.*: 22, 30]. It is worthy to note that he, when describing differences in various trees, comments on "the kinds which [local] people distinguish by special names" [*op.cit.*: 213]; thereby, Theophrastus anticipates the "nomenclatural" method of the future researchers of folk systematics (see [Berlin 1992; Pavlinov 2015] on the latter).

At the same time, in some places, Theophrastus applies the notions in question in partly metaphysical (Aristotelian) meaning, which is evident from the following passage. He points out that "it is the character of the flavor which becomes more complete and matures into something separate and distinct; in fact we must consider the one as 'matter', the other as 'form' or specific character" [Theophrastus 1916: 87–88].

Theophrastus' botanical encyclopedia will significantly influence the late Medieval and Renaissance development of ideas about the diversity of plants. The authors of herbals will be guided by it in their representation of the structure of this diversity, and scientific systematics will afterwards begin to grow from them (see Sections 4.3).

* * *

One of the important results of the development of the cognitive situation in classical Antiquity is the development of two general cognitive programs, which are called with some conditionality (using Linnaeus' terminology) "*methodical*" and "*collecting*" [Zuev 2002, 2015; Pavlinov and Lyubarsky 2011; Pavlinov 2018]. The first goes back to Aristotle and aims to develop methods for describing the diversity of natural objects; the second goes back to Theophrastus and aims at their description itself. These programs will have a significant impact on the prehistory and early conceptual history of systematics in the 15th–17th centuries (see Sections 4.3).

2.2 NEOPLATONISTS: PORPHYRY'S QUESTIONS

The most important contribution to the late antique (postclassical) development of European philosophical thought belongs to the doctrine of Neoplatonism, which is quite heterogeneous in its content [Losev 1960; Wallis 1972; Lloyd 1990; Remes 2008; Wildberg 2016]. From the point of view of the main subject of this book, a combination of Plato's natural philosophy and Aristotle's categories with the addition of Pythagorean numerology should be considered its principal achievement [Lloyd 1990; Sorabji 1990; Strange 1992; Evangeliou 1996; Hadot 2015]. From the first, an account of Nature as a hierarchy of ideas is taken, at the beginning of which lies The One as the root cause and the source of all things; from the second, the description of this hierarchy by means of predicates, including *genos* and *eidos*, is applied; and from Pythagoreanism, presentation of Nature as a harmony of numbers preceding *eidê* as forms is adopted. One of the important results of the intellectual endeavor of Neoplatonists is a fundamental rationalization of the entire antique onto-epistemology, which, after its refinement by scholasticism, will form the basis of early scientific systematics [Pavlinov 2018, 2021b].

The Neoplatonic stage of the development of late antique philosophy begins with the fundamental treaties (in Greek) *Enneads* by Plotinus (AD 204–270). His very mystical world picture is shaped by a fundamental hierarchically organized triad: Platonic One, Intellect or Mind (νοῦς, σοφία), and Soul (ψυχή), which are arranged in something like internested spheres [Bréhier 1958; Losev 1960; Corrigan 2005].

The most important to us are two treatises included in the *Enneads*, "On the genera of being" (VI. 1–3) and "On whether or not there are ideas of individuals" (V. 7).[4] In them, Plotinus expounds his understanding of Plato's greatest *genê*, treating them as intelligible beings and assigning them the status of basic categories irreducible to each other, and not so much epistemic as ontic [Evangeliou 1982; Strange 1987, 1989; Corazzon 2020]. On this account, he refines the general hierarchical subordination of *genê* and *eidê* proposed by Aristotle and acknowledges the ontic primacy of the former relative to the latter: *genê* (genera) generate *eidê* (forms) by means of *differentiae*, and not vice versa, and The Soul finds its intelligible embodiment in this hierarchy [Remes 2008; Chiaradonna and Galluzzo 2013]. With this, Plotinus identifies particular *eidê* (forms), partly in an understanding close to Aristotle's: they, eternally existing in The Soul, are embodied in individual things and by this become their "individual souls" (substances), or "inner *eidê*". To Plotinus, ontologically regarded "sensible substances are then more or less material—elements, plants, animals—their species, both individuals and universals" [Plotinus 2018: 645]. In his onto-epistemology, the things (individuals) are prior with respect to their perceptions, "but naturally prior are those which are more general, i.e., species" [*op.cit.*: 646].

The categories of our interest, as they are considered in *Enneads*, are basically partonomic in that they are identified with the essential properties (souls, forms, etc.) of both intelligible and sensible entities. Accordingly, Plotinus treats the hierarchy of categories as a result of partonomic division: "Intellect as a whole encompasses

[4] Throughout the *Enneads*, these notions should read *genê* and *eidê*.

all things just like a genus encompasses its species and just like a whole encompasses its parts" [Plotinus 2018: 632]. At the same time, when he considers *genos* and *eidos* epistemically as applied to the descriptions of the things, he seemingly ascribes taxonomic status to them and does not follow strictly to their Aristotelian hierarchy. For instance, he writes about "species of animal", among which there "is one genus of horses", whereas there are "all the other animals that can be said to form single genus", so that "we must again bring all these species back together under the one 'living being' as their genus" [*op.cit.*: 274]. And then, further developing this idea, he asks rhetorically, "are not living beings must be classed by their species, and then into the single genus of 'non-living beings'?" [*loc.cit.*]. Thus, it is evident that for Plotinus, *eidos* is a universal both ontically and epistemically: he writes that "straight line is [...] a species of line" and considers "change" as a species of "motion" [*op.cit.*: 646]. It is worthy to note also, that, when describing relations between *genê* and *eidê*, Plotinus applies the Aristotelian notion ὁμοειδῆ to identify *eidê* of the same *genos* [*op.cit.*: 388].

* * *

A detailed study of relations between the notions of *genos* and *eidos*, very suggestive for their further development, is provided by Plotinus' pupil Porphyry (AD 232–304) in his treatise (in Greek) *Introduction to the "Categories" of Aristotle* [Spade 1994; Karamanolis and Sheppard 2007]. Despite its title, it is not limited to an analysis of the content of Aristotle's *Categories*, but in fact is an attempt to harmonize two classical versions of onto-epistemilogy, viz., genology of Plato (and thus also of Plotinus) and ousilogy of Aristotle; it is called sometimes as an attempt of "platonizing" of *Categories* [Strange 1992; Hadot 2015]. The *Introduction* has a pronounced didactic character, and in accordance with this, its main sections are entitled to indicate the subjects they consider: "On genera", "On species", "On differences", etc. [Porphyry 2003].

Unlike Plotinus, Porphyry regards the categories of *genos* and *eidos* epistemically as predicates and begins his treatise with the following statement: "It being necessary, [...] for a schooling in Aristotle's predications, to know what is a genus and what a difference and what a species and what a property and what an accident" [Porphyry 2003: 3]. A little further on, he promises "avoiding deeper inquiries and aiming suitably at the more simple" and "decline to discuss" the ontology of these categories because of the complexity of the issue [*loc.cit.*]. However, paradoxically enough, his enumeration of these inquiries clearly outlines the main scholastic content of the genus-species problematics regarding "genera and species: whether they subsist, whether they actually depend on bare thoughts alone, whether if they actually subsist they are bodies or incorporeal and whether they are separable or are in perceptible items and subsist about them" [*loc.cit.*]. These questions are sometimes called "fateful", and their discussion will become one of the focal points of the entire scholastic onto-epistemology, with their discussion serving as the basis for an evident divergence of realism, nominalism, and conceptualism [Spade 1994; Neretina 2006; Chiaradonna 2008].

In the section "On genera", Porphyry designates three main accounts of this category, which can be represented as follows (in the terminology adopted here): *genos* can be understood (a) taxonomically as "the collection of things related somehow to one thing and to one another", (b) genealogically, meaning "the origin of each person's birth, whether the one who begot him or the place in which he was born", and (c) hierarchically as "that to which a species is subordinated", so it "seems also to *include* the whole multitude contained under it" ([Spade 1994: 1–2], italics in original). With this, "genera differ from species, even though they are predicated of several, are yet not predicated of what differ in species but only in number" [*op.cit.*: 2]. The contemporary commentators often emphasize that Porphyry understands *genos* as an aggregate, logical class [Sorabji 1990; Richards 2010; Shatalkin 2012], but by this, they omit the second account that clearly indicates a movement towards a deeper "generative" understanding of *genos*, initiated by Aristotle.

In the section "On species", Porphyry indicates the following main attributes of this category [Spade 1994: 3–4]. On the one hand, (a) "'Species' is said of each thing's form", i.e., it is considered partonomically. On the other hand, Porphyry emphasizes (b) "that what is under a given genus is also called a species" and notes their certain equivalence: "since both genus is the genus of something and species is the species of something, each of the other, both must be used in the definition of both". They are, however, in a strong subordination: (c) "Species is what is arranged under the genus and is what the genus is predicated of with respect to what the thing is". With this, he emphasizes on the correlative nature of *genos* and *eidos*, which are perceived not in isolation from each other, but "taken in relation to one or the other", and also that "the division of a genus is always into several species".

On this basis, one of the key conceptualizations is introduced in this section that goes far from concerning just *genos* and *eidos*: in fact, Porphyry outlines verbally the future genus-species scheme of division of notions as follows. There is the most general notion, which is "the greatest genus" exemplified by substance. There is the most specific notion, which is "the most specific species" exemplified by a thing. And between them, there is an intermediate multistep hierarchy: "What are between the two extremes [are] 'subordinate' genera and species. Each of them they hold to be both a genus and a species, taken in relation to the one extreme or the other" [Spade 1994: 3–4]. For instance, "Animate body is a species of body but a genus of animal. Again, animal is a species of animate body but a genus of rational animal" [*loc.cit.*]. Such character of the genus-species hierarchy ascends to Aristotle, and it is called *sliding* above. Subsequently, a tree-like scheme will be invented to illustrate graphically this scheme, which will be called the *Tree of Porphyry* in honor of the author of the *Comments* (see Section 3.5).

Thus, it is clear from the above quotations that Porphyry, while preserving the Aristotelian subordination of *genê* and *eidê*, puts in them a more formal and slightly different meaning. They are equivalent as predicates to the degree they are interchangeable within a rankless sliding hierarchy, so both their position in the latter and meaning are not fixed, and the entire hierarchy shaped this way is arbitrarily fractional. Because of this, some modern Aristotelians believe that Porphyry "demolished" species as a separate category [Pellegrin 1990; Lyubarsky 2018], and some others qualify Porphyry's position regarding these two categories as nominalistic [Lloyd

1990; Sorabji 1990]. Both these contemporary treatments are hardly true. Indeed, "relativity" of *genê* and *eidê* is already present in Aristotle's *Categories*, while Platonism of Porphyry, like that of Plotinus, implies a realistic understanding of their hierarchy. What is certainly non-Aristotelian and non-Platonic in the conceptualistic doctrine of Porphyry is that the highest position in that hierarchy is occupied by substance, which is treated as the *greatest genos*, and not the *primary ousia*, though, in agreement with Plato, the sensually perceived things are considered imperfect instantiations of the ideal forms [Chiaradonna and Galluzzo 2013].

* * *

The realistic account of Neoplatonic onto-epistemology is strengthened by Proclus Diadochus (AD 412–485), one of the last Neoplatonists of the late Antique period. In his *Commentaries on Plato's "Timaeus"*, he unambiguously deduces the logical subordination of *eidê* to *genê* from the ontic one [Taylor 1928]. According to the descending causal cascade in Plato's world picture, more general categories serve as the causes of more particular ones. Therefore, this hierarchically arranged sequence should be reflected in a corresponding sequence of a logical division of notions as universals of different levels of generality [Rosán 1949; Chiaradonna 2008; Martijn 2010]. This general consideration lays the ontological foundation of the epistemic genus-species scheme, with which mastering (let us recall once again) the development of biological systematics will begin.

3 The Middle Ages
Species as Universalia

The period of the development of European philosophical and cognitive traditions, following the Antiquity, is known as the Middle Ages, and it lasts about a thousand years. Its lower historical boundary is clearly marked by the collapse of the Western Roman Empire, with which the Antiquity ends as a certain type of culture in the fifth century AD. There is no such certainty with its upper historical limit: it is outlined by the beginning of either the Renaissance (the 14th century) or the Modern Times (the 16th century). According to the second variant, the Middle Ages are divided into three main stages—early (5th–11th centuries), classical (11th–14th centuries), and late (14th–16th centuries).

The onto-epistemic basis of medieval cognitive situation is shaped by the biblical world picture, according to which the source and cause of the universe is the primary supramaterial acting force, God the Creator. Everything created by him is originally prescribed in the Divine Mind containing Divine ideas (archetypes), they are the root causes (*rationes seminales*) of all things, and the Divine plan of creation serves as a means of their realization. In this cognitive situation, the attention of medieval thinkers is of "Platonic" kind by being aimed at a comprehension of Divine ideas as the cause of all things, with the Bible as an unconditional source of true knowledge about them. Continuity with the Antiquity in the early Middle Ages is based on Platonic eidology, and in the classical one, on Aristotelian ousiology. The continuity with the latter is framed by an acknowledgement that the created world begins with the interaction of passive matter (*genus*), containing all being in potentiality, with active forms (*species*) implementing Divine ideas to embody matter's potentiality into actual being.

This new cognitive intention of medieval philosophers is divided into two parallel and closely intertwining streams: *theology of revelation* focuses on God, guided mostly or totally by the faith in the unconditional truth of the Bible, while *natural theology* focuses on a rational cognition of what is created by God, based on the arguments of reason and also taking into consideration empirical data. Such motivation for cognitive activity leads to a loss of interest in the "collecting" program, i.e., in studying facts about things (the realm of natural science). Because of this, the Middle Ages are usually called the "dark time", meaning that it does not lead to a significant increase in the empirical knowledge about Nature. But with this, rationalist philosophers actively develop the "methodical" program aimed at finding the rational ways of inference of true statements about "nature of things", i.e., a method of reasoning in its general sense (as Aristotelian *Organon*). This intellectual activity frames a *scholastic* (from Greek σχολαστικός, school) cognitive program, based on the just outlined theo-philosophical rational onto-epistemology [Gaydenko 2003].[1] Therefore, all actors of this program are equally theologians and philosophers.

[1] Scholasticism is so named because the logic developed by it becomes an important part of the basic education course (*trivium*) in the schools and universities.

DOI: 10.1201/9781003346296-4

Because of the initially close connection with the bibleism, scholastic program and the method developed by it are usually treated by the materialistic philosophy of the Modern Times as unworthy of attention. However, it is this program that, with time, becomes an important prerequisite for the formation of scientific rationality and the logic of scientific research [Lektorsky 1995; Osler 1997; Gaydenko 2003]. Indeed, as the historian and philosopher of science Alexandre Koyré writes, "scholastic philosophy was something very significant. The scholastics carried out the philosophical education of Europe and created our terminology, which we still use; it was their works that allowed the West to perceive or, more precisely, establish contact with the philosophical heritage of Antiquity" [Koyré 1973; cited after Russian edition of 1985: 51].

In particular, with the main subject of our book in mind, scholasticism pays much attention to that part of the onto-epistemic frame of the new cognitive situation, in which a hierarchical genus-species scheme takes a focal position. Largely on the basis of this scheme, a classificatory cognitive program will arise, which will shape the foundations of early (scholastic) systematics, with its key cognitive motivation being the classification of the diversity of living nature. As it develops conceptually, systematics will get rid of much of the scholastic tradition—but this does not negate the historical fact that this discipline, with the problems it explores and the tasks it resolves, has its roots in scholasticism.

* * *

Numerous philosophical treatises of scholastics concerning genus-species problematics rotate mainly around two basic issues: the origin of genera and species and their onto-epistemic status as universals [Sokolov 1979; Gilson 1991; Spade 1994; Motroshilova 2000; McGrade 2003]. In both cases, the scholastics are primarily interested (in our terminology) in their partonomic, and not taxonomic aspect, i.e., in the ideas (forms) and essences (substances) embodied in the things, but not in the latter's aggregates. With this, their ontology presumes what is called here partonomic-taxonomic dualism, though in a peculiar fashion. Indeed, scholastics usually apply the notions of form and species in a very close or even identical sense; therefore, their unity will be referred to hereafter by a combined notion of *species-form*, wherever appropriate, to distinguish it from taxonomically understood *species* as such. Their focus is on the relation between ontic and epistemic accounts of genus and species as universals and their relations with singulars (*naturales*), which shapes a large part of the problematics characteristic of scholasticism. Its initial formulation goes back to Stoic philosophy of classical Antiquity, its advanced development occurs in late Antiquity, and it is regarded in the Middle Ages in the context of three main confronting worldviews—realism, conceptualism, and nominalism [Bronowski 2007; Klima 2017], outlined elsewhere (see Section 1.1.2).

The disagreements between them become a key part of the genus-species problem as it is developing in medieval scholasticism; a perpetual dispute between different philosophical schools lasts for several centuries and turns out to be very productive by contributing to the development of modern onto-epistemology [Koyré 1973; Marenbon 1988; Gilson 1991; Motroshilova 2000]. Together with the latter, these disagreements will move into biological systematics, in which the question of reality *vs.* nominality of taxa will remain one of the most fundamental and actively discussed until now [Simpson 1961; Sokal and Sneath 1963; Mayr 1982, 1988; Pavlinov 2018, 2021b].

* * *

The main trends of the development of scholastic cognitive situation largely coincide with the abovementioned historical stages of the Middle Ages, spanning to the Renaissance and partly the early Modern Times [Sokolov 1979; Motroshilova 2000]. The onto-epistemology of early scholasticism (before the 11th century) is formed under the strong influence of Plato's ideas (through Neoplatonism), in which realistically oriented theology and philosophy are closely interconnected. In the middle (mature) scholasticism (the 12th–13th centuries), the influence of Aristotle is more evident, whose principal works become known after their translations of Arabic philosophers (see below). At this time, a divergence of mystic theology and rational philosophy begins, and a logical apparatus for describing the structure of the world is actively developed, including genus-species scheme. This development is completed by the late scholasticism (the 14th–16th centuries), in which a nominalistic worldview is most noticeable.

The conceptual history of scholasticism, encompassing the development of our problematics, in many ways proceeds as the development of antique tenets. This continuity is evident in that the problematics in question remains genus-species one, and it is manifested in two ways. On the one hand, partonomic-taxonomic dualism persists, meaning that genus and species are treated either partonomically (intensionally) or taxonomically (extensionally), depending on the cognitive contexts. On the other hand, they understood in a conjugacy, which means that (a) species is treated as but a detailing of genus and thus does not exist independently of it, (b) therefore genus is primary in relation to species, (c) they are interconnected by genus-species uncertainty within a rankless sliding hierarchy, and (d) their particular status is interchangeable and defined contextually. With this, in ontology, the questions posed by Porphyry about reality or nominality of genus and species as universals comes to the fore, while in epistemology, the substantiation of a deductive argumentation scheme with genus and species as its main elements attracts most attention. What is of importance to us is that it is this general understanding of the genus-species problematics that will be inherited by early systematics, which initial development will proceed under a strong influence of scholasticism (see Section 4.3.2).

The initial development of the genus-species problem during the transition from Antiquity to the Middle Ages involves the change of its linguistic "framing" following the transfer of the center of gravity from Greek to Roman culture [Jorge and Newton 2016]. Due to this, Latin becomes with time a principal language of medieval European philosophy, in which an exceptional role is played by the conjugate development of theology and philosophy under the auspices of Roman Catholic Church

[Koyré 1973; Gaydenko and Smirnov 1989]. This inevitably affects the basic thesaurus of our problematics: the most significant notions of *genus* and *species*, *essentia* and *substantia*, *definitio* and *differentia* are introduced into philosophical and scientific circulation. With this, while γένοσ is uniformly transliterated by Latin *genus*, έίδοσ appears either as *idea* or *forma* or *species* depending on the semantic contexts it is placed in. In the first case, it most closely corresponds to an original Platonic understanding of "idea", whereas in the second and third cases, it is closer to an Aristotelian understanding of "form", so *forma* and *species*, as mentioned above, often appear in the Latin of medieval philosophers as interchangeable synonyms [Gilson 1960; McGrade 2003].

A noteworthy interpretation of species-form by some important actors of scholasticism involves conception of *species multiplication*. It has two different meanings, ontological and epistemological, i.e., there are actually two such conceptions; they show how broad content configures the species problematics in the medieval and partly post-medieval period of its conceptual history. According to one of them (R. Bacon), numerous partonomically treated species-forms participate in the physical interactions of things, being their attributes [Lindberg 1983; Kedar 2017, 2021]. According to another (Thomas Aquinas, F. Suaréz), they are elements of a *cognitive chain* to participate in the cognitive processes, in which they are arrange sequentially and denoted by special terms (*species sensibiles*, *species in medio*, *species intelligibiles*, etc.) [Spruit 1994; McGrade 2003]. This interpretation is not considered here in detail, as it seems to have no impact on the conceptual development of early systematics. However, it is to be noted that only species are included in these considerations, while genera are discarded: this is probably the first instance of indirect disjoining of these two categories.

* * *

This chapter describes the development of the genus-species problematics by medieval European scholasticism. It is discussed through the prism of philosophical ideas of its key figures, whose views are grouped into three above-mentioned cognitive standpoints, viz., realism, conceptualism, and nominalism. Quite understandably, the main emphasis is placed on the hypostases in which the notions of genus and species are presented in these worldviews.

3.1 AN EVE OF SCHOLASTICISM: THE BEGINNING OF GENUS AND SPECIES

The formation of scholastic philosophy begins with the linking of Greek antique and Jewish biblical ontologies, based on interpretation of the notions of Neoplatonic philosophy in terms of *Old* and *New Testaments*. This results in a kind of *Christian Platonism*, which influences the philosophical ideas of early scholasticism about Nature and its causes are very significant [Neretina 2013; Hampton and Kenney 2021].

The first important step in this direction is made by the Jewish theologian Philo of Alexandria (ca. 25 BC–ca. AD 50), writing in Greek. In his doctrine, Plato's

The One and Aristotle's the First Root (Prime Mover) are identified with God the Creator (θεόσ), whose creative force is contained in the Word (*Logos*), while *eidê* in their general sense are conceived as incarnations of *Logos* and are treated as Divine archetypes (θεϊκά αρχέτυπα) [Sandmel 1979].

The close successor of Philo is one of the first Christian theologians Clement of Alexandria (150–ca. 215), who in his treatise *Stromata* (in Greek) emphasizes the necessity for Christian Church to study "true philosophy" (mainly the teachings of Plato and Pythagoras) as an instrument of cognizing God [Chiaradonna 2017]. He examines the nature and causes of subordination of *genos* and *eidos* and discusses three main ways of their divisions: the whole into parts (partonomy), set into subsets by essential features (taxonomy), and set into subsets by any of accidents (another taxonomy). Clement clarifies their interrelations by assuming that *genos* contains in potentiality all the *differentiae* distinguishing its particular *eidê*, while the latter actualize these potentialities and implicitly contain their *genos* undistorted. This tenet will be reproduced from time to time in various natural-philosophical concepts, up to some versions of typology of the 19th and 20th centuries [Pavlinov 2018].

* * *

For the development of the genus-species problem, of great importance are the ideas of Aurelius Augustine (354–430), one of the founders of doctrinal Christianity. His worldview is initially formed by Neoplatonism and Manichaeism and is subsequently supplemented by biblical ontology; he anticipates conceptualism by emphasizing that an *intention* is required for any cognitive act [Inge 1900; Gilson 1960; Van Fleteren 1992; Pasnau 1997]. In the treatise *The City of God*, Augustine introduces the notion of *essence* into the thesaurus of scholasticism, directly deriving it from the notion of *being* (Latin *esse*) and linking it with the Aristotelian notion of *ousia* [Saint Augustine 2000]. The writings of Augustine are originally written in Latin, and due to this, the Latin term *species* becomes fixed in the European philosophical literature. His teaching will be among the most influential in European scholasticism up to the 13th century.

Following our terminology, we can be sure that Augustine puts partonomic content into the considered categories: for him, species is a *species form* (*forma specie*), and not a particular aggregate of things. Discussing the processes of generation of these forms, Augustine categorizes them into lower-level (perceiving) and a higher-level (cogitation) ones and refers to some "rule of specific *vs.* generic knowledge" resulting in "secondary specific and generic notions" [McGrade 2003; Nawar 2021]. In his treatise *On the Trinity*, he distinguishes four categories of species forms: those existing in objects (*species corporea*), in their human perception (*species impressa*, or *sensible species*), held in human memory, and lastly formalized in the human mind, or rather, "inner gaze of soul" (*species expressa*, or *intelligible species*) [Saint Augustine 2011]. The intellectual activity results ultimately in a *notion* (*notio expressa*) as a generalized image of the things existing in the cognizing mind.

Reflecting on the unity of Divine essences that make up the New Testament Trinity, Augustine examines the relationship between *genus* and its *species* and argues that "no species extends beyond the definition of its genus" [Saint Augustine 2011: 292], and he proceeds with saying that "the definition of the general term must necessarily embrace all its species" [*op.cit.*: 516]. This provision is achieved by that species predicates about all individuals endowed with the corresponding essences (substances), and genus predicates about all species embodying the corresponding forms. This clause, after passing through all stages of its evolution in scholasticism, would eventually affect procedures of early biological systematics, including an initial version of taxonomic nomenclature [Pavlinov 2015, 2022b].

When discussing the relation between the unitarity of each archetypal idea in Divine Mind and multiplicity of species-forms embodying it in the things, Augustine focuses on their *similitude*, or likeness (*similitudo*) as a link connecting them. With this, it links partonomically treated species-form with its taxonomical treatment: indeed, "the likeness of the individuals is the basis of the relative unity of the species just as the likeness of the parts is the basis of the relative unity of the individual" [Gilson 1960: 212]. In its subsequent understanding by scholastics, who tend to nominalism and partly conceptualism, Augustinian similitude becomes a prototype of what would later appear as a phenetic idea (see Section 4.3.4).

The ideas of Aurelius Augustine are echoed by onto-epistemic considerations of Augustine's contemporary, the Platonic philosopher Calcidius (the fourth century), whose *Commentary on Timaeus* will become popular in mature scholasticism [Albertson 2014]. He means partonomic account of species-form, and for him, two parts of this notion are synonymous and interchangeable, so he prefers, like Augustine, to designate form as species. In some discourses, Calcidius distinguishes between archetypal, or main (*species archetypa*, *principalis*) and secondory (*species secunda*) species forms. In the latter, he recognizes two hypostases: one means *native* species-form (*species nativa*) embodied in the world of things, also designated as *corporeal* (*species corporeae*), another means *intelligible* species-form (*species intelligibilis*) existing in the mental world of notions [van Winden 1965; Gilson 2010].

* * *

One of the key features of the genus-species scheme, which is an important part of the scholastic method, constitutes the hierarchical ordering of relations between genera and species of different levels of generality. In this regard, one of the significant events in the conceptual history of the genus-species problem in the early Middle Ages is an invention of the very notion of *hierarchy* as a component of the basic thesaurus of scholasticism [Verdier 2005; Lyubarsky 2018].

This notion is developed in the context of an idea of the biblical ontology of "angelic ranks" that compose the Heavenly *Ladder of Perfection* directed to the Creator [Drobner 2000; Dorofeeva 2001]. Its topology is similar to the antique Great Chain (or Ladder) of Being, with both being arranges linearly (therefore the term "chain") [Lovejoy 1936]. The notion in question fixes such ordering terminologically as an element of the theological conceptual space, and it appears in the fifth-century treatises known as *Corpus Areopagiticum*, or *Corpus Dionysiacum* (originally in Greek) attributed to a semimythical personage Pseudo-Dionysius [Hathaway 1969; Rorem 1993; McGrade 2003]. This heavenly hierarchy developed by him reflects a sequence of incarnations of the ideas in the Divine Mind, which he calls *Divine names*: thus understood, the Pseudo-Dionysian neologism is etymologically deciphered as the origin (ἀρχή) of the order of sacred essences (ἱερω).

The early variants of the Pseudo-Dionysian ordering is linear to include a sequence of 9–11 angelic steps; in its more advanced variants (Pseudo-Clement, Aquinas, etc.), these steps are grouped into 3–5 fixed "orders" [Dorofeeva 2001]. By this, an initial linear hierarchy turns into an *encaptic* (inclusive, nested) one, in which more general notions and the corresponding aggregates of the heavenly host include less general ones. It is noteworthy that the latter hierarchy, in contrast to the rankless antique sequence of *eidê*, represents the first version of the ranked hierarchy, in which categories (ranks) are fixed and designated terminologically. It is equally applied to both partonomically and taxonomically accounted ranks and their order.

In the Modern Times, an idea of linear hierarchy, as a special form of ordering the manifestations of Nature according to their stages of perfection, will frame the basis of natural-philosophical concept of the Ladder of Nature [Lovejoy 1936]. The latter will become in the 18th century a serious challenger to the concept of System of Nature; disagreements between them will cause disagreements between supporters and opponents of species reality (see Section 4.3.5). An idea of encaptic hierarchy will frame the basis of the genus-species scheme of division of notions developed by scholastics, and it will be included in the natural method of systematics (see Section 4.3.2).

* * *

For conceptualization of the notions of genus and species by scholastics, of key importance are the ideas of Anicius Manlius Torquatus Severinus Boëthius (ca. 480–524). In this respect, he is best known for his Latin translations of and comments on the works by Aristotle and Porphyry on logic and partly metaphysics, which throw a kind of "conceptual bridge" between antique and medieval philosophical systems [Sorabji 1990; Asztalos 1993; Magee 1998; Marenbon 2003; McGrade 2003; Neretina 2006; Martin 2009], including an account of species [Wilkins 2009a]. His treatises are of particular significance for, starting with the comments on the classic and late antique treatises, he works out his own conceptual constructs, some quite nontrivial [Sorabji 1990; Neretina 2006; Martin 2009].

Cognitive position of Boëthius, as that of Augustine, can apparently be thought of as close to conceptualism: he regards the ontic status of genus and species given not by itself, but by what a cognizing subject is interested in [Maurer 1982; Sorabji 1990; Neretina 2006]. The main question considered by him is this: do there really (objectively, outside the human mind) certain universals exist in Nature, which correspond to certain general notions (concepts) developed by human mind? Proceeding from this, Boëthius contemplates the "right ways" of positing genera and species that would allow, by applying the deductive method of Porphyry, to answer this question correctly. This means that the development of his own method goes from epistemology to ontology, and genus and species are initially treated as predicates, for which it is necessary to find certain correspondences in Nature. Such predicates define universals intensionaly, so Boëthius' genera and species can be assumed to remain largely partonomic *eidê*.

In *Commentary on Porphyry's Isagoge*, Boëthius undertakes to answer the questions that Porphyry posed in his *Introduction to Aristotle's Categories* (see Section 2.2), somehow expanding and specifying them. He summarizes his position as follows:

> Genera, species, and others are to be found either in corporeal things, or in those whose existence is connected with things; and if the spirit finds them in disembodied things, he receives the disembodied concept of the genus. [...] We imagine genera and species by extracting from the individual objects, in which they are found, those traits that make these objects alike. [...] A certain sign of similarity, conceived and truly considered [by mind], becomes a species. Thus, [genera and species] exist only in the particular things (*in singularibus*). With this, [only] general (*universalia*) is thought of, and the species should be accounted nothing else than a thought derived from the substantial similarity of a multitude of dissimilar individuals; whereas genus is a thought derived from the similarity of species. [...] So, [genera and species] exist in the realm of the sensuous, but they are thought apart from things. [They] exist in one way and are thought of in another; they are incorporeal, but, being connected with the sensible [things], they exist [in the realm of] sensible.
>
> *[Boëthius 1906; quoted after Russian edition of 1990: 27–28]*

In this passage, an idea of the universal as a twofold essence, existing in reality and in thought, is detailed. The peculiarity of the universal as an attribute of reality is that it is substantial, but noncorporeal and thus has no particular space and time limits. According to this position, Boëthius in his *Commentary on Porphyry's* reinforces Aristotle's distinguishing between primary and secondary substances. Commenting on Aristotle's *Categories*, he emphasizes that "individual substances are the first substances, in the highest way and in the proper sense of the word", and therefore, "of the second substances, species is more substance than genus, for it is closer to the first substance" and, referring to their predicative function, adds that "genera predicate of species, but not vice versa, species of genera" [Boëthius 1891; quoted after Russian edition of 2001: 140–141]. He repeats the latter in his *Commentary on Porphyry's* as follows: "that one genus always has many species; that the genus predicates of species, and the species predicates of individuals; that all the superiors predicates of the inferiors, but in no way vice versa" [Boëthius 1906; quoted after Russian edition of 1990: 71]. Boëthius thinks it is possible to comprehend genus and to divide it into species in two ways: in some cases, species appear as *instantiae* corresponding to particular things (houses, people), whereas in others, they are *differentiae* corresponding to the general notions (beauty, animality). As can be assumed, in the first case, Boëthius presumes a mixed partonomic-taxonomic account of the genus and its species, while in the second, he means a purely partonomic one.

In his *Commentary on Porphyry's*, Boëthius pays great attention to species. At the beginning of this treatise, he emphasizes that

> if we do not realize what species is, nothing would secure us from delusion. [...] Knowledge of the nature of species helps not only to avoid confusion between species of different genera, but also to choose the closest species within any genus".
>
> *[Boëthius 1906; quoted after Russian edition of 1990: 12]*[2]

In the special section "On species", he examines the ways of positing species as form and emphasizes its polysemy, speaking of its "threefold meaning: firstly, about species as a substantial quality; secondly, as the own form of every individual thing; thirdly, [...] about species subordinate to genus" [Boëthius 1906; quoted after Russian edition of 1990: 51]. As can be assumed, in our terminology, the first understanding is related to the partonomic aspect of the diversity of things, and the third one probably to its taxonomic aspect, whereas the second aspect corresponds to an everyday understanding of form.

Referring to the third understanding and quoting Porphyry, Boëthius designates successive steps of the deductive definition of notions, thereby formalizing and structuring the genus-species scheme for the first time. In it, the upper and lower boundaries are set by the *highest genus* (*genus generalissimum*) and the *lowest species* (*species specialissimum*), respectively; the former can never be species, for there is no genus above them; the latter can never be genus, for it is not further divided into species (i.e., it is *species infima*). There are *subordinate genera* (*genera subalterna*) between them, which "are species associated with higher ones, and genera associated with lower ones"; a subordinate genus, which is divided into the lowest species, is *proximal genus* (*genus proximum*) [Boëthius 1906; quoted after Russian edition of 1990: 54–55, 63, 70]. As an illustration of his reasoning, Boëthius uses the already-mentioned Tree of Porphyry, which "shows an order of predicating from the highest genus to the individuals" [*op.cit.*: 56]; this is probably the first publication of the logical tree-like scheme, from which all its other variants originate [Gontier 2011; Verboon 2014] The categories (steps) of this scheme recognized by Boëthius, after their clarification by Albert the Great (see below), will become the basis of the genus-specific logical scheme of later scholastics and will be subsequently included in the basic methodology of scholastic systematics (see Section 4.3.2).

Of Boëthius' idea of three ways of division, as they are set out in his *Commentary on Porphyry's* and *On the division*, two are especially significant for us: genus into species and whole into parts [Magee 1998; Arlig 2011]. They are implicitly present in classical antique literature, but Boëthius designates and analyzes them explicitly for the first time, indicating critical distinctive features for them [Henry 1991; Magee 1998; Tonoyan 2008]. The division of genus into species in Boëthius is basically taxonomic and quite traditional, whereas the division of the whole into parts is

[2] Thus, all past and current (and probably future) researchers, one way or another involved in the species problem, in a certain sense are "Boëthians" [Pavlinov 2018, 2022a].

not quite trivial for its being supplemented with consideration of the distribution of the essence of the whole between its parts. On the one hand, it implies such trivial cases of partonomization as the division of a house into roof, walls, and foundation, or the division of a human into soul and body. On the other hand, Boëthius considers an unusual variant of this division: "the parts of man are Cato, Vergil, Cicero, and the single men who, although being particulars, nevertheless combine to make up the sum total of man" [Magee 1998: 9]. It can be assumed that this corresponds to the literal meaning of Aristotle's indivisible species (ἄτομοη έίδοσ, *species individua*): it is not divided into specific things, but as a whole is distributed among them.

An important issue regarding the problem of relations between forms, genera, and species, relevant to the understanding of their integral or composite nature, is the one of *univocality* or *equivocality* of the essences or substances that are embodied in the things and make them what they are "by their nature". This issue, which is directly related to the understanding of the essences of living organisms, will be analyzed in more detail by Duns Scotus (see Section 3.3).

3.2 REALISM: SPECIES REALLY EXISTS

A realistic standpoint of the scholastics goes back to the conviction of Platonists and Neoplatonists that intelligible eternal *eidê*, being universals, are as real as sensually perceived things. As shown in the previous section, justification of the reality of universals with reference to the biblical world picture places a specific accent on this standpoint. It presumes that everything existing in Nature is real to the extant it is created by God and represents an embodiment of the Divine ideas pre-existing in the Divine Mind (see also Section 5.1.1). Since *eidê* are identified in scholasticism with genera and species, the latter are also recognized as real, i.e., existing outside and apart from the human cognizing mind (extra-mentally). This is the foundational idea of medieval *realism*, which is addressed to ontology.

Modern researchers usually distinguish between two degrees of realism featured by scholastics. An *extreme realism* recognizes equal unconditional reality of universals at all levels of generality, existing outside and before things. A *moderate realism* recognizes different levels of reality of *eidê*, and it takes two forms: one goes back to Aristotle's ousiology, and considers those embodied in the things (as *primary ousiai*) most real; another tends to Platonism and regards the higher-order level *eidê* more real in the sense that they more fully embody the supreme (root) cause.

* * *

In early scholasticism, a realistic account of universals is developed by John Scot (Iohannes Scotus) Eriugena (ca. 810–ca. 870), who follows Augustine in many respects. He expounds his biblical worldview in five books *On Nature* (known as *Periphyseon*), which translations are often called *On the division of Nature* [O'Meara 1987; Motroshilova 2000; Gilson 2010]. The latter title focuses on the hierarchical structure of Nature, in which he recognizes the following fourfold basic division of "great species" (*species magnae*): (a) that which is not created, but creates (*Natura non creata creans*), this is God as the original cause of all things, (b) that which created and creates (*Natura creata creans*), these are Divine ideas as intermediaries between the Creator and being, (c) that which is created but does not create (*Natura creata non creans*), this is everything that exists as the embodiment of eternal Divine ideas into temporary materialized forms, and (d) that which is not created and does not create (*Natura non creata nec creans*), this is God as what all things strive for, including man with his cognitive intention.

In his division of created Nature, Eriugena proceeds from the heavenly hierarchy of Pseudo-Dionysius: it is generated by Divine archetypes, which are the root causes of all "lower" genera and species. Eriugena accounts this hierarchy in the Platonic spirit and in terms of the genus-species scheme, ascribing the following important features to the latter. He believes that at the beginning of the Divine creation lies a *supreme essence*, containing all the potential multitude of particular genera and species. When embodied in the material world, they exist both outside things as universals and in them as substantial incarnations. The ordering of all things into genera and species is inherent in being itself, and not a result of human cognitive activity. With this, genera and species, as ontological entities, are endowed with different degrees of reality: genera are more real than species, for they are closer to Divine ideas in the heavenly hierarchy [Erismann 2007; Gabbay and Woods 2008; Gilson 2010].

Eriugena emphasizes inextricable interconnection and correlative nature of successive steps of genus-species hierarchy: "the genus is the genus of species and species is the species of genus" [O'Meara 1987: 61]. With this, a sequential division of categories into subcategories does not deprive them of their essential integrity, which is predetermined by their divine origin:

> The genus even when it is subsists as a whole in each species, and all its separate species form a single whole in it. The same virtue is also seen in the species, which, even when it is divided among individuals, preserves undiminished the force of its own indivisible unity, and all the individuals into which it appears to be infinitely divided are in it finite and an indivisible unity.
>
> *[O'Meara 1987: 61–62]*

Therefore, each genus is contained in its entirety in its species: the genus "four-legged animal" is contained in the species "dog", "cat", "cow", etc.; accordingly, each species is completely contained in particular things, i.e., the species "cat" is contained in all living cats. Eriugena follows Proclus in believing that a necessary condition for comprehending the root causes of the universe is an understanding of what genera are and how they are divided into species. According to this, epistemically understood genus-species relations are put in correspondence with the ontic hierarchy, so that a complete (as far as possible) isomorphism should be observed between them [Petrov 2000; Erismann 2007]. In such an account, the general onto-epistemic foundations of the future logical genus-species scheme are reinforced, in which the genus is considered primary to its species.

* * *

An important step towards scholastic rationalization of the cognitive activity is undertaken by Anselm of Canterbury (Anselmus Cantuariensis, 1033–1109) based on realistic onto-epistemology. His main objective is a rational deduction of all biblical truths with the help of reason that begins with faith and descends to particular judgments following certain rational argumentation. Thus, his "faith seeking understanding" (*fides quaerens intellectum*), borrowed from Augustine, is a rational way of getting an "ontological argument" (how Kant would call it later) in favor of the existence of God, which Anselm discusses in his treatise *Proslogion* [Davies and Evans 1998].

In the context of our book, two ontological constructs of Anselm are most significant. First, everything existing in the universe is an embodiment of Divine ideas (prototypes) into all concrete universals; therefore, the latter are as real and eternal as the very Divine ideas. Second, the Divine nature is a whole and single; therefore, its incarnations into universals and then into things must also be singles; thus, each thing is also characterized by a solid "nature".

Based on the latter idea, Anselm introduces a notion of *principal essence* (*essentia principalis*) inherent in each genus and species and the things embodying it [Holopainen 1996]. This goes back to Aristotle's belief that the basic eidetic content of an animal is its soul, which integrates all its particular *ousiai* into a single harmonious whole. This thesis will become an important contribution of Anselm to the initial development of scholastic systematics, concerned with the essentialist justification of recognition of the natural groups of organisms and finding their true places in the System of Nature [Pavlinov 2018, 2022b] (see Section 4.3.2).

* * *

A pretty specific moderate-realistic natural philosophy is developed by Roger Bacon (ca. 1214–1292).[3] Of his extensive legacy, the most significant for us is the series of three *Essays* complemented by the *Common Nature* [Hackett 1992, 2020].

The natural-philosophical standpoint of R. Bacon largely tends to Aristotelian ousiology, with an addition of Christian Platonism. It is based on an idea of Nature as a generative principle (*Natura generans*) of everything in the universe, and the latter is organized hierarchically according to the succession of its Divine creation [Maurer 1982; Hackett 1992]. The process of creation occurs as an interaction of matter and form, with the latter being considered as an active principle (*pars agens*), which initiates the entire series of successive incarnations in motion. According to

[3] Curiously enough, various authors place him among nominalists or realists, and the latter is sometimes called "extreme" [Maloney 1985].

the tradition going back to Aristotle, R. Bacon identifies matter and form with genus and species, respectively: in the *Third Essay*, he unequivocally indicates that "matter and genus are the same [and] we call distinctive features and forms species" (cited after [Appolonov 2002: 88]). Such their status is not fixed but sliding following the route of incarnation: a species-form interacts with genus-matter at a certain stage of incarnation to turn into matter and interact with other forms at the next stage. Passing through successive stages of incarnations, particular universalia (genera and species) do not exist by themselves, but as embodied in the substances of things, which are ontologically more real than what they embody [Hackett 2020].

According to Bacon, things are mutually similar due to a certain quiddity inherited in them, which he calls *common nature* (*natura communis*). It represents a concrete embodiment of a form that exists within the things and conditions their "mutual agreement" (*conveniencia*) in their essential properties [Maloney 1985]. A mental universal (*conceptum*) corresponds to this real common nature; it is generated by cognitive activity, which is aimed at revealing the common natures of things, so that natural universals precede the respective mental ones [Hackett 1992, 2020].

Based on ontic identification of matter and form with genus and species, R. Bacon correlates with them mental universals that are also called by the same names (as defined by Boëthius). Accordingly, for the successive acts of incarnations of the forms to be comprehended, the entire cognitive procedure should pursue a deductive logical division of notions. Following this argumentation, R. Bacon applies the already recognized genus-species sliding hierarchy that begins with *genus generalissimum* and finalizes with *species specialissimum*, in which mental universals are identified as either genera or species contextually depending on the particular cognitive tasks being resolved.

* * *

A significant contribution to the development of the realistic ontology of Christian Platonism is made by Albert von Bolshtedt (Albert the Great, Albertus Magnus, ca. 1200–1280). Reflecting on the comprehension of God's providence as a principal objective of cognitive activity, he considers the study of all manifestations of the material world, both universals and singulars, the basic means of achieving this end. With this, he clearly divides theology, philosophy and "science" (*Scientia*): the first is based on the truth of *Scripture*, the second on logic, and the third on facts [Gilson 2010; Tkacz 2012]. Starting from theology and turning to science, Albert concludes "that it is possible to become a complete philosopher only by learning both the philosophical teachings of Aristotle and Plato" [Gilson 2010: 386]. Observations of facts provide a principal instrument of discovering "a unity in multiplicity" and "common in different"; he treats in this vein onto-epistemic status of universals as the sought "commons" of different manifestations of being.

The philosophical basis of Albert the Great's worldview is shaped by a kind of synthesis of Christian Platonic and Aristotle's ontologies. According to Albert, in matter (genus) as a passive principle, everything is contained as *potential species*, while form, which is an active principle, concretizes plethora of potencies into substances, according to which the things acquire their *actual species* [Twetten et al. 2012]. With this, the latter, after their creation by God, remain eternal and unchangeable—and it is this unchangeability that provides an indispensable condition for their comprehension by means of rational science.

In the treatise *On the Six Principles*, Albert divides all universals into three basic categories: before things (*ante rem*), in things (*in re*), and after things (*post rem*). The universals *ante rem* are eternal Divine ideas existing in the Divine Mind as prototypes (primordial forms, archetypes) of all being, similar to Platonic *eidê*. The universals *in re* are embodiments of the ideas in the things, so they are Aristotelian *ousiai*, or substances. The universals *post rem* are concepts that emerge in the rational human mind as a result of the analyses of mutual similarities of things according to their substances. Thus, the first two categories of universals belong to the real (material) world, and the third one does to an ideal (mental) world [Tremblay 2008; Gilson 2010; Tkacz 2012].

Reflecting on the categories of our interest and trying to answers the fundamental questions of Porphyry (see Section 2.2), Albert emphasizes the subordinate position of species relative to genus and individual relative to species, and he extends to them an idea of threefold existence of universals [Tkacz 2012]. Accordingly, genera and species *ante rem* are real (extramental) principles of being, whereas they are ideal (mental) constructs, i.e., predicates, when considered *post rem* [Hackett 1994; Tremblay 2008]. This division echoes Arab kalamist Avicenna's (Abu Ali al-Husayn ibn Sina, ca. 980–1037) distinction between two hypostases of genus as a universal: *natural genus* is embodied in the essences of "one-kind" things, while *logical genus* corresponds to the former as a predicate, i.e., it tells of these things [Bäck 2008].[4] Regarded ontically, species *in re* makes individuals what they are "by their nature" by causing their qualitative certainty, while their quantitative differences depend on the peculiarities of the matter underlying them [Hackett 1994].

In outlining history of the natural-philosophical idea of the Great Chain of Being, the historian of science Arthur Oncken Lovejoy (1873–1962) notes that "Albertus Magnus, writing *De animalibus*, had already laid it down that 'nature does not make [animal] kinds separate without making something intermediate between them; for nature does not pass from extreme to extreme *nisi per medium*'" [Lovejoy 1936: 79]. In quoting this, Lovejoy refers to Kurt Ufermann's dissertation *Untersuchungen über das Gesetz der Kontinuität bei Leibniz*, 1927; he means the absence of gaps between different steps of the embodiment of genera-matter into particular species-forms.

Albert accepts the logical genus-species scheme as a basis of the rational way of describing the qualitative structure of the universe. He is perhaps the first among

[4] When outlining antique history of the genus-species problematics, this distinction is designated as natural-classificatory dualism. A few centuries later, a similar idea will be expressed by G. Leibniz (see Section 4.2).

scholastics to elaborate the method of division of notions in sufficient detail in his treatise *Questions Concerning Animals* (Aristotle's *On the Parts of Animals* is meant) and in the encyclopedia *On Animals* [Kitchell and Resnick 1999; Resnick and Kitchell 2008; Tkacz 2012]. Albert pays special attention to genus, whereas species is conceived by him as something secondary to it, as its particular. He adopts Boëthian sequence of generic steps in this scheme, including the highest (common) genus, subordinate (intermediate) genus, and proximal genus. Considering it an argumentation scheme, he emphasizes the following points: the need to proceed deductively from common to particular by a sequential division of notions; the insufficiency of dichotomous division; the importance of proper characters as *differentiae* for distinguishing genera and ordering them into an overall hierarchy; and the necessity of finding a "true" *differentia* for the correct setting of a "true" hierarchy of genera, which becomes a prototype of the *principle of the single basis of division*. Albert ensures that the adoption of these rules guarantees recognition and arrangement of genera according to their essential characters and naturally brings the entire hierarchical scheme to the final species. His verbose arguments, which occupy many pages of Book XI in the *On Animals* [Albertus Magnus 1999; Tkacz 2012], will be refined by logicians to become the most important elements of the genus-species scheme to constitute the basis for classifications in early systematics.

The treatise *On Animals* represents a multivolume encyclopedia of natural history, as it includes descriptions of all three "Kingdoms of Nature", viz., animals (26 books), plants (17 books), and minerals (5 books). It is basically a compilation focused (as Aristotle's) on the essences and how they are manifested in different living and inanimate things [Weisheipl 1980]; thus, this is (in our terminology) primarily a partonomic consideration of Nature, rather than a purely taxonomic one. The most voluminous zoological part of this encyclopedia is called by its first commentators *Summa zoologica*, by analogy with Aquinas' *Summa theologica* (see below). It consists of three main parts [Wallace 1999]: (1) a Latin translation of Arabic texts of Aristotle's works on animals prepared by the philosopher and mathematician Michael Scotus (1175–1232), (2) an abridged summary of these works by Albert himself, and (3) ample descriptions of the main representatives of the animal kingdom, compiled by Albert's disciple Thomas Cantimpratensis (1201–1272).

Anticipating a detailed outlining of the genus-species scheme in the Book XI of the *On Animals*, Albert promises in its Book I to present descriptions of all animals "according to [their] genera and species" [Albertus Magnus 1999, Vol. I: 47]. Indeed, at the very beginning of this Book, it is said about "those who walk according to their species, about those who fly according to the whole nature and species of flying creatures", etc. (*loc.cit.*). Elsewhere, he writes that:

> diversity or differences in the parts [of animals] is a matter of [their] increasing or decreasing, and in fact [they] are not understood as members of the same species, but of the same genus. I use "genus" in the sense of "genus of birds" or "genus of fish", since each of them contains many smaller genera and species.
>
> *[Albertus Magnus 1999, Vol. I: 52]*

It is noteworthy that Albert, when describing the particular animal species in their meaning close to the modern one, refers to the generative idea according to what the *Genesis* says. So he writes in the Book I that

> Species [...] participate in an eternal existence by reproducing one individual from another, and this can only be so if the force forming the animal is transmitted in the seed from the one who reproduces.
>
> *[Albertus Magnus 1999, Vol. I: 72]*

By this, Albert the Great certainly deserves to be counted as a direct forerunner of what is now known as the generative account of species.

Descriptive sections of Albert's encyclopedia are in many ways similar to the then popular bestiaries and herbals: categories of genus and species are often interchangeable as mixed partonomic-taxonomic units of various levels of generality, though they are mentioned terminologically only in few instances [Stannard 1980]. For instance, a generic category is indicated in the titles of some sections of the zoological series (e.g., XXIV. *De calais duobus generibus falconum*), and generic or species ranks are occasionally mentioned in the descriptions of particular groups [Albertus Magnus 1999, Vol. II]. In the "Synopsis" of the botanical series [Albertus Magnus 1867], there is a section "*Divisiones genera species plantarum*", in which the higher genera (*Genera summa*) are distinguished, which are trees, shrubs, herbs, algae, fungi, and one of the sections is entitled "*Genera et species secundariae*".

3.3 CONCEPTUALISM: SPECIES CONCEPT REFERS TO SOMETHING REAL

Some important provisions of conceptualistic onto-epistemology are noticeable in the works of proto-scholasticism (Augustine, Boëthius), whereas medieval conceptualism in its mature form is formed in the 11th–12th centuries based on a peculiar combination of certain fundamental ideas of realism and nominalism. From the former, borrowed is an idea that genera and species as universals are real to the extent that they are rooted in the eternal archetypal ideas of Divine Mind. The influence of the second lies in that these universals arise as concepts in the human cognizing mind due to rational insight of the results of perception and comparison of individual things. For this reason, conceptualism is often, but not quite correctly, presented as a "moderate nominalism". One of its salient features is recognition of an active role of the cognizing subject in the cognitive process: subjective intention focuses attention on certain manifestations of being, which makes them concrete objects of cognition.

* * *

The beginning of the transformation of realism of Neoplatonic kind into conceptualism tending to Aristotle is put by works of Pierre Abelard (Petrus Abaelardus, 1079–1142) [Neretina 1994; Saint-Andre 2002]; however, many researchers attribute his position to nominalism [Gaydenko and Smirnov 1989; Marenbon 1997; Shatalkin 2012; King 2004]. His teachers were philosophers with opposite views, the Platonic realist Guillaume de Champeaux (1068–1121) and the nominalist John Roscelin, briefly characterized below [Marenbon 1997; King 2004]. In respect of our interest, Abelard's *Dialectica* and *A Treatise on Understanding*, along with his commentaries on Aristotle, Porphyry, Boëthius, etc., are most significant [Neretina 1994; Marenbon 1997; King 2004; Ernst 2011].

Considering universals and genus-species relations, Abelard proceeds from Porphyry's and Boëthius' treatment of these categories and begins with exposing their fateful questions in his own interpretation: (a) do genus and species exist independently of mind, as they are laid as primordially given at the foundation of being, (b) if they exist independently, are they corporeal or incorporeal, (c) whether they are separated from sensible things or embodied in them [Maurer 1982; Neretina 1994]. Answering them, Abelard turns to the relation of Divine Logos to human concepts: he believes that all universals are initially present in Divine Mind as archetypes, while due to their embodiments in things, they are found in the latter as corporeal substances.[5] Man seeks the common in particulars by comparing the latter with each other and likening them (Augustine's *similitudo*) to Divine ideas that generate them. It is this similitude that yields genera and species as universals, belonging to reality in two ways—due to both their causal connection with Divine archetypes and their incarnations in things. This is the first connection that determines the reality and naturalness of universals and makes the things naturally similar in their generic and specific belonging; were it not so, the things would be mutually similar only accidentally [King 2004; Panchuk 2016]. Therefore, as Abelard's *Glosses on Porphyry* say, "the word 'man' names men on the basis of a common cause [...]. It is called a 'universal' on account of this cause. It also *constitutes a certain understanding*, a common one, not a proper one" (cited after [Spade, 1994: 41], italics in original). According to this interpretation, species exist in things as incorporeal substances, and genera likewise exist in species [King 2004].

In his *Dialectics*, Abelard points out that the connection through likening is established by the cognizing human mind, which generates the meanings of general notions (concepts, including genera and species) in two ways: by referring to the Creator's archetypal ideas and to their embodiments in things. Objecting to their purely nominalistic account, he emphasizes that had the universals no their own being immersed in reality, an insurmountable gap would arise between reality and the thoughts of it: by this presumption, conceptualism contrasts to nominalism [Saint-Andre 2002; Ernst 2011].

Thus, according to Abelard, genera and species as universals depend on the things and on their nature, not being these things in themselves; they exist in different fashions, being nevertheless interconnected. They can be thought of as *equivocations*, i.e., different meanings of the same, which is their common cause, namely Divine archetypes. With the latter's actualization, they "split" into *substances* in the things

[5] On this basis, the philosopher Wilhelm Windelband (1848–1915) attributes to Abelard an anticipation of the formula "*ante rem – in re – post rem*" [Windelband 1892].

themselves and *concepts* corresponding to them, in which both individuality and commonality are enclosed. Each genus and species has its own being as a special "conceivable entity" endowed with its own meanings, which is not identical to an actual thing, but a kind of its mental similitude. The choice of a particular meaning of the multifold being is set by the cognizing human mind, which makes cognitive act conscious and purposeful, and the grasping respective meaning becomes *intentional* (*conceptio intentionalis*) [Neretina 1994; Cameron 2007].

In this regard, Abelard introduces the notion of *intention* as an active agent of cognitive activity, directing it to a certain object or its aspect. As a result of the selective action of particular intentions, each universal as a general notion is mentally divided into an array of particular conceptions generated by different intentions of a cognizing mind. This is a cornerstone of Abelard's conceptualism: there is no separate ontology and epistemology in it, as in realism and nominalism, but there is instead their interrelation, making them a combined onto-epistemology, in which universals (genera and species) are endowed with a twofold mutually conditioned being (see also Section 1.1.1 on this).

In his *Dialectics*, Abelard touches on the relation, previously discussed by Boëthius, between divisions of genus into species and whole into parts, and he also finds significant parallels between them [King 2004]. He believes that, just as a meaningful division of a whole into parts cannot be accidental, but should discern its natural constituent elements ("carve at its joints" according to Plato), a meaningful division of genus should lead not to indefinite whatsoever species, but to the certain *proximal species* (*speciei proxima*); the latter roughly correspond to the *proximal genera* of Boëthius. Abelard admits in both cases a possibility of multiple variants of division, but considers the only one to be "natural" and therefore true; perhaps this idea correlates with the abovementioned principal essence of Anselm of Canterbury.

In this treatise, Abelard pays much attention to the relations between genus and species as logical categories, with which Aristotelian syllogistics could be built consistently. As with his predecessors, whom Abelard comments at length, these categories are understood conjugately and relatively: what is a species with respect to a certain inclusive genus, turns out to be a genus for the species, into which it is divided. At the same time, what is predicated by a species about things is also predicated about them by the genus that predicates about this species; as a result, encaptic sliding hierarchy of deductive definitions is framed, beginning with the highest genus and ending with the lowest species. For this, his *Dialectics* is counted as the beginning of the formation of a full-fledged scholastic logic [Marenbon 1997; Brower and Guilfoy 2004; Wilks 2008], including the genus-species scheme of division of notions, which will be incorporated by scholastic systematics.

* * *

During the period of mature scholasticism, the greatest contribution to the development of a conceptualistic view of Nature and the ways of its cognition, including understanding of genus and species, is made by Tommaso d'Aquino (Thomas Aquinas, ca. 1225–1274). His position is largely shaped by the influence of one of his teachers, Albert the Great, although the latter and his disciple disagree in many ways (Plato and Aristotle are recalled) [Gilson 2010]. Aquinas sets out his main ideas, which are of our interest, in his *On the Truth of the Catholic faith* (the early 1260s) and more complete *Summary of Theology* (the early 1270s), supplemented by commentaries on Aristotle, Boëthius, Pseudo-Dionysius, etc.

The basis of Aquinas' ontology (in an aspect most relevant here) constitutes an idea of the sequence of acts of interaction between primary matter and substantial forms [Wippel 2000; Gaydenko 2001]. In each of these acts, an infinite potentiality, contained in matter and determining everything that possible (*ens*), is actualized through an action of a concrete form as an accident, which acquires a finite genus and species embodied in a thing that really exists (*esse*) [Summa 1981]. The totality of such acts results in a linear hierarchy (as of Pseudo-Dionysius) of all manifestations of being according to their perfection due to their proximity to Divine Mind. This hierarchy supposed by Aquinas, like that of Albert the Great, is continuous in the sense that there are no gaps between particular species: "the lowest in any higher genus always comes into contact with the highest in the lower genus" [Thomas Aquinas 2001: 335].

Thomas Aquinas interprets the problem of universals following Albert the Great and recognizes their threefold existence: *ante rem* (before things in Divine Mind), *in re* (in things), and *post rem* (after things in human minds). When analyzing genus-species ontology, he distinguishes between *universal* and *common* (corresponds to *common nature* of R. Bacon, see above): the former refers to the general notions as such (beauty, goodness, etc.) and the latter refers to the quiddities of concrete things (animality, humanity, etc.). In addition, Aquinas demonstrates his Aristotelianism by a statement in his *Disputed Questions on Truth* that "individual things are beings, and more than universals: after all, universals do not subsist by themselves, but reside only in individuals" [Sancti Thomae de Aquino 1970: 115].

However, Aquinas is not quite consistent in his judgments. In some places, he rejects the reality of both common nature and universals, considering them predicative attributes of cognizing mind. He presumes that all what the latter observes is but a similarity of things, which is close to Ockham's nominalism. In other places, he argues that acknowledging the reality of individual (accidental) differences between things is impossible without acknowledging reality of something common to them (i.e., their common nature), viz., those universals that embrace accidents individualizing things: this corresponds to a realistic position. Such discrepancies seem to be explained by that Aquinas recognizes the reality of common nature with regard to some accidents and denies it with regard to others, with the former accidents being intentionally fixed by a cognizing mind as different aspects of essences [Vdovina 2009]. This conclusion is evident from Aquinas' thesis that "we must [...] admit the existence of an active power if we are to explain human knowledge" [Gilson 2002: 242], which fits the conceptualism of Abelard.

The last thesis of Aquinas receives a particularly accentuated interpretation by Giovanni di Fidanza (Bonaventure, ca. 1218–1274). His general view of essences is most consistent with that of Aristotle: in his treatise *Mind's Road to God*, he writes that "every thing presupposes as many forms as it has properties. Consequently, in every thing there are many forms arranged hierarchically in such a way that they form a unity" [Gilson 2010: 339]. This leads (in modern terms) to a pluralistic position regarding species classifications, which multitude is a result of fixation of certain forms (essences) for particular cognitive purposes.

This means that the standpoints of Aquinas and Bonaventure, rooted in Pierre Abelard's ideas, should be regarded conceptualistic, according to which knowledge about an object is its selective representation in a cognizing mind [D'Onofrio 2008; Vdovina 2009]. Therefore, it is hardly by chance that some recent authors ascribe to Aquinas' species, thought of as an aggregate of actualized things united by a common nature, an ontic status of *natural kind* in its modern conceptualistic understanding [DeYoung et al. 2009; Brower 2016] (see Section 6.2).

It is noteworthy how Aquinas, in his *On the Truth*, closely associates the existence of species with the forms understood as Divine archetypes (he means angelic forms and species, but it does not matter to us). He believes that

> Any distinction in [things] based on a difference of form divides them into separate species, while any distinction based on matter makes them numerically different individuals. [...] With incorruptible beings, there is only one individual to each species; in other words, there is neither numerical distinction nor matter. This is because the individual is incorruptible and therefore sufficient to assure the conservation and differentiation of the species. With beings capable of generation and corruption, many individuals are necessary in order to insure the conservation of the species. Accordingly, individuals only exist as numerically distinct within a species in order that the species might exist as formally distinct from other species. [...] Matter provides the basis for numerical distinction in each species. [...] Beings of the same species but differing in number as distinct individuals have a similar form united to different matters. [...] Accordingly, the multiplication of species adds more nobility and perfection to the whole universe than the multiplication of individuals in one and the same species.[...] The species and forms determining them are differentiated by the greater or lesser degree of the act of existing they share. Species are like numbers. Add or subtract a unit and the species is changed.
>
> *[Gilson 2002: 182, 195, 244]*

From all of the above, ending with an extensive quotation, it follows a conclusion, important for a proper understanding of the history of the species problematics, that Thomas Aquinas seems to separate ontologically the notions of form and species, which are put very close in their meaning by many scholastics. This distinction can be understood, in our terminology, in such a way that he refers form (essence) to partonomy, and species to taxonomy. The connection between them is provided for by that

> abstraction is first of all the agent intellect's apprehending in every material thing what places it in its own species, leaving aside the individuating principles belonging to its matter. Just as we can consider separately the color of a fruit without taking account of its other properties, so our intellect can consider separately, in the phantasms of the imagination, what constitutes the essence of man, horse or stone, without taking into account what differentiates the individuals in these species.
>
> *[Gilson 2002: 242]*

It is important to notice that in both interpretations, these categories are not specific to any particular manifestation of being. Indeed, Aquinas' description of the hierarchy of Nature, as that of Eriugena's, is based on the overall genus-species scheme as an all-encompassing representation of the diversity of things and ideas. With this, he follows the established tradition and considers species indivisible and in this sense finite (*species individua et infima*).[6] Following provisions of this scheme, he examines in the *Summary of Theology* a fundamental question of the existence of God and asks, whether can He relate as a species to a certain genus, and answers it negatively [Gaydenko 2001]. Also, Aquinas regards two forms of movement towards truth, sensuous and intellectual, to be different species, or steps of the same genus of action, and in another place, he writes about genera and species of motion, meaning by the latter any effect of some primordial cause (form) on matter [Gilson 2002].

* * *

A different line of the development of medieval conceptualism is marked by John (Ioann) Duns Scotus (1266–1308): in the foundations of his onto-epistemology, set forth in his treatises *Ordination* (also *Opus Oxoniense*) and *Treatise on the First Principle*, he reveals himself to be a Christian Aristotelian. His standpoint is interpreted by various researchers as conceptualism, moderate realism or nominalism for the same reasons as that of Abelard: because of its rather complex configuration [Motroshilova 2000; Appolonov 2001; Gilson 2010, 2018].

The onto-epistemology of Duns Scotus is based on the following basic assumptions [Appolonov 2001; McGinley 2008; Gilson 2018]. The supreme reality of all being is a *potentiality* of Divine ideas, whereas the material world consists of their final incarnations, i.e., the things (individuals) endowed with an *actual reality.* A moving force of the process of cognition is its aiming at the mystical insight of God. Thus conceptually framed, the mind turns to concrete things and, based on rational thinking of their sensory perceptions, formulates an initial rational truth (approximately corresponds to *species expressa*), which then ascends to higher-order universals. Each step of this intellectual ascent corresponds to a certain notion (species or genus) endowed with a *formal reality.* These notions, according to Scotus, are not "free" (abstract) creations of reason, for they have certain correlates in being and are real, albeit in a different meaning than things. Such understanding of the concept is born by his conviction that "universal is an existent (*universale est ens*), since non-existent cannot be cognized by reasoning" (cited after [Kurantov and Styazhkin 1978: 114]).

Duns, as other scholastics, ascribes genera and species a universal character: say, categories of different colors and of different organisms are equivalent ontic entities. Characteristic of his ontology is an idea of the endowment of each genus and species with a *common nature* inherent to it (*natura communalis*; almost the same as R. Bacon's *natura communis*). It results from a specific combination of matter and form; it is this common nature that Duns considers an intelligible essence. It is

[6] Obviously, the individual status of species thus understood has no relevance to its modern account (see Section 6.2).

incorporeal and in this sense "indefinite"; its embodiment into each genus/species, and ultimately into things belonging to the latter, is achieved by its *individuation*, due to which a thing acquires its own specific form (*forma individualis*) and its specific *thisness* (*haecceitas*), corresponding to Aristotelian whatness. The degree of individuation (definiteness, commonality of nature) of forms increases from the highest genus through intermediate genera to the lowest species and finds full completion in the individuals. With this, Duns asserts in the *Ordination* that each genus and species is shaped by "a commonality that belongs to Nature independently of mind, which is also true for its unitness; with this, commonality belongs to Nature by its very essence, while unitness belongs to Nature through something that lies outside of Nature" (cited after [Noone 2003: 110]). According to this metaphysics, "if species would not already possess some unitness—however, of a lower order as compared to numerical unitness of a particular individual,—then our concepts would not correspond to anything at all. Conversely, commonality of species, which is fragmentary found in various individuals, always carries a certain sign of unitness" [Gilson 2010: 454].

A peculiar realistic nature of Duns' account of universals is manifested in his regarding a possibility to interpret concrete things as parts of their species, and species as parts of their genus [King 1992]. However, their relations are not completely partonomic; in fact, they are partly taxonomical, as is seen from the following. According to him, a genus is embodied in its species and remains the same genus in this embodiment (which is Eriugena's idea), so it differs, as a universal, in this respect from a thing, which wholeness is not embodied in its parts [*op.cit.*]. This position of Duns Scotus seems to be relevant to the modern dilemma of consideration of species either natural kind or individual (see Section 6.2).

Analyzing common nature as actually existing, Duns Scotus argues that it is neither pure universality nor pure individuality (*ens reale, nec universale nec particulare*). Its comprehension as a universal faces a contradiction: it is impossible to consistently deduce a singular from a general, and, conversely, a general from a singular without certain provisions. To remove this contradiction, Duns introduces an onto-epistemic assumption of *univocation* as an agreement in the meaning of the judgments about God and all being. It is founded on a belief that the latter results from the embodiments of the integral Divine nature in a multitude of common natures of genera and species and thisnesses of things belonging to them. This assumption, which echoes an idea of similitude of Augustine, Bacon, etc., allows for a possibility to judge about a common in particular things, and to substantiate knowledge of things by knowledge of their common [Pini 1999, 2005].

Following Aristotle, Duns writes about an endowment of each thing with different actual substances (equivocation of Abelard). Transferring this multiplicity to genus and species and considering it potential, Duns in *Questions on Aristotle's Categories* states that "*concepts* of the same *species* can be ascribed to different realities, because *concepts* are ascribed to these realities as far as they are conceived by the intellect" (cited after [Vos 2006: 286], italics in original). To substantiate the validity of such a pluralistic view of common natures of genera and species, Duns appeals to a specific active cause contained in the cognizing mind. This is Abelard's *intentionality*, i.e., an aspiration to cognition of certain aspects of common nature of species, which is actualized in different ways in thisnesses of its individuals. Accordingly, every particular conceptualization is a result of the selective work of intellect, which "snatches" from common nature what is most significant for it at the moment. As said above about Abelard's position, this general idea is one of the key ones in conceptualism.

In a more particular aspect important to us, Duns' methodology fits into the genus-species scheme of the hierarchy of predicates developed by Boëthius. All its sequential divisions are substantive, i.e., involve the essences as such (animality, coloration, etc.) [Paasch 2014], so for Duns, as for other scholastics, this scheme is primarily partonomic. On the other hand, thinking of individuals, Duns invariably points out that they are combined in certain aggregates (*collectiones*) by common nature of respective genera and species [McGinley 2008]. This obviously means a taxonomic account of these categories, which is secondary to partonomic one.

* * *

A very peculiar worldview of Nicolaus Krebs (Nicolaus of Cusa, 1401–1464), which marks a transition from the Middle Ages to the Renaissance, is based on a mixture of Neoplatonism and biblical theism, seasoned with a significant amount of Hermetic mysticism [Hopkins 1978; Tazhurizina 1979]. One of his key ideas, set forth in the treatise *On Learned Ignorance*, rather popular in his times, is that being and knowledge about it are organized hierarchically. The universe resides in a descending hierarchy of etxra-mental universals of different levels of generality, its origin is the Creator as an "absolute universal" (*universalissimum*), and its completion is the embodiment of essences in individual things. The knowledge of being resides in an ascending hierarchy of mental universals, which origin are things as embodiments of the lowest universals (*specialissima*), and which ends with the comprehension of God.

Categories of genus, species, and form are important elements of this onto-epistemology. Cusanus conceives their interrelations quite complexly, with placing different context-dependent meanings on them. Indeed, on the one hand, in his *On Learned Ignorance*, he regards genus and species as steps of the sequential "outflow" of higher-order universals into lower ones (an evident echo of Plato's emanation). So, he believes that

> The universe, before its concrete incarnating in each particular individual, goes through three steps. The universe represents a universality of ten highest universals, then there are genera, and then there are species, [which] also represent universals at their steps, in a certain natural order existing before the things that constrict them into a concrete actuality. With this, in view of this concreteness, the universe is found only expanded in genera, genera are found only expanded in species, while individuals exist actually. In them, everything in the universe resides in a specific way. [...] With such consideration, it is clear that universals actually exist only in a concrete certainty, [...] in what actually exists.
>
> *[Tazhurizina 1979: 113–114]*

On the other hand, discussing the relation between these categories in a cognitive aspect in his *The Layman on Wisdom*, Cusanus refers form to ontology, while genus and species to epistemology, thus connecting them with the different aspects of being. He believes that intellect, "when perceiving [things] in the mode of matter, creates genera; perceiving in the mode of form, it creates differences; and perceiving in the mode composed of them, it creates species" [Tazhurizina 1979: 428]. With this, "genera and species [...] are rational essences that intellect creates for itself based on combination and separation of sensible things. Therefore [...] they exist by their nature later than sensible things, of which they are similitudes" [*op.cit.*: 392]. Detailing the cognitive scheme developed by Thomas Aquinas, Cusanus asserts in his *On the Pursuit of Wisdom* that the source of knowledge is an essential form (*forma essentialis*) contained in things and making them what they are "by their nature". He regards intelligible species (*species intelligibilis*) an active force of cognition, and the main element of knowledge is *species formalis* as a mental image of the essential form [Spruit 1995].

It is to be added that Nicholas of Cusa, in some of his contemplations, follows Duns Scotus in treating genera and species taxonomically in their epistemic consideration. This is evident from his statement that "intellect creates universals by comparison" [Tazhurizina 1979: 114], which results in that "dogs and other animals of the same species are united due to their common species nature, which is inherent in them" [*op.cit.*: 115]. But at the same time, Cusanus believes that such universals "cannot have an absolute being" [*op.cit.*: 114], existing actually in things.

One of the remarkable manifestations of an overall integrity of the universe of Cusanus is the continuity of all things, in which, like in Albert's and Aquinas', there are no gaps between different steps of the abovementioned "outflow". The continuity of genus-species relations implicitly refers to the antique Great Chain of Being in its specific biblical treatment [Lovejoy 1936; Wilkins 2003a, 2009a]. In implementing this idea, Cusanus presumes such order of genera and species, in which genera shape a "skeletal" frame of the hierarchy, and within each genus, there is its own gradation of species from higher to lower. Accordingly,

> between different genera, into which the universe is actualized, there is such connection of lower with higher that they coincide in an intermediate, and this order of combination, in which the highest species of lower genus coincides with the lowest species of higher genus directly adjacent to it, makes the cumulative universe a single continuum. [...] With this, the adjacent species of adjoining lower and higher genera are interconnected not by some intermediate link [...], but by a third species, whose individuals are scattered along their steps of differences, with none of them belonging to the two former species equally, as if it were composed of them, but concretizes a unified nature of its own [third] species, and only in comparison with others they look like being composed of the natures of lower and higher species.
>
> *[Tazhurizina 1979: 145]*

3.4 NOMINALISM: SPECIES CONCEPT REFERS TO NOTHING REAL

Nominalistic account of the categories of genus and species dates back to the antique philosophy of Stoicism [Bronowski 2007; Chiaradonna 2008]; in scholasticism, it begins to take shape in the 11th century [Neretina 1994; Spade 1994]. Its general standpoint is opposite to realistic: only particular things really (objectively) exist in Nature, while any general concepts (universals) are mental entities existing only in the cognizing mind in the form of notions and their names, and there is no reality (either material or ideal) behind them.

This idea, and in its extreme manifestation, is expressed by the Johannes Roscelin of Compiègne (ca. 1050–ca. 1122): he calls the names denoting general notions (genera and species) just an "air concussion" (*flatus vocis*) [Kluge 1976; Gilson 2010]. His extreme lies, for example, in that Roscelin considers a "word play" not only the division of genera into species but also wholes into their parts [Marenbon 1988]. Roscelin does not leave any written treatises; his only original text that will reach us is a polemic letter to his scholar Pierre Abelard. But it is the Roscelin's emphasis on names (*nomina*), as denoting nothing really existing, that becomes a motive to call this doctrine *nominalism*.

The most notable and influential actor of medieval nominalism is William of Ockham (Gulielmus Occamus, ca. 1300–1347/1349), a scholar of Duns Scotus. His views on being and ways of knowing it are set forth in voluminous treatises, of which the most significant for us is the *Sum of Logic*. In developing nominalistic world picture, Ockham regards an "accepted view" of universals as real entities an "absolutely false and absurd, [...] the worst error of philosophy" (cited after [McCord Adams, 1989: 13, 30]). To the contrary, he believes that there is nothing in common in things, but their similarity, that would correspond to the universals in their realistic understanding.

Ockham's nominalistic ontology is based on a deep theology, the core of which is his belief in the absolute free will (*potentia absoluta*) of God the Creator of all things [Boehner 1964; McCord Adams 1989; Copleston 1993b; Maurer 1999]. He criticizes realists and conceptualists for their constraining the Creator's will by a necessity to be guided in His thoughts and deeds by certain "binding rules" called Divine universals by philosophers. Ockham suggests something different: there are no general ideas (universals) in the Divine Mind that constrain His will in any way [Klocker 1992]. Instead, God creates all the things as such, i.e., without mediation of these universals, so they cannot exist in the created world as embodiments of Divine universals—neither in individual things (against Aristotle) nor outside them as independent essences (against Plato) [Maurer 1999; Motroshilova 2000]. In Ockhamian ontology, the Divine plan of creation contains ideas not of universals, but of individual things, so each of the latter exists "by itself" with its own thisness, and not as an embodiment (individuation) of a certain genus or species.

Ockham's denial of eternal general ideas and universals as "redundant essences" is aphoristically expressed by a formula "essences must not be multiplied beyond necessity" (*entia non sunt multiplicanda praeter necessitatem*). Its well-known "label" is *Ockham's* (or *Occam's*) *razor*, which means that everything unnecessary should be cut off in any reasoning about being by means of a "logical razor". The latter, formalized as the *principle of parsimony*, will make William Ockham very popular among modern philosophers and methodologists of natural science [Sober 2015]. However, in fact, he himself does not invent this "razor": the author of quite clear formulation of this principle is Duns Scotus, and the very term will appear at first in the works of the 17th–18th centuries as a metaphor of "blade of nominalism" (*novaculum nominalium*) [Thorburn 1918; Maurer 1999].

Rejecting genus and species as universals at the level of ontology, William of Ockham pays great attention to their epistemic function. The latter is based on an assumption that objectives of the knowing of the created world are the things and not universals. Considering the latter's genesis, Ockham details Albert's conception of intention: primary intention directs cognition towards the creation of primary conceptions (species, e.g., "cat"), and secondary intention—of secondary conceptions (genera, e.g., "animal"). Every general notion (*conceptum*) is formed in the "soul" of the cognizing subject as a kind of representation of multiplicity of real things, impelled by a certain cognitive intention. The things cannot be cognized based on a prior comprehension of universals, as they do not exist in reality: instead, knowing begins with a sensory perception of things, which precedes any conceptualization. As a result of perception, an unconscious primary intuitive image (*fictum*, close to Aquinas' *phantasma*) of each thing arises. Its formation is the first act of knowing (*actus intelligendi*); it is then transformed into a conscious mental image (*habitus*) of a thing, and there are no intermediate links between the latter and its image (*sine omni medio*). The terms denoting universals are but units of language, and their generality lies not in the universals' being (ontology), but in their unified use by people denoting certain things (epistemology) [McCord Adams 1989; Panaccio 2004; Vdovina 2009; Paasch 2014].

For Ockham, it is only the similarity of things that explains and justifies the use of generic and specific notions to describe their diversity. For example, the statement "human is a species" refers not to a certain imaginary human essence, but just to a similarity in some of the men's features that is grasped from their comparison. In this regard, it is of interest how Ockham deals with the "Aristotle's problem" (see Section 2.1.3 on it). Indeed, the sensually perceived things are similar to each other in various features and with various degrees, so it is possible to group them differently accordingly to cognitive intensions based on specific similarity patterns.[7] And since similarities between things are quantitative, it is possible to fix various degrees of their "agreement" (*assensio*) as various steps of genus-species division. Conceptualization of species may be based on the mental image of even a single thing, whereas for developing a more abstract notion of genus, it is necessary to acquire images of several things [McCord Adams 1989; Panaccio 2004].

3.5 GENUS-SPECIES SCHEME

One of the important (for our consideration) results of the rationalization of cognitive activity by scholastics is the development of a sufficiently formalized apparatus of logic as a system of consistent inference of true ideas about cognizable world. A purposeful movement in this direction is started by Stoics and Aristotle, continues by Neoplatonists (Porphyry, Boëthius) and early scholasticism, to reach its apogee in the period of mature scholasticism [Makovelsky 2004; Gabbay and Woods 2008]. This is evidenced by treatises on logic with an exposition of its foundations,

[7] Such is a position of modern phenetism in systematics, supported by positivist philosophy of science [Pavlinov 2018].

conceptual apparatus, and procedures authored by Abelard, R. Bacon, Ockham, Buridan, etc. In addition to the works of these luminaries of medieval philosophy, who develop conjointly metaphysics and logic, special manuals on the latter are regularly appearing since the 13th century [Shuman 2001; Martin 2009]. In them, the notions of genus and species are defined and standard schemes of their application are outlined.

An important part of the medieval logical apparatus is the *genus-species scheme*, with which the structure of the worlds of both ideas and things is analyzed. From our perspective, its elaboration seems one of the most significant results of the conceptual development of scholasticism. Based on this scheme, a classificatory cognitive program will take shape in the 15th–16th centuries, and one of its applications will give rise to early biological systematics (see Section 4.3.2). That is why the latter's conceptual framework will be shaped basically by the notions of genus and species developed by scholasticism as the focal elements of the scheme in question.

It is important to emphasize that the genus-species scheme, as an epistemic tool, becomes developing initially on a profound ontic background, which includes an idea of the hierarchy of genera and species as extra-mental universals [Erismann 2007; Gilson 2010]. According to Neoplatonists (Porphyry, Boëthius), this hierarchy appears due to a descending cascade of causality, in which genus is a cause of its species. Early Christian theologians (Eriugena, Augustine, etc.) refine this tenet based on the biblical *Genesis*, so they consider the hierarchy in question as a result of Divine creation of genera and their subsequent fragmentation into species [Marenbon 1988].

From this general world picture, a thorough cognitive system is derived, which is based on the hierarchy of genera and species as mental universals; sometimes it is called *scala praedicamentalis* [Dardağan 2018]. Eriugena is one of the first to formalize it in his treatise *On Nature* and to represent the logic as "that art which divides the genera into species and resolves the species into genera, which is called *dialectic*" ([Erismann 2007: 79], italics in original). He substantiates it by emphasizing that

> dialectic follows the ordered and hierarchical structure of reality. [...] The ontological ordering of beings into genera, species and individuals is not an *artificial* conjecture of the human mind, but a transposition of the structure of reality: the logical and ontological classifications are similar and express an identical reality.
>
> *([Erismann 2007: 79], italics in original)*

This dividing procedure constitutes the basis of a deductive argumentation scheme, in which particular judgments are derived from more general ones by their "divisions"—that is, species from genera as their detailing. Along with this, Eriugena considers an inductive scheme, according to which particular judgments constitute the basis for the formulation of general ones; however, he also puts it into the system of genus-species relations, building it in an ascending order. It should be added that nominalistic scholastics (like Ockham), denying reality of universals at the level of ontology, actively use a deductive argumentation scheme to arrange genera and species as mental universals into a single hierarchy.

* * *

Viewed from an epistemic perspective, the genus-species scheme is a logical (mental) procedure: it deals not with the real universals themselves, but with the corresponding notions and their definitions. In it, as it is initially outlined by Aristotle, the relations between genus and species are firmly set: genus predicates of species, and the reverse course of thought (from species to genus) is prohibited. In this sense, species always occupies a subordinate position in relation to genus by being its detailing; therefore, in scholasticism, the main attention is paid to genus and its divisions. The treatment of a universal, fixed at some step of division, as a genus or a species, is quite simple and formal, depending on whether it is further divided or not: if it is divided, it is a genus, otherwise it is a species.

The upper and lower boundaries of this hierarchy are designated by Boëthius as the *highest genus* (*genus generalissimum*) and the *lowest species* (*species specialissimum*); in later versions, *genus generalissimum* is replaced by *genus summum* (sometimes also *supremum*), and *species specialissimum* by *species infima.* Successive steps of generic division are designated as *subordinate genera* (*genera subalterna*); the last subordinate genus, which is divided into species, is designated as *proximal* (*genus proximum*). In some versions of this scheme, additional lower steps are distinguished, which are designated as *subordinate species* (*speciei subalterna*). Whatever genera and species are identified by applying this logical scheme, they are formally treated as classification units.

Logical division begins with the fixation of the highest genus, which represents a certain essence (substance) in its all-inclusive understanding, and ends with a species that is not divided by respective essence, so Aristotle calls it *indivisible* (ἄτομοη). Neither particular substantive meanings nor the number of division steps are fixed: all is determined within the framework of a particular cognitive task. Thus, in practical application, the genus-species hierarchy turns out to be *rankless* and *sliding*, so that what in one context is a genus, in another may become a species, and vice versa. What is important for us is that the particular task also determines the step, at which the sequence of logical divisions of generic notions terminates, and this level is conventionally designated by species notion.

The terms for the universals according to their position in the genus-species hierarchy change over time. The most popular terminological guise of this scheme, which will be mastered by early systematics, looks like that:

Genus summum
  Genus intermedium
    Genus proximum
      Species infima

The genus-species scheme, with its sequential definitions of notions, is universal in the sense that it is applied to the diversity of any kind of objects. The possible substantive understanding of this scheme is of two kinds; in the terminology adopted here, it can be partonomic or taxonomic (see Section 1.1.2). In the first case, the scheme describes the diversity of attributes and properties of things (including organisms), so it is a partonomic division, in which genera and species identified correspond to the parts (*genera subalterna* or *species infima*) of a whole (*genus summum*). In the second case, the scheme provides a taxonomic division of certain aggregates of things, starting from the broadest (i.e., all organisms) and ending with the narrowest (i.e., human tribes). An essentialist vision of Nature presumes that the partonomic scheme is primary with respect to taxonomic one: first, a hierarchy of essences (partons) is revealed, and then a hierarchy of aggregates of things (taxa) characterized by these essences is elaborated based on it.

* * *

Some formal aspects of the genus-species hierarchy in its refinement by logicians look as follows [Gorsky 1983]. The descending from one level to another must always involve *division*: a certain genus is "detailed" at the next step by at least two subordinate genera or species. This division should be carried out from top to bottom according to a *single basis of division*, which refers to the same essential property throughout the whole hierarchy: simultaneous use of different bases is prohibited, because it can lead to mutually exclusive divisions that violate the logical *principle of the excluded third*. The latter principle requires that the logical division be *dichotomous*: every judgment about a universal may be either true or false. It should provide *discrete* notions (they should mutually exclude each other), be *exhaustive* (the genus is divided into its species without remainder), and *continuous* (the hierarchy should be without semantic breaks).

Each notion fixed at a certain division step is provided with a definition following a certain logical rule: it is defined *by generic common and species particular* (*per genus proximum et differentiam specificam*). The generic common indicates a property that corresponds to the higher-level notion (as a logical "genus"), in which context the one being defined is individuated, and the species particular corresponds to a property that separates this defined notion (as a logical "species") from others, recognized in the same "genus" along with this one. This means that the species are defined not as such, in an "absolute" sense, but relative to each other within a certain genus. The properties considered in this regard should ideally be not accidental, but essential, i.e., referring to those forms (*eidê*) that structure Nature and allow it to be "carved at its joints". Aristotle exemplifies this concern by two definitions of man: he may be defined either accidentally as a "biped without feathers" or essentially as a "reasonable animal". In both variants, the first part of a definition is generic common, and the second part is species particular.

Observing the principle of a single basis of division faces a serious problem of potential multiplicity of these bases. It is a consequence (according to Aristotle) of a composite nature of the substantial quiddity of each thing that integrates several essences (*ousiai*). As a result, for a certain set of things, it is possible to elaborate several genus-species hierarchies following different division bases. From a nominalistic standpoint, this is not a problem at all: since there are no real universals in Nature, any genus-species divisions are arbitrary and therefore equally admittable. From a realistic standpoint, the problem is ontological: since the general hierarchical structure of Nature is provided for by a single cause (The One, Divine Mind, etc.), there should be the only one true (natural) hierarchy of universals. One of its decisions is suggested by Aristotle, who divides animals using several essential properties: this leads to the violation of some important formal provisions (including dichotomy) but allows distinguishing the sought natural groups. Another is proposed by Anselm of Canterbury in the form of principal essence: it implements an idea of Aristotle of an organismal soul that can be represented by a certain principal essence to serve a unique division basis. Both these ideas will be mastered by early systematics.

* * *

The sequence of deductive genus-species divisions is most usually represented graphically by a tree-like scheme, in which each step of division is corresponded by a transition from one branching point (node) to another. The main task of this scheme is to show both the sequence of divisions itself and the hierarchical structure of subordination of the universals generated by it. Such tree-like configuration first appears in Boëthius' *Commentaries on Porphyry* of the fifth century, and it is termed *Tree of Porphyry* in the logical treatises of the 13th century [Archibald 2014; Dardağan 2018]. This personification refers to the Neoplatonist Porphyry, who is the first to expound verbally an Aristotelian idea of genus-species division (see Section 2.2).

The variants of outlining logical tree are quite diverse [Archibald 2014]. The most simple and formalized tree represents only sequences of the nodes connected by the edges, with recognized universals (notions) of different levels of generality indicated in it (preserved to nowadays). An opposite is a very picturesque tree, a reminiscent of stylized real trees. Regarding its orientation with respect to a book page, this tree in its initial configuration is vertical: this is a paradigmatic Tree of Porphyry, which can be either descending or ascending. A horizontal orientation will be added later to this in the form of *Tree of Ramus* (see Section 4.1 on it).

This logical tree admits a possibility of both partonomic and taxonomic treatments of the sequential divisions illustrated by it. In the first case, it, like a genealogical tree, is *connective*: it connects partonomic genera and species resulted from a sequential division of a whole into its multiple parts, which corresponds to their original Platonic meaning. In the second case, it means *dividing* (classification) tree that is exemplified by the sequence of the division of larger aggregates into smaller ones, including taxonomic genera into species. It is the second meaning, especially illustrated by the *Tree of Ramus* (see Section 5.1), that will become most popular in scholastic systematics, whereas the first one is inherent in contemporary phylogenetic systematics [Pavlinov 2018, 2021b].

4 The Renaissance and the Early Modern Times

Species as Classification Unit

The Renaissance cultural revolution of the 14th century involves an important shift in the conceptual history of West European cognitive program, partly refusing its scholastic tradition. Its key motivation becomes focusing mainly on the "Book of Nature" (the metaphor of Augustine) instead of the "Book of Revelation" (Bible), and on the diversity of things and not on universals describing them [Ogurtsov 1993; Motroshilova 2000; Del Soldato 2020]. However, due to a noticeable mystical bias toward Hermeticism, going back to the preantique cult of Hermes Trismegistus [Evola 1995], the Renaissance program exhausts its intellectual potential in the 16th century and gives way to a new rationality, which begins developing in two main streams. One of these is the "second" (postmedieval) scholasticism, and another is the natural science of a mechanistic (Cartesian) kind [Gaydenko 2003].

The movement within this general integration trend of a new rationality, previously designated by Albert the Great as *Scientia*, does not at first lead to its institutionalization, which would correspond to the modern understanding of natural science. Instead, its branches get distinctly separated based on two different epistemes, quantitative mathematical and qualitative classificatory, which are designated by F. Bacon as *natural philosophy* and *natural history*, respectively. The first one continues the Pythagorean tradition and reduces the variety of natural phenomena to mathematical formulas that link with each other different variables expressing certain properties of things, e.g., body free fall acceleration with its mass, the rate of a chemical reaction with the temperature of a solution, etc. The second continues the Aristotelian tradition and represents the diversity of natural phenomena by classifications that link different things into certain categories, viz., genera and species, based on their mutual resemblance in some features, i.e., flowers or limbs (partons), flowering plants or quadrupeds (taxa). The philosopher and historian of science William Whewell (1794–1866) suggests calling *classificatory* the sciences of the second kind, to which the species problematic belongs [Whewell 1859].

The active development of natural history in such vein, which is most significant for us, marks the rise of a specific *classificatory program* of cognitive activity [Rozova 1980, 1986; Ereshefsky 2001; Breidbach and Ghiselin 2006; Ereshefsky and Reydon 2015]. This program partly continues and partly refines (due to new formalizations) the scholastic method of describing Nature by the hierarchical genus-species scheme. The legacy of scholasticism manifests itself in natural history in two ways. At the level of ontology, it is evident from the preserving biblical world picture, according to which the universe is a result of Divine creation, and at the level of

DOI: 10.1201/9781003346296-5

epistemology, from preserving understanding of genus and species as classification universals, equally relevant to all manifestations of the universe.

An expansion of the classificatory program to the section of natural history dealing with the study of the living leads to the origin of biological systematics. Its principal task is the classification of organisms by genera and species, so the development of the species (more precisely, at first still genus-species) problematics during this period occurs almost exclusively within the framework of this program. With this, the treatment of these categories as referring to the aggregates of organisms (taxa) and not to their properties (partons) becomes predominate, which yields *taxonomization* of the problematics in question. This shift seems significant enough to be considered a local scientific revolution named after A. Cesalpino (see Section 4.3.2).

An important innovation in the development of the classificatory program by early systematics is an evident *structuring* of the genus-species hierarchy by inventing a scale of fixed ranks. One of the outcomes of the latter, most significant to our history, becomes a "disconnection" of the categories in question. This results in *autonomization* of the species category, now taking a fixed position in the classification hierarchy, which yields the removal of genus-species uncertainty. Moreover, a partial transfer of cognitive emphasis (unlike in scholasticism) from genus to species occurs, and genus becomes logically secondary relative to species as the latter's aggregate. Due to this, the general notion of species gradually gets a central position in the considered cognitive situation, and the whole genus-species problematics inherent in scholasticism transforms gradually into a simpler species. This shift in the treatment of the species notion marks the next scientific revolution in the conceptual history of the species problem, which is called after J. Pitton de Tournefort and C. Linnaeus (see Section 4.3.3).

This transformed cognitive situation develops toward strengthening of a specific dual understanding of species that goes back to Aristotle, viz., its *natural-classificatory dualism*. The latter is born by an old confrontation between realistic and nominalistic standpoints to become an important part of the genus-species and then species problematics at this stage of its conceptual history. On the one hand, species continues to be treated epistemically as a universal classification unit, equally convenient for the description of the diversity of any branch of Nature. On the other hand, it becomes understood ontically as a natural unit in a significantly different sense, as compared to scholasticism: its natural status is determined by a specific biological mechanism (reproduction) that provides for both discreteness and persistence of species. This new natural treatment of species, supported by its autonomization, is especially significant in that it becomes a prerequisite for the subsequent *biologization* of the entire species problem.

This chapter describes the development of the classificatory program during the Renaissance and the early Modern Times promoted by the initial formation of natural history. The first part of the chapter is devoted to the same philosophical issues that were discussed in the preceding one. Its second part characterizes the development of the genus-species problematics and its initial transformation into biologically meaningful species within the nascent systematics as a biological discipline. Herbal epoch begins this development, scholastic systematics continues it, and early postscholastic systematics modifies it in its own way.

4.1 THE RENAISSANCE TURN AND THE "SECOND" SCHOLASTICISM

The Renaissance owes its name to the revival of interest in the antique cultural heritage, which was previously pushed into the background by Christian scholasticism. In its philosophical part, it involves new translations of antique authors and their Arabic interpreters, and comments on them for discovering in their works some arguments in favor of new ideas [Di Liscia et al. 1997]. The latter not only mature in a polemic with the old ones but also use some of them as a cognitive tool, including genus-species categorization. The reason is that scholasticism still retains a fairly strong position in university courses on philosophy [Pedersen 1997]. In the development of its program, one of the principal motivations becomes further elaboration of a universal logical apparatus, which leads to the "second" (postmedieval) scholasticism of the 16th–18th centuries [Copleston 1993b; Spruit 1995; Shmonin 2006; Heider 2014].

* * *

A brief review of the period in question is pertinently to begin with the ideas of Marsilio Ficino (1433–1499), who works out in his multivolume *Platonic Theology* realistic ontology of Neoplatonic kind, including a peculiar interpretation of genus and species [Kristeller 1964; Allen 1995; Farndell 2010]. According to Ficino, the Creator is "the first among species" (*princeps specierum*), and an ideal world created by Him is shaped by primary matter, considered as a "universal genus", and *divine species* (*species divinae*). The latter are ordered into four steps, or circles, called ideas (*ideas*), causes (*rationes*), seeds (*semina*), and forms (*formae*), which contain prototypes of all actual (natural) species. The primary matter interacts with divine species to produce the secondary matter with a plethora of "actual genera", which in turn, by interaction with forms, give rise to natural species.

Discussing actual genera and species, Ficino assumes their linear arrangement according to the ontological principle "*first of its genus*" (*primum in aliquo genere*), which is rooted in antique natural philosophy [Kristeller 1964]. For instance, in Aristotle's *Metaphysics X*, it is stated that "in every genus there is one [that comes] first, that is the measure of all those that come after" (cited after [Ashworth 2013: 239]). This means that in a genus, its species are ordered according to the completeness of embodiment of the generic essence, which is transmitted to others in a descending order, beginning with the first species (*species primum*) and becoming less and less complete in subsequent embodiments by other species. Before Ficino, something similar was said by Nicolaus of Cusa when he considered gradation of species within a genus (see Section 3.3). As examples, Ficino considers genera of movement and luminosity: in each of them, some species embody movement and light in full, while others are far from generic perfection. From this, it is evident that Ficino implies (in our terminology) partonomic account of both genera and species, i.e., as essences of things, and not as their aggregates, and in this principle, the elements of the Ladder of Perfection are clearly visible.

> The principle of "*first of its genus*" will subsequently turn into a specific method of systematic description of the diversity of organisms. It first characterizes the most typical member of a certain group (species within a genus, genus within a family, etc.), while others are described by various additions to this characteristic. W. Whewell will call this, with the reference to G. Cuvier, the *method of type* [Whewell 1859]. J. Wilkins considers this linearity a specific manifestation of the generative species conception [Wilkins 2003a, 2009a, 2010], but this treatment is hardly correct.

Unlike detailed ontological considerations of being, Ficino does not pay much attention to the epistemic issues [Kristeller 1964; Allen 1995]. The reason is that for him, the main subject of philosophical knowledge is the comprehension of God, which is achieved not rationally, but by illumination of human mind by Divine light (as in Augustine). Along with this, Ficino implies something like an inductive analysis of universals: one must ascend from lower species to higher genera, and from actual species as embodied forms to divine species as embodied ideas of Divine Mind. In this regard, he follows Aquinas in arranging cognitive forms (Ficino's *species intellectualis*) in a cognitive chain that leads from sensory experience to cognition of universals [Spruit 1995].

* * *

The development of renaissance rationalism in a nonscholastic vein, including significant refinements of the notions of genus and species, is remarkably demonstrated by the ideas of the humanist Lorenzo Valla (1407–1457). In the treatise *Correction of Dialectics and Philosophy*, he attacks the system of Aristotelian categories [Trinkaus 1993; Nauta 2021]. According to his ontolo-epistemology, Nature consists of things (*res*) and is described by three basic categories, which are *substance*, *quiddity*, and *action*. When describing Nature with these categories, he also uses (as auxiliary) the categories of genus and species to analyze how substances and quiddities relate to each other and to the things. For instance, describing the structure of a "wet thing", he writes that "*humor* (liquid) is not a quality, as *calor* (heat) is, but body, and water is contained under it as the species of a genus; [...] and '*humor*' differs from '*humiditas*' (humidity) and '*calor*' from '*caliditas*' (hotness) as genus from species" (cited after [Trinkaus, 1993: 292–293], italics in original). In another place, he considers every concrete thing a "genus" with two "species", which are color and shape (*figura*), with both having many gradations called *subspeciei*. By the latter term, Valla also denotes subdivisions of species (aggregates) of vegetative and nonvegetative things. In a similar fashion, he divides genus "movement" into species "change" and "displacement" with respective subspecies.

Thus, Valla's standpoint is realistic, although not precisely in the Platonic sense: he considers any essences and phenomena real (natural) entities, and not abstract ideas existing in cognizing mind only. With this, though depriving genus and species of the status of categories, Valla constantly uses them to denote the divisions of his basic categories into respective subcategories. In some cases, his species correspond to essences (humidity, hotness) and mean partons; in others, they denote divisions of natural phenomena (as movement) and occasionally aggregates of things (such as plant species), which are taxa.

* * *

A remarkable understanding of genus-species relations is developed by Francisco Suárez (1548–1617), whose theo-philosophy, intermediate between nominalism and conceptualism, is set out in the *Metaphysical Disputations* [Spruit 1995; Shmonin 2006; Heider 2014; Vdovina 2019].

An important hallmark of Suaréz's onto-epistemology is that he violates a tradition coming from Aristotle and Neoplatonists to be fixed by scholastics and separates the notions of category and universal. In his conceptual system, categories serve as a cognitive means of ordering universals into a single inclusive hierarchy, with genus and species being members of the latter [Heider 2014; Jorge and Newton 2016]. His nominalism manifests itself in the acknowledgement of real existence (nonmental being) of things only, whereas universals (including genera and species) do not exist outside the cognizing mind and therefore are mental entities. According to Suarez's *principle of individuation*, one of the fundamental properties of what really exists is its indivisibility, so the actual things (including living organisms) are literally *individuals*. This radically distinguishes them from mental universals (including genera and species), which are semantically integral, but logically divisible. However, by developing a conception of *objective precision* (*praecisio obiectiva*), Suaréz tends to conceptualism: he postulates a rather strict similitude of a properly elaborated system of universals to the real structure of diversity of things.

Bearing in mind the future development of the species problematics, Suaréz's *Treatise on the Labors of Six Days* is of interest. In it, he insists on a literal account of the biblical *Genesis*, including an assertion that God "personally" created each plant and animal "according to its kind". Based on this, some historians of biology emphasize his significant input in the propagation of the biblical picture of the origin of species and, by this, detaining the development of an evolutionary idea [Mivart 1871; Osborn 1913; Locy 1915].

* * *

An important novelty in the comprehension of genus-species relations, which has significant consequences for future systematics, is introduced by an onto-epistemic system of Pierre de la Ramée (Petrus Ramus, 1515–1572). The core of his anti-Aristotelian position is a special understanding of the general system of constructing knowledge based on a specific unified logic [Ong 1958]. The main provisions of the latter are set out in *Training in Dialectic* and *Remarks on Aristotle* [Ong 1958; Walton 1970]. These propositions are as follows: the ideas about forms (*speciei*) rooted in intuition are false; Aristotelian general categories (*universaliae*) as predicates should be replaced by direct judgments (*inventiones*) about things; the persuasiveness of judgments should follow from observations of nature, and not from proofs based on formal syllogistics; these judgments are not discrete (true or false), but probabilistic.

In the unified mental and physical being studied by Ramus, the highest level is set by Aristotle's *topoi*: he calls them *arguments* and distinguishes several dozen of most general ones. The lowest level of being is formed by elementary corpuscles (*simples*): the things themselves (individuals) are made of material corpuscles, and the primary judgments about them are made of mental ones.[1] Individuals, for Ramus, are species, their grouping gives certain aggregates (clusters), which he calls genera. At higher levels of generalizations, the clusters are considered *composite individuals*, whose associations yield clusters of a higher level of generality, which are composite genera. Thus, Ramus decisively rejects an identification of genus with matter and species with form, dating back to Aristotle and accepted by many medieval scholastics, and ascribe them completely different uniform meaning—basically (in our terms) taxonomic. With this, he retains the scholastic treatment of genus and species as interchangeable entities within a sliding hierarchy.

The method of Ramist dialectics is basically inductive by going from particular primary judgments about things to the corollaries that constitute more general secondary judgments. However, the general framework of the entire cognitive procedure is shaped by the most general *topoi*. It seems to follow from this that Ramean onto-epistemology is a something between nominalism (denial of the reality of universals) and conceptualism (a prior introduction of high-level *topoi*).

Notwithstanding that genera and species of Ramus are clusters (aggregates of corpuscles and things), his logic of structuring their overall hierarchy has not taxonomic, but seemingly partonomic sound. This means that "the relation of genus to species cannot even be that of common to proper, but is always the simple relation of whole to part" (cited after [Ong 1958: 204]). Therefore, for him, these categories are not universals that predicate many, but something relating to a certain thing, an individual. For example, for Ramus, "the genus 'animal' would 'answer' not concerning man but only concerning this particular individual animal" [*loc.cit.*].

From the main theme of our book, it is interesting how Ramus identifies categories of genus and species as *topoi*, allocating them to rather low levels in his hierarchy. He uses these categories very formally, without ascribing any special substantial account to them; however, he does not consider them purely logical predicates but sees some reality behind them. Therefore, his clusters are not so much mental as real units of the structure of being, revealed by the analytical means of his dialectic. Thus, detailing the *topos* "humanity", Ramus designates the cluster "humans" as a genus, and the clusters "man" and "woman" as its species.

A complex inductive-deductive nature of Ramean dialectics is manifested in the way he illustrates relations between *topoi* by a specifically shaped tree graph known as the *Tree of Ramus*. Like that of Porphyry (see Section 3.5), it represents a cognitive scheme, following which and generating more and more *topoi* of different contents and levels of generality, it is possible to explore any fragments or aspects of Nature [Ong 1958]. The Tree of Ramus differs from the canonical Tree of Porphyry in two aspects, indirectly indicating different meanings of

[1] In the 16th and especially the 17th century, an idea of corpuscularity becomes one of the most fundamental in understanding the basic structure of the universe.

underlying logical operations. First, it is oriented not vertically, but horizontally with respect to a book page, and second, it shows relations between universals of different levels of generality in a different manner. The Tree of Porphyry in its original "linear" outline indicates the subordination of *eidê* as parts of the whole (partons), so it implies a partonomic division, which corresponds to the priorities of medieval scholastics. The "bracket" outline of the Tree of Ramus indicates the subordination of genera and species as clusters (taxa) and, thus, implies a taxonomic division. Therefore, it is the Tree of Ramus, and not that of Porphyry, that should be called a *classification tree* rightfully in the full sense of this notion, and it is this tree that will become the principal grapheme used in the herbals and early systematic treatises (up to mid-19 century). However, in the contemporary literature, the notion of Tree of Porphyry is accepted in a very broad sense, including that of Ramus.

A remarkable application of the Tree of Ramus to represent a global hierarchical structure of the diversity of all things and ideas is invented by the philosopher and linguist John Wilkins (1614–1672) in his *An essay towards a real character* [Wilkins 1668]. For him, the "philosophy" is in many ways a "taxonomy", since both are engaged in the ordering of knowledge about the universe. To implement his idea, Wilkins elaborates a firmly ordered universal hierarchy of the System of Nature, represented as the "bracket" Tree of Ramus. In this System, which is taxonomic in its meaning, the whole of being and knowledge about it is divided into forty categories called *genera*, which are then divided into *differentiae* (usually nine in each genus), and these in turn into *speciei* (usually 15 in each *differentia*) [Slaughter 1982; Maat 2004]. The hierarchy of the Wilkins' scheme descends to the "things", which in our case correspond to the particular "species" of plants and animals. The scholastic systematician John Ray participates briefly in the Wilkins' project [Stearn 1986] (see Section 4.3.2 on him).

4.2 THE EARLY MODERN TIMES: NATURAL-CLASSIFICATORY DUALISM

The cognitive program in the early Modern Times (16th–17th centuries) is based on a mechanistic world picture, in which the laws of Nature (they are attributed to divine origin) are reduced principally to mechanical interactions between physical bodies and their parts [Russell 1967; Ogurtsov 1993; Gaydenko 2000, 2003]. At the level of onto-epistemology, it is marked by an obvious inclination to empiricism, represented by two basic cognitive models, largely the same as in scholasticism. Nominalistic model (J. Locke, F. Bacon, partly R. Descartes, etc.) acknowledges the meaningfulness of only sensory knowledge about particular things. The conceptualistic model (G. Leibniz, etc.) acknowledges the importance of global presupposed knowledge containing "*a priori* truths" (universals of a high level of generality), in which context-specific knowledge about things is inferred.

One of the initiators of this new rationality, the natural philosopher and mathematician René Descartes (1596–1650), in his *Principles of Philosophy*, asserts that universals have no existence outside the mind, being "only modes of thinking", and he indicates that "five universals are usually counted: genus, species, difference, property, and accident" [Descartes 1982, Pt. I: § 58, 59], which are those recognized by Porphyry. He adds to this that "these universals are created solely by the fact that we use one and the same idea in order to think of all individual things which are similar to one another: and as we also give one and the same name to all things represented by this idea; this name is universal" [*op.cit.*: § 59]. Thus, Descartes' position with respect to genus and species is mostly nominalistic. In this regard, it is noteworthy that direct references to him can be found in the works of the luminaries of scholastic systematics (Ray, Tournefort, Linnaeus).

* * *

Among the versions of onto-epistemology of the early Modern Times that most closely concern the genus-species problematics, of particular significance is the one developed by John Locke (1632–1704), which is set out in his fundamental treatise *An essay concerning human understanding.* It can be considered a combination of moderate nominalism and moderate realism. Indeed, on the one hand, he agrees with nominalists that "general and universal belong not to the real existence of things; but are the inventions and creatures of the understanding, made by it for its own use" [Locke 1999: 399]. On the other hand, he offers a very original realistic account of essences and substances embodied in things, which markedly differs from that of the nominalists [Ayers 1981, 1993; Look 2009; Jones 2020].

Locke believes, following a number of natural philosophers of his time, including P. Ramus and R. Descartes, that the fundamentum of the universe constitutes the smallest material corpuscles, various combinations of which produce various individual things. Each corpuscle is a carrier of a certain property that a thing is endowed with; some of these properties are essential, and others are accidental. Together, they make up the *real essence* of a thing, which "may be taken for the very being of anything, whereby it is what it is. And thus the real internal, but generally (in substances) unknown constitution of things, whereon their discoverable qualities depend" [Locke 1999: 402]. With this, for Locke, the "object of thinking" is not a thing, but an idea, and all ideas "come from sensation or reflection" [*op. cit.*: 86–87]. Among them, most important are *simple ideas* that "have a foundation in nature; such as have a conformity with the real being and existence of things, or with their archetypes" [*op.cit.*: 355]. They relate to the real essences of things, so Locke calls them *real ideas*, and their formation initiates cognition. A generalizing knowledge about things involves *complex ideas*, which are "combinations of simple ideas put together, and united under one general name" [*op.cit.*: 356], and they correspond to the *nominal essences*. These "complex ideas of substances are real, when they agree with the existence of things, [but] those are fantastical which are made up of such collections of simple ideas as were really never united" [*op.cit.*: 357–358]. Of the highest level are *general terms*, which denote *general ideas* about things, and they "become general, by separating from them the circumstances of time and place", so they correspond to "general natures [that] are nothing but abstract and partial ideas of more complex ones" [*op.cit.*: 396–397].

The Locke's notions of real and nominal essences seem to be analogous to those of the abovementioned actual and formal realities of Duns Scotus, respectively (see Section 3.3).

Arguing *in absentia* with scholastics about essences and therefore being forced to communicate with them in their conceptual space, Locke pays much attention to the questions: "what the species and genera of things are, wherein they consist, and how they come to be made" [*op.cit.*: 389]. According to Locke, the need to form generalizing notions about things in the process of cognition "is the reason that we collect things under comprehensive ideas, with names annexed to them, into genera and species; i.e. into kinds and sorts" [*op.cit.*: 370]. Accordingly, "this whole mystery of genera and species [...] is nothing else but abstract ideas, more or less comprehensive, with names annexed to them" [*op.cit.*: 398].

For genera and species to get deprived of their "mystery", Locke simplifies the conceptual apparatus rooted in Antiquity and refined by the scholasticism. In this regard, without abandoning the Latin terms *genus* and *species* altogether, he considers them just synonymous with the ordinary English *kind* and *sort* and emphasizes this in some places of his *An essay* [e.g., Locke 1999: 370, 389].[2] In one place, he describes "gathering things under ideas", also using ordinary words to denote the corresponding categorical terms: "general" instead of *genus* and "sortal" instead of *species* [*op.cit.*: 403]. Of importance for us is that this Lockean terminology seems to indicate his understanding of genera and species taxonomically: for him, they are groups of things (organisms, etc.), and not their substances [Kuklok 2013; Newton 2014]. Regarding this issue, he notices that in scholasticism, "the word essence has almost lost its primary signification: and, instead of the real constitution of things, has been almost wholly applied to the artificial constitution of genus and species" [Locke, 1999: 403]. Therefore, Locke's "anti-Aristotelian" treatment of these categories is sometimes called "classification theory" [Sloan 1970; Crane 2003], although this is hardly the case. As a matter of fact, he seems to be not interested in classifications at all, but his standpoint can be considered as a claim for taxonomization of the entire genus-species problematics, which is important for the development of systematics. Lockean terminology will be picked up by some logicians in the 19th century (such as J. Mill, see below).

J. Locke endows genus and species with nominal essences and emphasizes "that things are ranked under names into sorts or species, only as they agree to certain abstract ideas, to which we have annexed those names, the essence of each genus, or sort, comes to be nothing but that abstract idea which the general, or sortal [...], name stands for" [Locke 1999: 403]. Thus, regarding genera and species, Lockean position is definitely nominalistic: he deprives them of realistic ontology and leaves for them the epistemic function of mental universals [Cain 1997; Stamos 2003; Jones 2020]. At the same time, he partially breaks the genus-species linkage inherent in the scholastic argumentation scheme, following the position of Thomas Hobbes (1588–1679).

---

[2] It is possible that this replacement by J. Locke's is inspired by the King James English edition of the Latin canonical text of the *Old Testament*, in which the term *genus* is translated as *kind* (see Section 5.1.1).

The latter writes that "there is no such generic mental act to be designated 'affirmation' or 'judgement', of which intuitive knowledge, opinion and so forth are the species" (cited after [Ayers 1993: 73]). Accordingly, Locke focuses almost exclusively on species, leaving aside consideration of genus and, thereby, indirectly minimizing genus-species uncertainty in the cognitive situation, in which systematics begins its development at this time.

Locke reinforces his position by stating that the nominal essences and their corresponding species are "the workmanship of the understanding, but have their foundation in the similitude of things. [...] Nature, in the production of things, makes several of them alike. [But] sorting of them under names is the workmanship of the understanding, taking occasion, from the similitude it observes amongst them, to make abstract general ideas [...] so they come to be of that species" [Locke 1999: 400–401]. And he further says that "our ranking and distinguishing natural substances into species consists in the nominal essences the mind makes, and not in the real essences to be found in the things themselves" [*op.cit.*: 432]. With considering essences nominal, Locke believes that no real existence of things is necessary for recognizing species: "the [nominal] essence of a species rests safe and entire, without the existence of so much as one individual of that kind" [*op.cit.*: 405]. Thus, according to Locke, nominal essences and corresponding species are generated by the cognizing mind on the basis of the similarity of things, but they do not cease to exist in some mental space of ideas with disappearance of the things.

An emphasis on similarity makes Lockean nominalistic account of species purely phenomenological; in particular, as far as species of living organisms are concerned, there is nothing in this account that presumes a generative idea [Cain 1997]. With this, Lock continues nominalistic theme that goes back to Boëthius and refined by Albert the Great and William Ockham: they all concertedly assert that there are no real boundaries between different species, which are drawn by a cognizing mind. Developing this idea, Locke notes "that in all the visible corporeal world, we see no chasms or gaps" [Locke 1999: 405], and he exemplifies this consideration by the same facts from living nature that are lay down in the idea of Ladder of Being. For this reason, Locke "cannot see how it can be properly said, that Nature sets the boundaries of the species of things: or, if it be so, our boundaries of species are not exactly conformable to those in nature" [*op.cit.*: 446]. Thus, although Locke admits (by the reservation "if it be so") a possibility for real boundaries between species to exist, he doubts the latter's unambiguous delineation—not because it is difficult to find these boundaries, but simply because they are actually absent [Ayers 1981, 1993; Goodin 1999]. The particular decisions about demarcating species depend on the cognitive and other needs of men: indeed, no other but "men determine the sorts of substances, which may be sorted variously [and whereas] Nature makes the similitudes of substances [...]; it is men who [...] range them into sorts, in order to their naming, for the convenience of comprehensive signs [...]: and in this, consists the whole business of genus and species" [Locke 1999: 449–450]. This thesis of J. Locke marks the beginning of that version of nominalistic account of species, which is now commonly referred to as *conventionalism* [Crane 2003; Wilkins 2003a, 2009a; Newton 2014; Bird and Tobin 2018].

* * *

The philosopher and mathematician Gottfried Wilhelm von Leibniz (1646–1716) summarizes very interesting ideas about genus and species that are quite characteristic of a certain circle of natural philosophers of the 17th century. The basis of his picture of material world is constituted by *monadology*, a variant of natural-philosophical idea dating back to Antiquity, according to which all things consist of *monads* as invisible and indivisible elementary corpuscles. For Leibniz, they are clots of energy, which actively interact with each other and, by forming certain combinations, produce the whole variety of things of inert and living matter. From this, he infers the *principles of plenitude* (*lex completio*) and *continuity* (*lex continui*), expressed by an aphoristic formula *Nature does not make leaps* (*Natura non facit saltus*). The latter represents another antique idea of the Ladder of Being, which, according to Leibniz, is discrete at the micro level, but continual at the macro level: matter consists of particular monads, and their potentially infinite combinatorics makes up the whole of Nature without breaks. An important component of Leibniz's worldview is expressed by his claim that "essence is fundamentally nothing but the possibility of the thing under consideration" [Leibniz 1996: 294]. By this, he seems to mean "that individual essences or complete individual concepts are consistent with other sets of essences of individuals and are within the infinite and eternal mind of God" [Look 2020: 13].

G. Leibniz embodies these natural-philosophical constructs in his ideas about genera and species, discussing them in detail in his treatise *New Essays on Human Understanding*. This treatise is conceived by him as a correspondence polemic with J. Locke about real and nominal essences and, in this regard, about genera and species; therefore, it is sometimes called "Anti-Lockean". Leibniz sees the main reason for the divergence of the views in that his opponent is committed to Aristotelianism, while he himself is committed to Platonism [Leibniz 1996: 48]. In this polemic, Leibniz focuses on the reasons for the existence of genera and species, in what respect they are continuous and in what are discrete, what is their relevance to the description of the diversity of things, etc. [Goodin 1999; Look 2009; Smith 2011]. He understands these categories in a taxonomic sense: for him, genera and species are certain aggregates of things, but not the essences that abide in them: here, Platonist Leibniz agrees with Aristotelian Locke.

Leibniz begins consideration of this issue by pointing out that "one should distinguish between *physical* (real) genus and *logical* (ideal, also mathematical) genus" ([Leibniz, 1996: 64], italics in original).[3] And although Leibniz writes here about genus, in the subsequent text, he discusses almost exclusively species and treats genus as "species aggregate"; thanks to this, physical (natural) species actually becomes the principal subject of his ontological account. Leibniz distinguishes

[3] This distinction was evident in some of Aristotelian issues (see Section 2.1.3).

between physical and logical species by that the former is characterized by essential properties of bodies it includes, with which they are endowed with necessity, whereas the latter is characterized by accidental properties. According to this, "there are sorts or species such that if an individual has ever been of such a sort or species it cannot (naturally, at least) stop being of it, no matter what great events may occur in the natural realm" [Leibniz 1996: 306]. Thus, Leibniz's position regarding physical species is quite realistic.

Leibniz's clear idea of distinguishing between physical and logical species provides an obvious expression of their *natural-classificatory dualism.* It should be recalled that the latter arises in the works of Aristotle, and Avicenna and Albert the Great also separate two meanings of these categories similarly [Windelband 1892; Tweedale 2013]. This dualism will implicitly appear in late scholastic systematics and will be accentuated in early postscholastic systematics (see Sections 4.3).

Leibniz treats logical genera and species nominally in agreement with Locke and writes that they "are merely sorts" [Leibniz 1996: 305]. He regards them infinitely divisible: "the tiniest difference which stops two things from being alike in all respects makes them 'of different species'"; for their separation, "the least dissimilarity is enough, so that each different idea yields a new species, whether it has a name or not"; and "furthermore, a single individual will move from species to species, for it is never similar to itself for more than a moment" [*op.cit.*: 309, 326]. As regarding physical species, the differences between them, according to Leibniz and contrary to what Locke claims, involve essential properties of things, and the "boundaries of species are fixed by the nature of things" [*op.cit.*: 303]. Therefore, when individuating the physical species, one should proceed "with the presumption that they have some essential and unchangeable nature" [*op.cit.*: 326]. In this passage, Leibniz evidently means real essences, while Locke's nominal essences do not make sense for him: "a 'nominal essence' is understood to be a false and impossible one—something that appeared to be one but really is not" [*op.cit.*: 294]. Objecting to Locke's thesis (going back to Aristotle) that real species are characterized by multiple nominal essences, Leibniz suggests distinguishing between essence and definition and emphasizes "that although a thing has only one essence, this can be expressed by several definitions" [*op.cit.*: 295]. The latter admission yields a possibility to consider one thing in different aspects and seems to indicate a certain conceptualistic motive in the Leibniz's standpoint [Goodin 1999].

Leibniz's acknowledgement of discrete nature of physical (real) species looks paradoxical in the light of his principle of plenitude, but, in fact, it is not so [Smith 2011]. As noted above, this principle means that there are no breaks in the Ladder of Being, which is continuous at the macrolevel. Its continuity means that the gaps between "adjacent" species are imaginary: Leibniz believes that any close study should fill any seeming gaps between them with intermediate species.

For our discussion of species of animals and plants, it is important that Leibniz invariably counts them as physical (real) units, into which living nature is divided [Smith 2011]. Leibniz considers their constancy and discreteness in two ways, which are partially present in Aristotle and evident in early Biblical scholars of a realistic kind, and will be adopted by the naturalist systematicians (see the next section). One of them refers to the unchangeable essences of species, rooted in Divine ideas, each being concrete and therefore discrete; this implies an essentialist account of species. Another is a continuous sequence of generations uniting members of the same species into a single whole, just like links in a chain, which corresponds to a generative idea going back to Aristotle (discussed in detail in Section 5.1.2).

> It is noteworthy how Leibniz substantiates existence of physical species of living organisms, combining Aristotelian teleonomism and his own anthropocentrism [Smith 2011]. A physical species endowed with essence is the same "thing" as an organism. The particular animals and plants are designed and exist to perform certain functions, including their uses for the humans. It is reasonable to assume that each species was also created by God for something like this: the bees make honey and wax, the silkworm larvae make silk threads, the cows produce milk and meat, the horse serve humans, etc. And in general, different species were created by Creator so that people could comprehend His wisdom by studying their essences.

Seeing no "philosophical" sense in the multiplicity of nominal essences, Leibniz addresses a specific classification issue of more "practical" kind, which seriously concerned systematists since Antiquity and will concern them up to recent times [Goodin 1999; Smith 2011]. This is a contradiction between the multiplicity of possible ways of distribution of species among genera because of multiplicity of classifying traits, on the one hand, and a presumed unitarity of "true" arrangement of them corresponding to the "nature of things", on the other hand. Leibniz believes that "there is nothing wrong with employing multiple systems of classification. Indeed, the more the better, as each one may help to reveal something new about the structure and nature of the entities in question" [Smith 2011: 245]. On this account, Leibniz obviously anticipates an idea of taxonomic pluralism, which will be developed in different versions by the systematics of the 20th century.

Proceeding from this account, Leibniz considers reasonable a rather pragmatic approach to solving this problem, according to which the choice of one preferable classification among many possible is determined by its significance for researchers or users. One of these choices is the "Ariadne thread" of the scholastic systematicians (Ray, Linnaeus, etc.), with which the diversity of organisms is navigated; another choice is the "general purpose" classification of modern positivist systematicians (Gilmour, Sneath & Sokal, etc.) [Pavlinov 2018, 2021b]. However, if the fundamental task of elaboration of the single "true" classification is presumed, Leibniz thinks it necessary to reveal *real* similarities between species that reflect their *real* (natural) affinity. He, following the scholastic realists (Augustine, R. Bacon, etc.), understands the latter as a proximity of the physical species to their prototypes in the Divine plan of creation. As a naturalist, he believes that since the persistence of the species after their divine creation is ensured by the continual reproduction of organisms, in the case of plant species, the similarity of their flowers is the most reliable indicator of their grouping into real genera [Smith 2011].

Leibniz's suggestion to distinguish between physical (real) and logical species is supported by the philosopher Immanuel Kant (1724–1804). Impressed by G.-L. Buffon's ideas about "generative" nature of biospecies (see Section 5.1.2), Kant prepares in 1775 a lecture *Of the different races of human beings*, in which he expounds his view on (in modern terms) taxonomic diversity of humans [Kant 2007].

Although Buffon ranks the latter as species (*espèce*), whereas Kant writes about genus (*Gattung*), this does not change his account much [Sloan 1970]. In the introductory section of this lecture, Kant follows Leibniz in dividing genera into natural (*Naturgattung*) and logical (*Schulgattungen*), and he writes that "the unity of the [physical] species is nothing other than the unity of the generative power that is universally valid for a certain manifoldness of animals" [Kant 2007: 84]. He continues with asserting that "the scholar division concerns *classes*, which divide the animals according to *resemblances*, the natural division concerns phyla [so] the first only aims at bringing creatures under titles; the second aims at bringing them under laws" ([*loc.cit.*], italics in original).[4]

* * *

Following the route of the considered conceptual history and looking two centuries ahead, it seems to be a proper place here to reflect on a particular way of "kinding" of the diversity of things (see [Kendig 2016] on this notion), in a way similar to that of Leibniz, by the logician John Stuart Mill (1806–1873), who is famous after his repeatedly reprinted fundamental *A system of logic.* Some of his suggestions concerning our categories will have a significant impact on the ideas about their ontology in the 20th century, so they deserve a brief summary here for not to get back to the logic issues again on the next pages of this book.

Mill regards genus and species as universal classification interchangeable units, and he presumes their predominantly taxonomic interpretation. In fact, repeating Porphyry and Boëthius, he writes that "the words genus, species, etc., are [...] relative terms; [...] the same class which is a genus with reference to the sub-classes or species included in it, may be itself a species with reference to a more comprehensive, or, as it is often called, a superior, genus" [Mill 1843: 162–163]. Referring to examples from zoology, he points out that "genus and species are not only used by naturalists in a technical acceptation not precisely agreeing with their philosophical meaning, but have also acquired a popular acceptation, much more general than either" [*op.cit.*: 162]. Mill reproaches them for their following the norms of everyday language, so "the word species is used in a very different signification in logic and in natural history" [*op.cit.*: 169].

Mill recognizes and designates two categories of classes depending on their properties, characterizing them as follows. For him, "these classes, distinguished by unknown multitudes of properties, and not solely by a few determinate ones, are the only classes which, by the Aristotelian logicians, were considered as genera or species. Differences which extended to a certain property or properties, and there terminated, they considered as differences only in the *accidents* of things; but where any class differed from other things by an infinite series of differences, known and unknown, they considered the distinction as one of *kind*, and spoke of it as being an

[4] This passage from I. Kant makes him a real forerunner of the phylogenetic account of biodiversity [Pavlinov 2018].

essential difference, which is also one of the usual meanings of that vague expression at the present day" ([Mill 1843: 167–168], italics in original). For fixing these categories more rigidly, he borrows the respective terms not from logic but from everyday language: so, those classes that are distinguished based on "essential difference" are called by him *Kinds*, and sometimes *real Kinds*, whereas any other possible classes distinguished by accidents are just *kinds*. Considering them in another place, Mill says that "every class which is a real Kind, [...] is either a genus or a species" depending on whether it is divided further or not [*op.cit.*: 171]. Thus, though calling them *Kinds* or *kinds* instead of Porphyryan *category*, Mill considers them scholastic classification units and makes no rank-dependent substantive differences between them (see [Magnus 2015; Quinn 2020] for discussion). So his genus-species hierarchy remains sliding and retains genus-species uncertainty.[5]

The terminology used by Mill partly reproduces that proposed in the 17th century by J. Locke [Ayers 1981; Crane 2003; Look 2009; Kuklok 2013; Newton 2014]. With this, however, his approach largely represents a specific development of the ideas outlined in the *History of Inductive Sciences* by W. Whewell [Magnus 2015]. Considering classifications, Whewell routinely designates the units in them as *kinds*. The latter have a universal meaning by corresponding to any classes of things, be they chemical elements, minerals, or living organisms. However, he argues that the task of natural sciences is to develop law-like *natural classifications* reflecting the "nature of things", and such classifications (by tautology) should contain *natural classes* (kinds) of these things. Whewell's "nature of things" largely corresponds to their essential properties; in particular, regarding botanical classifications and referring to Linnaeus, he asserts that "the knowledge of the natural arrangement is the knowledge of the essential construction and vital mechanism of plants" [Whewell 1859: 397].

An important difference between the positions of Whewell and Mill is that the former defines a natural kind inductively through the "nature of things" common to it (position of Aristotle and other realists), whereas the latter does this deductively by referring to a certain set of any characters shared by a natural kind's members (Locke's position) [Wilkins 2013b]. So J. Mill's account of "natural kind" can hardly be thought of as anticipating somehow its modern conception (of Quine et al.), as it is applied to biospecies [Quinn 2020] (see Section 6.2).

Interpretation of genus and species as "classification kinds" based on the universal rules of logic can be found in the works of some biologists of the 19th century (e.g., [Bentham 1827; Usov 1888; Sanson 1900]). In particular, the future botanist George Bentham (1800–1884) argues in his early *Outline of a new system of logic* that "there is no such thing as an infima species, as any one, however low, is susceptible of still further subdivision", therefore, depending on the context, "the lowest species [...] must, for the purpose of exposition or ulterior division, be considered as summa genera" [Bentham 1827: 35, 117]. Naturalists of the 19th century inclined to such formalisms appear to be nominalists: for instance, the zoologist Sergei Usov

[5] This Mill's standpoint deserves highlighting, because systematic biologists of that time already undermine this uncertainty by ranked hierarchy (see Section 4.3.2).

(1827–1886), fascinated by Kantian philosophy, in his essay *Taxonomic units and groups* interprets species as "a transcendental ideal, *prototypon transendentale* [...] an abstract individual, an individual as *noumen.* [...] Species is an *ideal image.* [...] Therefore, we can call the species a *taxonomic unit* or, better, a *taxonomic individual*" ([Usov 1888: 315–317], italics in original). In the 20th century, the account of species and genus as classification units will be unanimously supported by logicians, ideologists of the so-called *classiology* as a general theory of classification [Pokrovsky 2014], as well as by some rationalist biologists [e.g., Lyubishchev 1972; Meyen and Shreyder 1976; Zarenkov 1988; Marradi 1990]. It is noteworthy that in the modern manuals on logic, one can find the use of these units to denote both essences (partons) and groups (taxa) of things (e.g., [Voyshvillo 1989]), i.e., the partonomic-taxonomic dualism is preserved.

4.3 THE ORIGIN OF SYSTEMATICS: SPECIES OF THE LIVING

As noted at the beginning of this chapter, a large portion of that branch of natural history, which deals with the diversity of living beings, masters more or less actively classificatory program during the 15th century and beyond. It takes the form of employing certain elements of the scholastic genus-species scheme, including its basic categories as universals, to the description of that diversity. With this, an "adaptation" of this scheme to the cognitive tasks of natural history yields the latter's fundamental onto-epistemic specificity, as compared to scholasticism. In particular, the relation between these universals is discussed without reference to the matter-form interaction, and they are interpreted phenomenologically as classification units.

This new trend in the considered conceptual history is ordered specifically by descriptive "collecting" and analytical "methodical" cognitive programs (see Section 2.1.4 on them) [Pavlinov 2018, 2021b]. The dominance of the first in the 15th–16th centuries gives rise to the renaissance herbalism of naturalistic kind, in which the main attention is paid to the description of particular groups of plants and animals in terms of their "uses" to humans. Accordingly, their classifications play an auxiliary role, and classification units are nearly formally applied to identify and arrange the groups of interest to herbalists. A cognitive shift to the second is initiated in the 16th century by a purposeful and consistent application of the scholastic genus-species scheme, and this leads to arise of scholastic systematics.

The classificatory activity, as applied to living organisms, affects both aspects of their diversity, partonomic and taxonomic, with developing mutually agreeing partonomic and taxonomic classifications. With their gradual conceptual disjoining, the implementation of the classificatory program leads to the separation of two departments within the overall cognitive situation: one of them is anatomy dealing with the partonomic issues, and another is systematics proper (taxonomy) analyzing taxonomic diversity of organisms. This causes *taxonomization* of the genus-species problematics, according to which taxonomic tasks become dominating, and the basic result of classificatory activity becomes not a partonomic arrangement of essences but a taxonomic classification of organisms. This means, first, a weakening and eventually elimination of partonomic-taxonomic dualism and, second, the change of logical basis of analysis of taxonomic diversity of organisms: an intensional logic of essences gets replaced by an extensional logic of classes [Cassirer 1950].

At the initial stage of this phase of the conceptual history under consideration, genus and species are accounted as classification units in a "mechanistic" manner, i.e., as the aggregates of specimens united by their similarity [Hatch 1941; Mayr 1949; Luckow 1995; Pozdnyakov 2015]. With this, they are still regarded, just as in scholasticism, as interdependent elements of a unified genus-species scheme, so that its problematics also remains mostly genus-species one. By the end of this phase, an initial rankless hierarchy of the genus-species scheme transforms into the ranked one by fixation of a few ranks with more or less firm position and designation. One of the most significant results becomes *autonomization* of species as a classification unit, which leads to the elimination of genus-species uncertainty. This becomes an important prerequisite for turning the genus-species problem into the species one and its subsequent biologization, which accentuates natural and eventually biological understanding of species.

4.3.1 Herbal Epoch as a Precursor

Naturalistic *herbalism*, which determines the main content of the *Herbal epoch*, emerges as a result of the increased naturalistic and medical interests of people in plants and animals, which is characteristic of the Renaissance. It stimulates a revival of the "collecting" program that was pushed into the background by scholasticism, and this results in the appearance of specific descriptive treaties with descriptions of plants (*herbariae*), which are most numerous (therefore the name of this epoch), and animals (*bestiariae*) [Arber 1938; Stannard and Stannard 1999; Baxter 1998; Janick 2003].

Considered from the perspective of the species problem, renaissance herbalism differs from the very ancient medical herbal tradition in that it emerges under a certain rationalistic influence set by scholasticism [Perfetti 2000; Pavlinov 2018]. Accordingly, the new herbals and bestiaries, unlike the classical ones (such as *De Materia medica* by Dioscorides, the first century of AD), represent the diversity of organisms by classifications ordered in a certain manner, so they are rightfully regarded as an eve of implementation of the classificatory program to the study of the living. This is because the main actors of the Herbal epoch are physicians educated from universities, and their interests are not only in the medicines and their uses but also in a more general phytography. Therefore, the early Renaissance herbals and bestiaries of our interest date back to the 13th–14th centuries, when the first universities appear in Europe, while their heyday fells at the beginning of book printing in Europe in the 16th century [Arber 1938].

As Herbal epoch develops, its pragmatic emphasis is gradually supplemented and then replaced by the natural-historical one. In this regard, the most advanced treatises are reminiscent of systematic catalogues by their adherence to classificatory terminology. Nevertheless, as noted above, an account of genera and species by herbalists remains mainly phenomenological: meaningfully, they are more similar to taxa distinguished in folk systematics than to those that will be identified in scientific systematics [Atran 1990; Pavlinov 2018; Minelli 2022].

The epoch in question begins with the growth of interest of the Italian philosophers in the antique works on natural history, which are repeatedly rewritten and commented on [Stannard and Stannard 1999; Ogilvie 2006]. More significant to us is the next phase of the history of herbalism, in which development the principal role is played by German, Flemish, and French phytographers [Sprengel 1807; Sachs 1906; Greene 1909]. Among their principal tasks is not commenting on the antique authors but rather preparing their own botanical encyclopedias. The latter's distinctive feature is the replacement of alphabetical listings of genera and species with hierarchically organized classifications. Due to this, the scattered data on diverse plants and animals become united into an appropriately structured pool of knowledge, which can be considered the beginning of natural history in general [Foucault 1970; Lyubarsky 2015] and systematics in particular.

The hierarchically arranged fragments, into which many books are divided, can be regarded as the first instances of encaptic hierarchy in natural history [Lyubarsky 2015, 2018]. For this, the conceptually advanced herbals and bestiaries consistently apply classification terminology. For example, in the *History of Plants* by Rembert Dodoens (1517–1585), one of the introductory sections is entitled "On the genera of plants" [Dodoens 1553]. The full title of the catalogue *Representation of Botany* by Gaspard Bauhin (1560–1624) includes an explanation of its main objective as the "methodical description of [plants] according to their genera and species" [Bauhin 1623]; it literally repeats the goal of systematics formulated by the "methodist" systematician Cesalpino (see the next section about him). In the most developed treatises, classification Trees of Ramus are provided, the examples are *The List of Aquatic Animals* by Conrad Gesner (1516–1565) and especially *Seven Books on the Insect Animals* by Ulisse Aldrovandi (1522–1605) [Gesner 1560; Aldrovandi 1602].

In applying the genus-species scheme, herbalists do not presume any substantial meaning of the respective categories, which simply designate certain levels of generality of the groupings. For this reason, the standard designations of classification units are frequently replaced by some others. The latter are most often borrowed from the book printing, and respective fragments are called Books (*Libres*), Chapters (*Capita*), Sections (*Sectio*), etc. Others are logical terms, such as Order (*Ordo*) in the abovementioned Gesner's *List of Aquatic Animals*, or Class (*Classis*) in the *History of Rare Plants* by Charles de l'Écluse (Carolus Clusius, 1526–1609) [Clusius 1601]. Along with this, an unusual terminology appears occasionally: for example, some sections and subsections in the *Images of Live Herbs* by Otto Brunfels (1489–1534) are designated by the musical terms, with generic-ranked sections called *Rhapsodiae*, while their species subsections called *Sententiae* or *Paraphrases* [Brunfels 1530–1536].

Hierarchically arranged genera and species in herbals may have additional gradations, with lower genera and higher species being often interchangeable in their application, so they should probably be considered "generic-species", as in folk systematics [Atran 1998; Ogilvie 2006; Pavlinov 2018]. This is instantiated by the *New List of Plants* by Pierre Petra (Petrus Pena, 1538–1605) and Matthias de L'Obel (Matthaeus Lobelius, 1538–1616). Thus, they distinguish in the higher genus *Narcissi* several intermediate genera, which are in turn divided into lower genera, and local species are named within the latter [Pena and Lobelius 1570]. In the abovementioned *Images* by Brunfels, many species and even intraspecific forms, as they are currently ranked, are assigned generic status: for example, two varieties of plantain are designated as *Plantaginis duo sunt genera, Maior & Minor.* In this regard, it is noteworthy that the naturalist Arthur Cain (1921–1999) designates classification categories in Bauhin's *Representation* by modified terms: *generoid* (as if genus) and *speciate* (as if species) instead of *genus* and *species*, respectively [Cain 1994a].

However, explicit designations of classification categories are not always provided in the herbals, so contemporary researchers usually distinguish them by the same lexical criteria, as in the folk classifications [Pavlinov 2018]. Accordingly, focusing on the recent "Linnaean" standards, they typically regard one-word names as generic, and multi-word ones as specific [Sachs 1906; Greene 1909; Larson 1971; Atran 1990]. But this rule does not always work, for different forms of generic names are occasionally used by herbalists: in some cases, they are one word, while in others verbose, examples could be found in the abovementioned *New List of Plants.*

It can be concluded from the above that the character of descriptions of "genera" and "species" in herbals and bestiaries indicates their taxonomic rather than partonomic account. Since a cognitive horizon of herbalism is limited to the living, this yields an impression that these categories are treated in a "biological" fashion. However, this is hardly true: they are employed by herbalists as universals intended not so much for expressing their ontic status as specific natural units as for representing the diversity of living organisms in the form of hierarchical classifications of a scholastic kind.

4.3.2 From Cesalpino to Ray: A Scholastic Legacy

The first phase of the conceptual history of scientific systematics, which begins to move toward the biological account of species, coincides with the rise of natural science, marking the transition from the Renaissance to the Modern Times at the turn of the 16th–17th centuries. Its relations to the previous philosophical ideas are rather complicated, and they may be summarized as follows. The ontic component of its cognitive situation is rather specific as compared to that of scholasticism. On the one hand, early systematicians, with a few exceptions, seem to have little interest in the interpretation of genus and species as related to matter and form (*eidos*). On the other hand, the natural-philosophical idea of the System of Nature becomes of paramount importance: it is perceived as an expression of the universal law embodying the Divine plan of creation [Larson 1971; Greene 1992; Breidbach and Ghiselin 2006; Smith 2009]. The epistemic component is shaped basically by just the scholastic program with an addition of certain elements of new Cartesian rationality. Scholasticism provides the method of description of the diversity of plants and animals based on the genus-species scheme, including elements of essentialist interpretation of genera [Pavlinov and Lyubarsky 2011; Pavlinov 2018]. The impact of Cartesianism is mostly expressed in the mechanistic account of an organism as an array of its constituent parts:[6] this becomes a prerequisite for its understanding as a

[6] As stated by G. Leibniz in his *Monadology*, "the organic body of each living being is a kind of divine machine or natural automaton, which infinitely surpasses all artificial automata" [Leibniz 2018: 11].

"sum" of firmly fixed diagnostic traits [Slaughter 1982; Smith 2011; Lyubarsky 2015; Pozdnyakov 2015].

On this general onto-epistemic foundation, a natural method of early systematics is being elaborated, which fits entirely into the classificatory program. Its main goal is to ensure the development of taxonomic system (classification), in which organisms take their places according to their nature (*quidditas*). Such accentuation shifts the focus of systematics to a global hierarchy of Nature, and the main cognitive task becomes elaboration of natural classifications most adequately representing it. Accordingly, both genus and species are considered classification universals, with the former being primary to the latter and the relations among genera of various levels of generality being in focus. With this, species gradually ceases to be treated interchangeable with genus in the overall taxonomic hierarchy, and the both are assigned more or less stable places in the hierarchy of Nature. This yields transformation of scholastic sliding genus-species hierarchy into a fixed one and the elimination of genus-species uncertainty.

* * *

The first among logicians who gravitate toward natural history and get seriously interested in classifying plants is the philosopher and naturalist Andrea Cesalpino (Andreas Caesalpinus, 1519–1603) [Whewell 1859; Sachs 1906; Bremekamp 1953; Pratt 1982; Atran 1987; Pavlinov and Lyubarsky 2011; Pavlinov 2018, 2021b], whom C. Linnaeus will call "the first true systematician" [Linnaeus 1751: § 54]. Cesalpino's philosophical views are a mixture of Neoplatonism with Aristotelianism, supplemented by something like biblical pantheism [Buhle 1816]; they are set out in his treatise *Peripatetic questions* [Cesalpino 1571]. Discussing Aristotle's *Metaphysics* in it, he regards an "absolute pure form", as an attribute of Creator, the only and the highest reality. It is a source of concrete forms that, when interacting with passive matter, generate substances embodied in things. With this, however, Cesalpino departs significantly from the canonical ontology of Aristotle: he does not identify either matter with genus or form with species. It can be assumed that (in our terminology) Cesalpino interprets matter and form partonomically as primary attributes of being, whereas genus and species are treated taxonomically as secondary attributes of being, as well as predicates of its cognition; this standpoint partly reproduces the one of Avicenna and Albert the Great.

The scholastic character of Cesalpino's onto-epistemology is evident from the fact that in the *Peripatetic Questions*, he does not rigorously distinguish between genera and species, introduces gradations in them (first and other genera, great and other species), and correlates them equally with substances. In Book I of this treatise, he discusses the multiplicity of species and writes about "species of species" (*specierum species*), while regarding genera, he suggests taking this account for "analogy, since the genera themselves are related to the first genus" [Cesalpino 1571: 10]. In addition, in one of his "peripatetic questions", Cesalpino divides substances into genera, with identifying each genus with a domain of particular science (*Scientias distingui secundum genera suibstantiae*) [Prefatio, Quest. IIII].

The content of Cesalpino's classification method and, with it, entire emerging systematics is quite clearly defined in his treatise *16 Books on Plants*: it is briefly outlined in its Dedication and detailed in Book I. Many of his theses look today quite trivial and even banal, but it must be understood that this is the first summation of the onto-epistemic foundations of classificatory program, as they are understood at that time. According to Cesalpino, "every science consists in uniting similar and separating dissimilar things into genera and species according to *their nature*" ([Cesalpino 1583: (4), italics added). However, since the things differ in many respects, this simple task becomes more complicated when the question arises on what grounds (properties, traits) to unite and separate the things to reflect their presumed "nature". By meaningful formulation of this question and the search for rational answers to it, Cesalpino draws the boundary between herbalism and systematics, in which the *principle of differential weighing of characters* becomes one of the most fundamental [Zarenkov 1988; Pavlinov 2018, 2021b].

The main classification unit in the Cesalpino's method is genus. Quite in the style of Boëthius, who wrote about species in the sixth century (see Section 3.1), Cesalpino asserts that "if you do not know the true genus, no description, even if precise [...], proves authenticity, but usually misleads: as a result, all things will be confused if the genera are confused, [so] if genera are not distinguished, species would necessarily be confused" [Cesalpino 1583: (4), 25]. But there seems to be no deep philosophy in this declaration, as it is justified pragmatically: all the vast variety of plants should be reduced to a reasonable number of genera in such a way that "every plant that has never been seen before can be placed in an appropriate genus and get its name" [*op. cit.*: (5)].

It should be highlighted that the categories of genus and species in Cesalpino, as in later scholastic systematicians, are accounted mainly taxonomically and only rarely partonomically depending on context. Due to this, partonomic-taxonomic dualism becomes clearly weakened, which means initial taxonomization of the genus-species problematics. It is accompanied by a partial fixation of the sliding hierarchy of genera and species, which means an initial undermine of genus-species uncertainty: according to Cesalpino, the "last species" (*species ultima*) does not mark the last arbitrary step of logical division, but is always allocated to the lowest level of the taxonomic hierarchy, as it is understood at his time. This important turn in the early conceptual history of systematics, which is associated with the name of Cesalpino [Whewell 1859; Pavlinov 2018, 2021b], can be appropriately designated as the *Cesalpinean Revolution*.

According to Cesalpino, genera are characterized by few essentially treated anatomical structures that embody the "soul" of plant organism. Cesalpino's main emphasis is on the organs of reproduction, since they provide the potential infinity of species (*specierum aeternitatem*) due to chains of generations. This appeal to reproduction makes Scott Atran believing that Cesalpino is the first to introduce in botanical systematics "a taxonomic notion of the species as a perpetually self-reproducing unity from a common seed" [Atran 1987: 202]; however, A. Cesalpino is not mentioned among the forerunners of generative concept (see Section 5.1.2). The structure of reproductive organs constitutes a single basis of the sequential generic division, which means the scholastic character of the Cesalpino's method [Sloan 1970, 1972; Pavlinov 2018, 2021b]. The species are delineated by him based on the combinations of secondary traits: this can be considered an anticipation of future phenetic idea (see Section 4.3.4 on it).

* * *

The "logical" classifications of plants by genera and species in the middle and second half of the 17th century become a real craze among "methodist" phytographers. Manuals are published one after another, and in their introductory sections, the principles of genus-species division are outlined. The classifications produced by these principles take the form of taxonomic catalogues, containing the lists of genera and species with their diagnostic features. All this indicates an active taxonomization of the incipient systematics and, together with it, the genus-species problematics. And yet, the guiding idea of the implementation of the classificatory program remains largely scholastic: genera and species retain their relative "sliding" status as classification universals (predicates), which preserve in part genus-species uncertainty.

An important contribution to the initial development of the method of scholastic systematics at this time belongs to Joachim Jungius (1587–1657). His *Hamburg Logic* will be highly appreciated by such recognized authorities as R. Descartes and G. Leibniz [Kangro 1969; Clucas 2010]. His epistemology can be defined as a kind of Cartesian rational empiricism: knowledge is based not on sensory perceptions as such, but on their rational comprehension. Jungius applies this approach to the description of diversity of plants: he develops a partonomic method of individuating and coding their anatomical elements and representing the latter's diversity as an ordered system of standardized combinations, superficially similar to mathematical formulas. This makes taxonomic classification of plants a formalized procedure of circumscribing, diagnosing, and arranging of genera according to these "formulas". His ideas are implemented into a taxonomic system in the *Brief Botanical Physics*: as this title indicates, Jungius tries to build descriptive botany in the image and likeness of physical science. This treatise will be published posthumously in the mid-18th century [Jungius 1747], but its handwritten text becomes available immediately and will have a noticeable impact on the development of systematics of a scholastic kind.

An overview section of the *Botanical Physics* begins with the statement about the need to reduce the entire diversity of plants to an ordered system of genera and species. This general thesis is justified by an appeal to the epistemic limits of cognizing mind, which is incapable of comprehending an unordered infinity; therefore, it is necessary to operate with a finite number of properly defined and arranged genera and species. To achieve this end, he suggests to "divide [plants] by generic and specific differences according to the laws of logic" [Jungius 1747: 69]. Thus, Jungius counts systematization of plants as a purely logical endeavor, his genera and species are classification universals, and their natural (realistic) account does not matter much to him.

A scholastic nature of the classification studies by the early "methodist" phytographers is evident from splitting each of the basic categories following scholastic terminology. For example, in the *Introduction of Botany* and the *New Distribution of Umbelliferous Plants* by Robert Morison (1620–1683), the terms *primum* and *minimum* can be found in sequential divisions of genera [Morison 1669, 1672]. August Bachmann (Augustus-Quirinus Rivinus, 1652–1723) applies the terms *generalissimum, summum, subalternum,* and *proximum* in this regard [Rivinus 1696], while Pierre Magnol (1638–1715) applies the category of *subaltern* to divide species [Magnol 1689]. Rivinus in his *General Introduction to the Plant Kingdom* summarizes this predominant intention by posing, with the reference to Cesalpino, the question of "how, based on the structure of fruits and flowers, [plants] can be divided into genera and species" [Rivinus 1690: 20]. In this regard, he poses a purely scholastic question, "whether it is necessary that what is told by a genus should also be told by each of its species?" [*op.cit.*: 10].

An important innovation appears in R. Morison's *New Distribution*, which marks a movement from scholastic classification to natural (and partly biological) arrangement of genera and species. He borrows the notion of *affinity* (*affinitatis*) from alchemy to bind these units into a single network of relations, which resembles more natural rather than logical links. This is illustrated by tree-like schemes of "cognitions & affinitatis" (*cognationes & affinitatis*), akin to genealogical ones, which will be actively mastered by systematics only 100 years later due to the development of evolutionary idea.

* * *

One of the most notable scholastic systematicians of the second half of the 17th century is John Ray (Joannis Raius, 1627–1705). He begins as a herbalist: in his first botanical catalog of 1660, plants are ordered alphabetically by generic names [Raven 1950; Sloan 1970]. But after getting acquainted with the classification ideas of Cesalpino and Jungius, whom he repeatedly mentions in his main works, as well as due to his direct contacts with John Wilkins in developing the universal "philosophical" language [Slaughter 1982] (see Section 4.1), he gets transfigured into an advanced "methodist" focused on the development of the natural system of plants and partly animals [Raven 1950; Sloan 1970, 1972; Bryan 2005]. Ray's attention is also attracted by the philosophical ideas of J. Locke that guide his thought toward empiricism [Sloan 1970, 1972; Sucker 1978; Cain 1996]. Besides, in connection with the justification of his method, Ray plunges into natural theology and reflects on *The Wisdom of God* [Sokolov 2008].

J. Ray's general onto-epistemic standpoint, largely based on the ideas of Locke [Sloan 1970, 1972; Pratt 1982], is evident from his following passages:

> Since Nature (as it is commonly said) does not make leaps, it proceeds from one extreme to another not otherwise than through intermediates. It always produces species intermediate between the highest and the lowest, species of dubious position that connect one type and another and have something in common for both.
>
> The essences of things are not known to us. Since all our knowledge comes from perceptions, we do not know anything about the things, existing outside us, except for the effects they have on our senses […] and thinking about them.
>
> *[Ray 1686, Prefatio: (2); Ray 1696: 5; Ray 1733, Prefatio ad Lectorem: (2)]*

The "general method" (*Methodus generalis*) of Ray, as of his immediate predecessors, is based on the scholastic genus-species scheme. In the Preface to the *History of Plant Species*, he emphasizes that such method, which for him is equal to classification, "is both natural and philosophical (*naturalis & philosophica*), being a convenient means of brief presenting botany" [Ray 1686, Prefatio: (2)]. It serves as an "Ariadne's thread (*Ariadneo filo*) that allow you easily navigating among intermediate genera and reaching the final species" [*loc. cit.*].[7] At the same time, Ray warns readers that he

> does not hope for a perfect and absolute method [...] that would arrange all plant species into genera according to characters, without leaving some of them common to many genera. The nature of things does not allow this.
>
> *[Ray 1686, Prefatio: (2)]*

The basic classification units of the Ray's method are genera and, to a less degree, species: as in other scholastics, they are universals applicable to any manifestations of Nature. Depending on the context, these units may be partonomic or taxonomic. In the first case, Ray, in the style of Aristotle's *Parts of Animal*, writes about species of seeds, flower parts, etc.; in the second case, he refers to certain aggregates of organisms divided by genera and species according to certain characters.

Ray's synoptic classifications, in which genera and species are always taxa, are illustrated by numerous Trees of Ramus (Ray calls them "tables"), with the diagnostic traits indicated at their nodes that distinguish respective genera. With this, Ray emphasizes in several places of his treaties *On Different Plant Systems* that "according to the rules of logic [...] the complete definition of [species] consists of proximal genus and essential difference" ([Ray 1696: 23], also [Ray 1733: 6]). But this scholastic thesis is partly opposed by an antischolastic one, which goes back to Locke:

> Since the same essences provide for the same quiddities, functions, as well as other secondary features [of organisms], there can be no more reliable indicator of an essential, and thereby generic unity, than agreement on many common features, i.e., on similarity in many parts and properties [of organisms].
>
> *[Ray 1733: 6]*

Two important corollaries can be drawn from this passage: first, Ray relies on essences in elaborating generic classifications; second, he believes that these essences are manifested "in many parts" of organisms. Obviously, this position is fundamentally essentialist and at the same time quite Aristotelian, if we keep in mind the previously mentioned "Aristotle's problem" (see Section 2.1.3). Accordingly, the species are gathered in genera, "whose appearance akin to each other [and] they have many attributes in common" [Ray 1696, Prefatio: (8)]. Simultaneously with Ray and with reference to him and for a similar reason, viz., due to Locke's influence [Sloan 1972], P. Magnol expresses a similar idea: he asserts that "those parts that do not serve

---

7 This metaphor is usually attributed to C. Linnaeus, but its author is J. Ray.

the fruit are no more secondary [to plant] than the limbs are secondary to animal" [Magnol 1689, Prefatio: (8)].

In Ray's taxonomic systems, classification units of various levels of generality form a multistep hierarchy, and when discussed in the introductory sections of his treatises, they are designated scholastically according to those levels. These are primary or higher (*primum*, *summa*; denoted as *classes* occasionally), secondary (*secundum*, *subaltern*), tertiary, or proximal (*tertium*, *proximum*) genera; sometimes species are mentioned as *infima* or *subaltern*. In *The Wisdom of God* (1690), he writes about four "major genera, or orders" of animate bodies: beasts (including snakes), birds, fish, and insects [Ray 1714: 21]. However, in the *History of Plant Species*, when discussing the foundations of botanical classification, Ray mentions zoophytes as *speciem* intermediate between plants and animals [Ray 1686].

However, in his synoptic classifications, Ray usually does without scholastic formalisms to denote successive steps of division of the main groups. For example, in the *History of Plant Species* they are designated as follows (listed in descending order): Part (*Pars*), Section (*Sectio*), Chapter (*Caput*), with the last division corresponding to *genus proximum*, since it is further divided into species [Ray 1696]. In the *Synopsis of Quadruped Animals*, the higher steps are not termed but indicated by descriptive names of the respective taxa (e.g., *Quadrupeda Bisulca*; *Animalia Quadrupedia Pede ungulato Anomala*). With this, the lower steps are designated as genera (e.g., *Bovinum genus*, *Ovinum genus*), which are divided into the lowest genera and species (e.g., *Urus*, *Bonasus*, *Ovis domestica*) [Ray 1693]. In a short *Discourse on the specific differences in plants* (1674), Ray criticizes herbalists for their splitting species: they pay great attention to accidents and take them for true species differences, so their "species" are delineated not "agreeably to nature" [Ray 1928a: 78].

Despite the strong influence of Locke's philosophy on Ray's general attitude, he acknowledges the real status of genus and species as natural units, as can be seen from the following statements in his essay *On Different Plant Systems*. Regarding genera, he writes that "to belong to Nature and to belong to genera is the same. For us, they are the nature or essence of things" [Ray 1696: 13]. Referring to species, Ray writes that they "are founded in things, [and] the individuals of the same species seem as samples (*exemplar*) of idea in Divine mind" [*op.cit.*, Prefatio: (6)]. In addition, in the *History of Plant Species* and in *The Wisdom of God*, Ray indicates one more criterion for the species integrity, which goes back to an antique-medieval generative idea: this is "special origin from the seed" (*distincta propagano ex semine*).

Thus, Ray seems to implicitly "divorce" logical and natural accounts of species.[8] In the first case, species is treated as a mental "unit of reasoning" (*entia rationes*) [Ray 1696: 1] and corresponds to the final step of the logical genus-species division. In the second case, with reference to Divine Mind, Ray acknowledges species as a natural unit of the structure of Nature. Of course, many realist theologians made this reference before him (Aurelius Augustine, Anselm of Canterbury, Albert the Great, etc.), since their biblical world picture obliged to this, but it is J. Ray who becomes

[8] The same will be done explicitly by G. Leibniz several decades later (see Section 4.2).

the first to incorporate this justification of reality of species in the context of systematics (see also Section 5.1.1). Accordingly, the logical genus-species uncertainty recedes into the background, and the natural-classification dualism comes to the fore in the cognitive situation, as it is outlined by the Rayan general method.

However, careful examination of *The Wisdom of God* unambiguously witnesses that Ray, when characterizing the results of divine creation, implies species of all material things. Inspired by the then popular corpuscular idea, he believes that "there are many *species* of bodies which the *Perepateticks* call mixed, which are as simple as the *elements* themselves, as metals, salts, and fume sorts of stones", so he "rather attributes the various *species* of inanimate bodies to the divers figures of the minute particles of which they are made up" ([Ray 1714: 60–61], italics in original). In agreement with this, he writes about "the *species* of Principles or indivisible particles", meaning some "elementary bricks", of which these bodies are composed [*loc.cit.*], and compares the number of species of inanimate and animate bodies by counting them as commensurable units.

And yet Ray's mentioning generation chains as an important condition for the unity and persistence of species of living organisms indicates that he distinguishes them in some way from "species" of inanimate matter. This brings a significantly new emphasis to the species problem, directing it toward the modern general idea of species as a biological unit (see Section 5.1.2).

4.3.3 Tournefort and Linnaeus: Disconnection of Genus and Species

The most important innovation in the implementation of the classificatory program to the studies on the diversity of plants and animals during the 16th–18th centuries becomes a rigid *structuring* of the genus-species hierarchy. This is achieved by introducing a limited number of fixed ranks that are not presumed by its scholastic rankless construction. It involves a taxonomic aspect of the diversity of organisms and does not concern partonomic one; as a result, a *ranked* taxonomic hierarchy gradually develops.

This trend partly begins with the herbalists (Brunfels, Gesner, Bauhin, Lobelius), the first naturalist systematicians (Rivinus, Ray) continue it, and the key figures of the last phase of the conceptual history of scholastic systematics, Tournefort and Linnaeus, complete it at the first approximation. Most often, the names of these categories are borrowed from book business (*Pars*, *Sectio*, *Capitum*), not rarely from logic (*Classis*, *Ordo*), as well as from social (*Familia*, *Tribus*) or military (*Cohors*, *Division*) terminology [Pavlinov 2015, 2018; Lyubarsky 2018].

Initially, these categories do not have a strictly fixed status: like scholastic genera of different levels (*subalternum*, *intermedium*, etc.), they simply denote succeeding steps of division; therefore, they are comparable neither in rank nor content in different classifications. At the final stage of structuring taxonomic hierarchy, unambiguous fixation of ranks occurs: a small number of them (4 to 6) is recognized, they are organized in a strictly defined sequence, definite terms are assigned to them, and this makes them comparable to a certain extent in different groups of organisms. However, in the 18th and in the first half of the 19th centuries, the complete unification of the ranked hierarchy is not yet worked out: it differs in different authors, and sometimes in different books of the same author. It is only in the middle and second half of the 19th century when its stabilization and unification will occur due to the initial codification of taxonomic nomenclature [Pavlinov 2015, 2022b].

It should be taken into account that the structuring of the taxonomic hierarchy, being an obvious departure from the scholastic canonical version of the genus-species scheme, is just as obviously based on it. Thus, Rivinus identifies his *Ordo* with *Genus summum* [Rivinus 1696]; Ray also identifies *Classis* with it [Ray 1696]; Linnaeus identifies *Classis* with *Genus summum* and *Order* with *Genus intermedium* [Linnaeus 1737a]. As a result, in the second half of the 18th century, largely due to the authority of C. Linnaeus, a certain correspondence between the old steps of genus-species division and the new fixed ranks of taxonomic hierarchy becomes acknowledged and partially canonized. From the perspective of the considered problem, the most important part of such structuring of taxonomic hierarchy is the appearance of a fixed gap at its lowest level: in its new ranked version, *Genus proximum* becomes just *Genus*, and *Species infima* becomes just *Species*. In such milieu, the latter notion finds its place among the most fundamental ones in emerging biology.

* * *

The phytographer Joseph Pitton de Tournefort (1656–1708) becomes the first among systematicians to modify seriously the genus-species classificatory scheme of scholastics. His general "methodical" attitude is outlined in his *Elements of Botany* (Latin edition is entitled *Institutiones rei herbariae*), and it is clearly visible from his statement that "without applying a method to the things as different as in botany, it is impossible to get a clear idea of Nature" [Tournefort 1719: 54]. With this, he is sure that the method is not arbitrary and "those who think that the correct classification and naming of species depends on someone's imagination are mistaken" [*loc.cit.*].

The classification method of Tournefort is based on similarity regarded not in itself, but as an indicator of some kind of affinity in its general natural-philosophical sense. According to him, "this similarity should be deduced from the signs of the closest affinity, i.e., from the structure of one part of the plant, and do not pay attention to the signs of a more distant affinity" [Tournefort 1694: 13]. Wishing to emphasize the fundamental soundness of this approach, Tournefort points out that none other than "Creator of things inscribed essential signs in the plants themselves, on which basis one can observe the similarity inherent in the species of the same genus" [*loc.cit.*]. When choosing these "signs", Tournefort rejects the natural-philosophical approach underlying Cesalpino's method: "there is no question about the goals of Nature or about the exclusive qualities of the parts; rather, it is a matter of finding the means to differentiate the plants as clearly as possible" [*op.cit.*: 6]. Thus, the classification method of Tournefort, including the separation of particular genera and species, cannot be qualified as essentialist and even more Aristotelian: contrary to a popular opinion [Cain 1958; Hull 1965; Mayr 1969, 1976, 1982; Panchen 1992; Ereshefsky 2001; Abruzzi 2015], he assesses the significance of the parts of plants not for their vital functions, but as the "signs" of their affinity.

Another important departure from scholastic canons involves certain elements of inductive argumentation: taxa are not divided into subtaxa, but are grouped of them. This is evident from the following definitions in the Dictionary of his *Elements of Botany*: "A class of plants is the cluster of several genera of plants, all of which are suitable in that they have certain common marks which essentially distinguish them from all other genera of plants"; "A genus of plants is the cluster of several plants which have a common character established on the structure of certain parts, which essentially distinguishes these plants from all others" [Tournefort 1694: 525, 542].

Tournefort is the first to introduce and consistently apply a simplified taxonomic hierarchy including only four ranks with fixed names and subordination: Class, Section, Genus, and Species. In this ranking scale, the categories of genus and species become clearly separated, which is of paramount importance. By this, his method excludes interchangeable "sliding" status of these categories inherent in scholasticism: it yields autonomization of species and signifies an initiation of a conceptual revolution that will be finalized soon by Linnaeus (see below).

With this, however, the category of genus remains basic for Tournefort: focusing on it, he is little interested in species. In the *Elements of Botany*, he writes that the species constitute "plants of the same genus [that] differ from each other in some features [but] have a [common] generic trait" [Tournefort 1694: 13]; in the Dictionary, species, unlike class and genus, is not defined in any way [*op.cit.*: 531]. Because of this emphasis, Tournefort tends to ascribe generic rank to the morphologically distinct species while treating many varieties as species. This results in the appearance of monotypic genera in his synoptic classification, which, though seeming technical, represents an important classification innovation as compared to the scholastic method: in the latter, nondivided monotypic "lower genera" are impossible; they should be treated as "final species".

The physician and naturalist Karl Nikolaus Lang (1670–1741) transfers the classification method of Tournefort to zoology [Bachmann 1896, 1906]. In his *New System of Seashells*, he applies the Tournefortian four-ranked taxonomic hierarchy with the same subordination, and these categories are listed in the full title of this book [Lang 1722].

* * *

The conceptual history of scholastic systematics is completed by the works of the great naturalist Carl von Linné (Linnaeus, 1707–1778). University education in Lund, where scholastic influence is strong at his time, makes him an advanced "methodical" systematician, not alien to an idea of "philosophical" language of science [Bobrov 1957, 1970; Stafleu 1971 Koerner 1996; Breidbach and Ghiselin 2006]. He publishes several dozens of repeatedly reprinted fundamental treatises arranged in two principal series. One of them, the most significant for us, begins with the *Foundations of Botany* (1736) of about 40 pages, continues with the *Analysis of Botany* (1737), and is completed by the *Philosophy of Botany* (1751) of more than 300 pages. The second series consists of taxonomic accounts: the *System of Nature*, which survives 12 editions during the author's lifetime (the first of 1735 is of 12 pages, the last of 1766–8 is of about 2300 pages), the *Genera of plants* (1737), the *Species of plants* (1753), etc. The significance of these works is emphasized by the fact that the whole systematics is often divided into "pre-Linnaean" and "Linnaean" major stages, though, from a conceptualistic perspective of its history, such division is more than doubtful [Pavlinov 2018, 2021b].

Linnaeus' ontology is based on the biblical world view: he believes that "Nature is the behest of God" (*Natura est jussum Dei*), or "law of God, endowed in the creation" (*lex Dei, in ipsa creatione rebus indita*) [Linnaeus 1749: 546]; therefore, to reveal the System of Nature through classifying its entities means to comprehend the providence of God [Ramsbottom 1938; Hofsten 1958; Larson 1968, 1971; Lepenies 1985; Petry 2001; Grene and Depew 2004; Breidbach and Ghiselin 2006; Harrison 2009). His epistemic position is distinctly conceptualistic: he assures that none other than "philosophers, by demonstration from rational principles, have reduced botanical knowledge to the form of a science"; he calls them "theoreticians" and thinks that botany is obliged to them for its "rules and regulations" [§ 19].[9]

For Linnaeus, the System (called by him "Arrangement") and the Method are the same: "the arrangement of vegetables [...] is commonly called *method*" [§ 153], so his statement that "the natural method is the ultimate purpose of botany" [§ 161] should be understood in this sense. The emphasis on the "method" indicates a noticeable scholastic influence on Linnaeus [Grene and Depew 2004; Pavlinov 2018], which is evident from his active use of the Tree of Ramus to illustrate his arrangements, e.g., in the introductory section of the *Genera of Plants*. However, contrary to an "accepted view" [Cain 1958; Hull 1965, 1985; Mayr 1982], this influence is not so comprehensive, which is emphasized by many authors [Müller-Wille 2001, 2007; Winsor 2001, 2003, 2006a,b; Vogt 2008; Müller-Wille and Scharf 2009; Pavlinov 2018, 2021b].

In the *Philosophy of Botany*, Linnaeus follows Cesalpino and Jungius in developing a fairly advanced partonomic classification of the parts of plants. This allows him, in accordance to the canons of Cartesian rationality, to turn a plant as a living organism into a certain cognizable object, mechanically divided into characters for individuating and naming taxa [Svenson 1953; Eriksson 1983; Cain 1994b; Pavlinov and Lyubarsky 2011; Lyubarsky 2015]. In this regard, being committed to the scholastic tradition, he occasionally designates partonomic units as "species" [§ 82, 85].

* * *

One of the main cognitive objectives of Linnaeus is the comprehension of System of Nature. This natural-philosophical notion in its overall account is well known to the philosophers of this time and underlies their ideas of a unified, rationally organized knowledge about Nature (Descartes, Leibniz, Wilkins, etc.). Linnaeus interprets an arrangement of things reflecting this System in a taxonomic rather than a partonomic sense and distinguishes two general categories of such

[9] Hereafter, unless otherwise specified, quotations are given from the last English edition of the *Philosophy of Botany* (Linnaeus, 2003), sometimes in a different wording; italics are always in original.

arrangement, *natural* and *artificial* [Linnaeus 1735, 1736, 1737a]. The meaning of this distinction is that the natural system (or rather arrangement), according to an initial presumption, is the only one as a manifestation of Divine providence, and it is an unattainable ideal, while artificial arrangements, termed by Linnaeus *synopses* [§ 154], are developed as pragmatic aids. Following J. Ray, he calls them allegorically "Ariadne's thread" allowing one to navigate the labyrinth of the diversity of Nature [§ 156]; therefore, "artificial systems are absolutely necessary" [Linnaeus 1735: § 12]. This distinction will later appear very important for the development of systematics [Smith 1822; Daudin 1927; Müller-Wille 2007].

For the taxonomic system, Linnaeus adopts its ranked hierarchy that includes five major fixed "divisions: *classes*, *orders*, *genera*, *species*, and *varieties*" [§ 155]. The first four are borrowed from Tournefort in a slightly different designation, the last one is added as new. As noted above, this hierarchy has an obvious scholastic background: Classis (Ordo Tournef.) corresponds to *Genus summum*, and Ordo (Sectio Tournef.) does to *Genus subalternum* [§ 200]. With this, Linnaeus clearly indicates their Tournefortian subordination: "Genera consist of Species, Orders of Genera, and Classes of Orders" [Linnaeus 1737a: § 251]. However, his taxonomic categories are hardly ultimately settled: for example, the main plant groups (fungi, algae, etc.) are called "species" in the *Foundations of Botany* [§ 78]; in the *Philosophy of Botany*, they are "families", while they are mentioned as "tribes" in the initial section [§ 3] of the 12th edition of the *System of Nature* [Linnaeus 1736, 1751, 1766–8]. Besides, in a brief introduction (*Delineatio Plantae*) to the *System of Nature*, he distinguishes only two categories, viz., genus and species, and includes class and order in the genus in its widest (almost scholastic) account. It is to be noted that Linnaeus applies these categories in a unified meaning when describing the diversity of all three "Kingdoms of Nature", which are animals, plants, and minerals. From this, it is evident that the taxa of these "Kingdoms", including species, are basically classification universals of scholastics [Mayr 1982; Pavlinov 2018].[10]

In the classification method of Linnaeus, the generic category seems to take a basic position [Sharp 1873; Cain 1958; Stearn 1959; Larson 1967, 1971], which also reflects his scholastic attitude. He reproduces several times Cesalpino's declaration that "if the genera are confused, it is inevitable that everything will be confused" [Linnaeus 1737a,b, Ratio operis: § 6; Linnaeus 1751: § 159], and adds that "without knowledge of genus, species is devoid of certainty" [Linnaeus 1737a, Ratio operis: § 6], which is repeated in other words in the *Philosophy of Botany* [§ 256]. At the same time, he writes in the *Species of Plants* that although "systematization as an Ariadnine thread is usually limited to genera [he] tries to extend it also to species, assuming occurrence of true (*letgitimis*) differences between them, since all *true*

[10] However, Linnaeus' occasional mentioning of generative idea in regard to the plants motivates some of his admirers to declare him an originator of the reproductive (biological) species conception [Eriksson 1983; Müller-Wille and Orel 2007].

knowledge depends on the knowledge of species" ([Linnaeus 1753, Lectori aequo: (2)], italics in original).

It is of particular importance how Linnaeus considers relation between species and varieties in the special section of the *Philosophy of Botany*: its "legitimization" is an obvious departure from the scholastic canons. In the latter, the final step of the genus-species division is denoted as *species infima*, and it is indirectly observed in the Tournefort's method, who ranks many varieties as species.[11] In Linnaeus, the species can be divided into varieties without losing its species status, so his innovation fixes ultimately a distinctive status of species initiated by Tournefort. Therefore, their coherent intellectual efforts yield a significant advancement in our conceptual history, which deserves to be qualified as *Tournefortian–Linnaean revolution*.

The general idea of intraspecific category becomes soon acknowledged by other systematicians who begin separating and designating them [Vinarskiy 2015; Brower 2021]. In particular, a Linnaeus' student Eugen Johann Christoph Esper (1742–1810), in his dissertation *On Species Variations*, introduces the category of *subspecies* into taxonomic hierarchy to distinguish by these essential subdivisions of species from accidental *varieties* [Esper 1781];[12] and it will receive full acknowledgement at the beginning of the 19th century [Persoon 1805; de Candolle 1819]. Eventually, these intraspecific categories will play a key role in the formation of Darwin's evolutionary conception and will significantly influence the understanding of the nature of species in the second half of the 19th and early 20th centuries (see Section 5.3).

Linnaeus' understanding of the ontological status of taxa initially depends on their position in the ranked hierarchy. In the *Genera of Plants*, Linnaeus asserts realistically that "Genera and Species are all natural" and "without adopting this principle, it is impossible to comprehend the art" of systematization [Linnaeus 1737b, Ratio operis: § 6]. Therefore, in the botanical section of the *Systems of Nature*, he emphasizes that "the basis of botany lies in the systematic arrangement and naming [of plants] by genera and species", whereas orders and classes are not mentioned [§ 2]. He details this observation in the *Philosophy of Botany* by noting that the genus and species are "always works of Nature, the variety is quite often the work of Cultivation, class and order are the work of Nature and Art" [§ 162].

Justifying the naturalness of species, Linnaeus follows J. Ray with referring to their divine creation and asserts repeatedly that "there are as many species as there were different forms produced by the Infinite Being in the beginning" [Linnaeus 1737b, *Ratio operis*: § 5; Linnaeus 1751: § 157]. This means that Linnaeus regards species as ontologically primary relative to other taxonomic categories. As for the genera, Linnaeus treats their origin controversially. On the one hand, he asserts that "there are as many genera [...] as there are common, proximate attributes of different species" [Linnaeus 1737b, Ratio operis: § 6] or (in other words) "as there are similarly constructed fruit-bodies produced by different natural species" [§ 159]:

[11] This gives Linnaeus a reason to note that "the introduction of varieties has done more to contaminate botany than any other thing" [§ 259].

[12] It should be reminded that this notion is already in use by L. Valla in the 15th century (see Section 4.1).

in this case, genus is secondary to species, which agrees with the just above primary treatment of species. On the other hand, he states in the same paragraph, referring to his *Systema Naturae*, that "every genus is natural, made in the first place such as it is, for this reason it is not to be capriciously split or stuck" [§ 159]. However, all these reflections seem to be nullified toward the end of his life, when Linnaeus develops a quite different account of divine creation of the hierarchical diversity of living nature, which we will discuss elsewhere (see Section 5.1.1).

* * *

An initially scholastic attitude of C. Linnaeus as a systematician is followed by his disciples and closest colleagues, the "Linnaeans" [Stafleu 1971]. They treat basic categories as universal classification units, and their principal task is to recognize natural genera (*genera naturales*). Thus, a disciple of Linnaeus, Peter Artedi (Petrus Arctaedius, 1705–1735), in the introductory section of his *Ichthyologia* (published posthumously by Linnaeus), discusses genera and species of minerals, plants, and animals [Artedi 1738]. These three kingdoms are indicated in the full title of the *Introduction to Natural History* by Giovanni (Johannes) Antonio Scopoli (1723–1788), and he mentions in its introductory section "natural genera" and "true species" [Scopoli 1777]. Another instance is the *System of Fossils* by Johann Gotthelf Fischer von Waldheim (1771–1853), who reproduces the Linnaean *System of Nature* by its structure: all its three kingdoms are divided uniformly into classes, orders, families, genera, and species [Fischer 1811].

Preservation of scholastic tradition in the botanical classifications of the turn of the 18th–19th centuries is exemplified by the *Principles of Botany* by Carl Ludwig Wildenow (1765–1812). He argues that "the system is primarily divided into *classes* and *orders*, [...] the last division is called *species*; [...] no system should be divided according to any other attribute than the one that is originally chosen" ([Willdenow, 1792: 132–133], italics in original).

However, in scholastic systematics of this epoch, classification account of the main categories does not always follow Linnaeus' standards. An example is the *Philosophical phytozoology* by Noël Martin Joseph De Necker (1730–1793), who strictly adheres to the logical interpretation of taxonomic categories as universals. In his version of the genus-species scheme, he distinguishes two categories, viz., *true genera* and *natural species*: the former correspond to the supra-generic categories and the latter to genus [De Necker 1790]. Regarding infima species, they are not recognized by Necker, who mentions them as "races" (*proles*).

4.3.4 Adanson, Jussieu, etc.: The Birth of Natural Systematics

In the second half of the 18th century, one of the most significant events in the conceptual history of systematics involves a transition from scholastic to postscholastic research program [Pavlinov and Lyubarsky 2011; Pavlinov 2018, 2021b]. The classic of descriptive biology at the beginning of the 19th century, A.-P. de Candolle, equates this shift with the scientific revolution in chemistry, which was committed by A. L. de Lavoisier at the end of the 18th century [de Candolle 1819]. In many ways, it represents a delayed (one might say, belated) antischolastic turn of systematics, which in other sections of natural science happened two centuries ago and marked the beginning of the Modern Times. In the 1760s–1780s, a new biophilosophy begins being actively shaped, it is based first of all on a substantive disjointing of living and inert matter and on a new understanding of organism—not as a certain "thing" characterized by a set of classifying traits, but as an actively functioning living creature interacting in a complex way with its environment. With this, regarding the species problem, this turn transfers the cognitive focus of systematicians from genus to species: it is acknowledged now that these are the species that "constitute the foundation of the science of botany" [de Jussieu 1789: xx] and "form the basis of all classification" [Lindley 1832: 307].

On this foundation, biologization of the explorations of biodiversity intensifies [Wilkins 2003a, 2009a; Pavlinov and Lyubarsky 2011; Pavlinov 2018, 2021b]. According to the new research program, classification activity should not be limited to artificial "Ariadne's threads", but strives to a truly natural system of plants and animals. For this, two principles of classification are to be followed that are opposite to the scholastic genus-species scheme. First, systematization should proceed following not deductive but inductive argumentation scheme, i.e., in an ascending manner from lower to higher categories. Second, organisms are to be compared not by a few diagnostic, but by all available characters, which yields an *overall similarity* as a means of revealing their natural affinity and results in their grouping into really natural genera and species. The second condition means that a *phenetic idea* becomes an important part of the epistemology of early postscholastic systematics, which differs significantly by this hallmark from scholastic essentialism. Taken together, these two principles refer to a growing impact of the Lockean empiricism developed in the 17th century.

Regarded from a broad conceptual perspective, the phenetic idea of early systematics appears not *de novo* but develops those insights that were being formed gradually by medieval scholasticism and its successor, scholastic systematics. A scholastic concept of similitude, formulated by A. Augustine and further elaborated by R. Bacon, can be considered a kind of its prototype. In systematics of the 16th century, A. Cesalpino expresses something similar to this idea with regard to species, and the same is done in the 17th century by J. Ray and P. Magnol with regard to the natural groups in general. The maturation of this idea at this time is evidenced, for example, by an opinion of the phytographer Alexander Christian Hackenholz, who, setting out his "combinatorial method" in his letter to Leibniz, emphasizes that "the comparisons of plants themselves should be instituted not only on the basis of the flowers, but also on the basis of most other major parts" (cited after [Smith, 2011: 307]). Among the naturalists of the 18th century, who champion an idea of the overall similarity of organisms for delineating their groups, the first to be mentioned is Georges-Louis Leclerc de Buffon (1707–1788), famous for his multivolume *Natural History*. Inspired by the ideas of Locke and Leibniz [Sloan 1970], he emphasizes in his *Premier Discours*, opening this serial, that botanists err when "wanting to judge about a whole, or a combination of several wholes, by a single part and by comparing the differences of this single part" [Buffon 1749: 15]. Developing this idea, Buffon writes:

> If the individuals have a perfect resemblance, or the differences so small that they can only be seen with difficulty, these individuals will be of the same species; if the differences begin to be noticeable, and at the same time there are always many more similarities than differences, the individuals will be of another species, but of the same genus as the first; and if these differences are even more marked, without, however, exceeding the resemblances, then the individuals will not only be of another species, but even of another genus than the former and the latter, and yet they will still be of the same class, because they resemble each other more than they differ: but if, on the contrary, the number of differences exceeds that of the resemblances, then the individuals are not even of the same class.
>
> *[Buffon 1749: 21]*

As a result, the "new systematics" emerges in the second half of the 18th century, which can be called with a sufficient reason not just postscholastic, but *natural* in the broadest sense [Stafleu 1969; Pavlinov and Lyubarsky 2011; Pavlinov 2018, 2021b]. Within its framework, partonomic-taxonomic dualism and genus-species uncertainty are eliminated from the cognitive situation, taxonomization and autonomization of species are finally established, and the genus-species problematics gets reduced to the species one. New understanding of the mechanism ensuring integrity and persistence of species as a natural unit comes to fore, viz., self-reproduction in the chains of generations, according to which its previously dominating interpretation as a universal classification unit is complemented with a biological one. As a consequence, the significance of natural-classificatory dualism increases, meaning a twofold accounting of species as both a natural and a classification unit, with the first treatment becoming a priority.

It is to be reminded that the scholastic Albert the Great wrote about this dualism (of course, without using this term) in the 13th century based on the ideas of Avicenna, and G. Leibniz similarly distinguishes between physical (natural) and logical (classification) species in the 18th century. Although the "founding fathers" of postscholastic systematics limit themselves by mentioning Leibniz and Locke, this does not mean an absence of a deeper historical continuity going back to the ideas of kalam and mature scholasticism.

* * *

The beginning of postscholastic revolution in systematics dates by the publication of the two-volume treatise *Families of Plants* by Michel Adanson (1727–1806), an active maker and advocate of the natural method in its new guise [Stafleu 1963, 1966; Guedes 1967; Pavlinov 2018, 2021b]. In the introductory theoretical section of this work, he argues that the natural method "should [...] be based on the nature of beings, which nature includes the totality of their properties and structures. There is no doubt that the natural method in botany can be achieved by considering the totality of all plant structures" [Adanson 1763: clv]. One of the sections of *Families*, entitled "How to distinguish classes, genera, species, individuals, and varieties", is specifically devoted to this issue, which Adanson calls "the most difficult and important". His realistic onto-epistemology is clearly expressed by his assurance that "before establishing principles for genera, species, etc., it is necessary to prove that they really exist in nature" [*op.cit.*: clviij].

In his historical review, Adanson discusses classes, genera, and species at length, with analyzing the methods of their delineation by his predecessors, but his main attention is occupied by the justification of the natural families, which category is borrowed from P. Magnol. It is for their recognition that Adanson develops his natural method based on the analysis of the multiplicity of features without their prior division into significant and insignificant; this original approach is most similar to the modern characters compatibility analysis [Nelson 1979; Pavlinov 2018]. In the systematic part of the *Families*, the genera are regarded the "carriers" of the familial characters, while species are mentioned only occasionally by necessity. Nevertheless, in the abovementioned theoretical section, Adanson devotes quite a lot of space to the species category.

Discussing the latter, Adanson argues (partly following Buffon) that it "consists of a constant and continuous sequence of individuals generated by crossing two sexes" [Adanson 1763: clx]. On the other hand, he treats species as an assemblage of similar individuals: for, "there are as many species as there are [aggregates] of individuals that differ from each other in one or more traits [characteristic] of the given family" [*op.cit.*: clxviij]. By this twofold consideration, natural-classificatory dualism of species is affirmed at the very beginning of postscholastic systematics. Besides, Adanson says positively about varieties as classification units: for him, they "indicate existing or possible transitions between very similar species" [*op.cit.*: clxix].

* * *

Due to an excessive novelty of the Adanson's classification method, his key ideas, including the assessment of overall similarity, remain almost without attention by his contemporaries; in fact, for example, Scopoli and de Candolle will refer to him only several decades later in their reviews. A more influential figure in initial formation of postscholastic systematics is Antoine Laurent de Jussieu (1748–1836) [Sachs 1906; Carr 1923; Daudin 1927; Kozo-Polyansky 1937; Uranov 1979; Stevens 1994]: this happens due to the fact that his method is more conservative and retains some important characters of its essentialist predecessors [Pavlinov 2018]. Assessing Jussieu's contribution to the development of systematics, one of the "founding fathers" of typology, Jean (Georges) Leopold Nicolas Frédéric Cuvier (1769–1832), writes that "his natural method [...] forms perhaps as important an epoch in the sciences of observation, as the *Chimie* of Lavoisier does in the sciences of experiment" (cited after [Carr 1923: 63]).

Jussieu's understanding of the natural method, which is largely inductive, is set out in the introductory section of his book with the quite Linnaean title *Genera of*

Plants [de Jussieu 1789]. His later detailed article *Principles of the Natural Method* is especially devoted to this issue [de Jussieu 1824]. In this article, Jussieu justifies and describes his inductive method of developing the botanical natural system. He observes that

> in order to have a complete knowledge of [true] nature, one cannot be content with studying a few organs, but that science must embrace them all. [...] The natural order, towards which our study must be directed, must be based on fixed, invariable principles, [and] it is only in nature itself that these principles can be found, observing its course in the formation of groups generally recognized as most natural. By studying these groups, we recognize that the beings who are gathered in each one are similar in the greatest number of their parts or characters, and that among these characters there are some that seem more constant, more important than others.
>
> *[de Jussieu 1824: 8–9]*

Jussieu believes that if we distinguish and designate species and genera by a few diagnostic traits, then it turns out simply "the science of nomenclature". In order to follow Nature in the system, it is necessary to "carefully examine and reveal the entire organization of plants, [...] all the characters [...] and not miss anything that would reveal the mutual affinity of all plants and achieve full knowledge of them" [de Jussieu 1789: xxxv]. However, when discussing orders, Jussieu focuses on the essentialist interpretation of characters for their substantiation. He ranks characters as primary, secondary, and tertiary based on an assumption that the "common and invariable characters cannot be obtained except from the organs that are most essential for life, namely for the reproduction of species", and "*one constant character turns out to be equal or even superior to many non-constant ones*" ([de Jussieu 1824: 27], italics in original).

His general account of species combines phenetic and generative ideas: in the introductory section of the *Genera of Plants*, Jussieu writes that "the plants that are consistent in all parts or in a common character, born from a similar [plant], and of which there are quite a lot, form a so-called species that was previously poorly defined, but now correctly determined through a continuous sequence of reproducing similar individuals. These aggregates and sequences are usually unchangeable and perpetual. [...] Such native species, precisely and strictly defined, constitute the foundation of the science of botany" [de Jussieu 1789: xix–xx]. According to this, Jussieu offers the following combined "phenetic-generative" definition of species:

> the plants that are similar to each other in all parts are individuals of the same *species*: they are born from similar individuals and must, in turn, produce the ones similar to themselves; thus, *species must be defined as a sequence of completely similar individuals, continuous due to reproduction*"
>
> *([de Jussieu, 1824: 18], italics in original).*

And yet, when Jussieu discusses the grounds for elaborating species classifications, he focuses on the selection (weighing) of characters based on the following observation. "Plants, like animals, also have reproductive organs that contribute to the renewal of individuals by creation of new beings similar to the preceding, and this constitutes the maintenance of the life of the species" [de Jussieu, 1824: 10]. Therefore, it is from these organs that "the characters that serve to distinguish species" should be taken [*op.cit.*: 12]. With this, he notices that "those who rely on the *similarity of organisms in all parts* [of plants] to unite them into species never experience a contradiction" ([*op.cit.*: 24], italics in original).

In a partly similar way, Jussieu defines genus. According to him, species are "united in genera on the basis of their affinity [...]. They include species that agree in many characters [...], selected not arbitrarily, but on the basis of rigid principles, so as to unite true relatives" [de Jussieu 1789: xx]. And further, "just as similar individuals are united into one species, the same law determines the union of similar species, [and] species should be united according to a greater number of characteristics" [op. cit.: xxxvij]. The natural method "requires *approximation of similar species according to the greatest number of different characters* for the formation of genera" ([de Jussieu, 1824: 24], italics in original).

* * *

Thus, an emphasis on the need to use numerous traits for revealing the affinity of organisms and delineating their taxa becomes immediately a dominant idea of early postscholastic systematics. Indeed, the already mentioned G. Scopoli in his *Foundations of Botany* indicates that "natural classes are distinguished based on a careful and accurate comparison of *all properties* of individuals, whereas artificial ones are based on one or a few properties" ([Scopoli 1783: 45], italics added). The Buffon's close colleague Louis-Jean-Marie Daubenton (1716–1800), in his article "Botany" for the *Encyclopedia* issued by D. Diderot, argues that the "natural order" should be "grounded on a complete description of 'all the relations' of resemblance, rather than on one concentrating on similarities in a single part" (cited after [Sloan 1979: 121]). Johann Friedrich Blumenbach (1752–1840) in his popular *Manual of Natural History* writes that "we should consider not just a few selected traits, but all external characteristics, [therefore] animals that are similar in 19 structures and differ only in the twentieth, should [...] be grouped together" [Blumenbach 1782: iii–iv].

This phenetic idea soon acquires a terminological frame. The botanist Charles-François Brisseau de Mirbel (1776–1854), a proponent of natural systematics, in his *Natural History of Plants* designates the former one-trait method of the scholastics as *monotypic*, and the new natural method based on many traits as *polytypical* [Mirbel 1806]. This terminology will later receive another interpretation by referring to the extensional account of taxa, and the terms of Mirbel will be replaced by *mono-* and *polythetic*, respectively [Sokal and Sneath 1963; Panchen 1992; Pavlinov 2018].

Due to attention to both phenetic and generative ideas, a combinatorial account of species reflecting its natural-classificatory dualism soon becomes characteristic of early postscholastic epoch. It is being actively developed by Augustin Pyramus (Pyrame) de Candolle (1778–1841), another "founding father" of natural systematics [Whewell 1847; Sachs 1906; Drouin 2001; Kamelin 2004; Pavlinov 2018].

Of his works, the most important for us is the *Primary Theory of Botany* [de Candolle, 1819], which revised English version is published in collaboration with the phytographer Kurt Polykarp Joachim Sprengel (1766–1833) as *Elements of the Philosophy of Plants* [DeCandolle and Sprengel 1821].

> Candolle begins outlining his taxonomic theory with separation of two main categories of classifications, partly coinciding with Linnaeus' artificial and natural ones. According to Candolle, *empirical* classifications are not related to the nature of plants, whereas *rational*, or *scientific* classifications reflect this nature, or more precisely, natural affinities of plants.

Candolle begins the presentation of the principles of the rational (scientific) classification, based on inductive methodology, with an "Idea of Species": this is the title of the corresponding subsection in the *Elements.* He defines species at first in a rather detailed way as "the collection of all individuals who are more similar to each other than they are to others; who can, by mutual fertilization, produce fertile individuals; and who reproduce by generation in such a way that by analogy they can be called all originally descended from a single individual" [de Candolle 1819: 193–194], and then noticeably simpler as "a number of plants, which agree with one another in invariable marks" [DeCandolle and Sprengel 1821: 95]. The latter mean, after Jussieu, diagnostic traits that remain unchanged, Candolle calls them *specific.*

According to an "Idea of Genus" (the next subsection of the *Elements*), it is defined as "the collection of species that have a striking resemblance to each other in the whole of their organs" [de Candolle 1819: 216] or as "the sum of the species which agree in certain constant properties of the essential parts" [DeCandolle and Sprengel 1821: 98]. With this, natural genus is the one that is "founded on an agreement in the properties of most of the parts", whereas artificial genus "only in the marks of a few essential organs" [*op.cit.*: 99]. The higher categories in Candolle's natural method are designated as tribes and families, with the latter being defined as "the sum of all the genera, which agree in one or more essential parts" [*op.cit.*: 103].

As can be seen, Candolle's definitions of species and higher categories differ in one significant respect. The former is defined with reference to both similarity and generation, so species is treated conjointly as both a classification and natural unit. The latter are defined as "sums" with reference to similarity in "essential parts", which means their treatment as classification units.

A similar account of species can be found in many treaties of this time. For example, the zoologist Johann Karl Wilhelm Illiger (1775–1813) considers this issue in a special work devoted to systematic terminology, and he asserts first that "the species originates through the extraction of common characters shared by several individuals", while in another place, he concludes that "it is established with certainty that information about fertile reproduction is the only arbiter of the validity of species" [Illiger 1800: xxvi, xxxii]. Essentially the same position is exposed in *The Philosophy of Zoology*, whose author John Fleming (1785–1857) writes that species is "a group, consisting of individuals possessing the greatest number of common properties, and producing, without constraint, a fertile progeny" [Fleming 1822: 148]. Close to this is a definition of species in Cuvier's *Animal Kingdom*: "The assemblage of individuals descended one from the other, or from common parents, or from such as resemble them as strongly as they resemble each other" [Cuvier 1834: 8–9].

* * *

The works of John Lindley (1799–1865), a classic of botanical natural systematics of the first half of the 19th century, are very significant for the development of the species problematics based on the combinatorial idea. In *An Introduction to Botany*, he examines the general principles of elaboration of the natural system and emphasizes that "its divisions are framed from a careful consideration of every, even the minutest, character that is appreciable; and consist of species, not arbitrarily collected by a few common signs, but agreeing with each other as far as possible in every material point of structure" [Lindley 1832: 318]. Similarly, in the introduction to *The vegetable kingdom*, Lindley notes that, when developing a truly natural system, it is necessary "that all points of resemblance between the various parts, properties, and qualities of plants shall be taken into consideration" [Lindley 1836: vii]. However, in the essay *A key to botany*, Lindley (as Jussieu before him) combines phenetic idea with certain elements of essentialism. Indeed, he assures that the "affinity is an accordance in all essential characters", and "the characters by which natural affinities are ascertained, are valuable in proportion to their importance to the existence of a plant" [Lindley 1835: 40]. Thus, we have here a very characteristic example of a complex definition of species, including references to three basic ideas—phenetic, essentialist, and generative. However, this is not essentialism in its Aristotelian sense: for Lindley, as before for Tournefort, "the characters of plants are merely the signs by which we judge of affinity" [*loc.cit.*].

Turning to the species, Lindley emphasizes that they "are created by Nature herself and remain always the same, in whatever manner they may be combined; they form the basis of all classification" [Lindley 1832: 307]. At a theoretical level, Lindley defines species comprehensively as "a union of individuals agreeing with each other in all essential characters of vegetation and fructification, capable of reproduction by seed without change, breeding freely together, and producing perfect seed from which a fertile progeny can be reared" [*op.cit.*: 365]. Based on this account, Lindley identifies "two sorts of species: the one, called *natural* species, determined by the definition given above; and the other, called *botanical* species, depending only upon the external characters of the plant" ([*op.cit.*: 366], italics added). This distinction repeats evidently Leibnizean idea of distinguishing between "physical" and "logical" species (see Section 4.2.).[13] It is noteworthy that, according to Lindley's account, "the former have been ascertained to a very limited extent; of the latter nearly the whole of sytematic botany consists" [*loc.cit.*]. And he adds, regarding the "botanical" species, that due to variability of their characters, their determination is "in all

[13] His colleague Hewett Cottrell Watson (1804–1881) calls them, respectively, "natural" and "book" species [Watson 1843].

respects, arbitrary, and must depend upon the discretion or experience of the botanist" [*loc.cit.*].

By acknowledging the real status of species as a natural unit, J. Lindley, partly following Linnaeus (and ignoring Adanson), distinguishes it fundamentally from higher categories, which he treats nominalistically. He regards genus as an "artificial means of condensing our ideas of the forms of plants, [and] orders like genera, are a contrivance for analysing and simplifying our ideas, by reducing their number" [Lindley 1832: 368]. He finally infers that these categories "have no real existence in nature, [so] they have no fixed limits, and consequently it is impossible to define them" [Lindley 1835: 40]. This distinction is rather common among the then systematic botanists [Stamos 2007]; in particular, just-mentioned H. Watson argues that genera are "purely conventional groups, but species are commonly believed to have a distinct and permanent existence in Nature" [Watson 1843: 613].

* * *

In the second half of the 18th and the early 19th centuries, an issue of distinguishing between species and intraspecific categories (subspecies, races, etc.) becomes an important part of the species problem (this is a future "species question", see Section 5.3). It arises due to an obvious contradiction between the theoretical idea and practical knowledge about the diversity of organisms. Its theoretical part, inherited from realistic scholasticism, is based on account of species as a discrete unit with constant diagnostic (essential) characters and distinct boundaries. Its practical part results from the numerous explorations that testify against this "theory": the species are variable, and the boundaries between them are blurred.

Of course, there is nothing new in this issue, as many naturalists wrote about it, albeit casually and incidentally, since Aristotle and Theophrastus times [Franklin 1986]. It was most often interpreted in the context of Aristotle's division of the properties of organisms into essential (*propria*) and accidental (*accidentia*), according to which any deviations from the species "norm" are insignificant. For example, J. Ray, focusing on the generative idea, writes in the *History of Plant Species* that "no matter what variations individuals or species exhibit, if they originate from the seed of one plant, they are accidental and do not violate the boundaries of species" [Ray 1686: 40]. Such understanding of accidental variations is expressed in a concentrated form in the *Philosophy of Botany* by aphorism "that a character does not make a genus, but the genus makes the character" [Linnaeus 2003: § 169]. Regarding species, such view becomes widespread in the early 19th century, as exemplified by the following statements: "varieties are accidental subdivisions of species" [Cuvier 1834: 8]; "some of these differences [...] are accidental, and constitute what are termed varieties" [Fleming 1822: 148]. With this, however, the attentive naturalists see certain regularities in these variations. Thus, Buffon observes that "if we still confide each species in different climates, we will find varieties sensitive for the size and for the form; all take a more or less kind of climate imprint" [Buffon 1756: 59]. Moreover, in the section "On degeneration of animals" of the 14th volume of *Natural History*, after reviewing the variations ("degenerations") of animals as deviations from original prototype, Buffon points to their "more close consideration, the view of which is much more extensive; it is that of the change of the species themselves" [Buffon 1766: 335].

A close attention to species variations in the early 19th century grows from two sources. One of them is the natural-philosophical idea of the continuity of Nature ("Nature does not make leaps"), encouraging naturalists to question the discreteness of the natural species and to look for all possible transitions between them. Such "ladderists" as Buffon and Lamarck, who object to the discreteness of classification units and their ranks, are especially keen on this idea [Lovejoy 1936; Sloan 1970, 1986] (see the next section). The second source is an introduction of the category of variety into the taxonomic hierarchy by Linnaeus; it is listed in the introductory part of the 13th edition of *System of Nature* equally with other categories [Gmelin 1792: 22]. This obliges "Linnaeans" to list the varieties along with the species in their synoptic classifications, which promotes the growing popularity of this "noncanonical" category. In this regard, a solid distinction between species and intraspecific categories becomes quite relevant for the early postscholastic research program.

A.P. de Candolle is one of the first to focus on this issue: how to distinguish between species and intraspecific forms based on combined generative and phenetic ideas, which he specifically examines in the *Primary Theory of Botany.* Candolle "calls *Variety* (*Varietas*) any deviation from the ordinary state of a species" ([de Candolle 1819: 196], italics in original), identifies several of their categories caused by particular causes (climate, soil, hybridization, etc.), points out the need to take into account their stability, and provides some practical rules for distinguishing species from intraspecific categories. In the *Elements*, he differentiates two principal kinds of such "deviations", which can be "either a Subspecies (*subspecies*), or a Variety (*varietas*). [They] continue indeed during some reproductions, but at last, by a greater difference of soil, of climate, and of treatment, are either lost or changed. [...] Scientific Botanist must therefore be particularly attentive to distinguish permanent species from the variable subspecies, degenerate plants, and varieties" [DeCandolle and Sprengel 1821: 97].

In the middle and second half of the 19th century, intraspecific forms (races, subspecies, etc.) will be acknowledged as full-fledged units of the diversity of organisms [Schlegel 1844; de Candolle 1867; Planchon 1874] to find their place in the taxonomic nomenclature [Pavlinov 2015, 2022b]. Such their status will be reinforced by the Darwinian evolutionary model, and subsequent equalization of species and subspecies (races) with respect to their status as natural units will cause "species crisis" (see Sections 5.2.1 and 5.3).

* * *

As can be seen from the above, the natural-classificatory dualism of species presumes two significantly different interpretations of this general notion. The natural species is defined through a relation by origin; this is a theoretical definition referring to ontology, whereas the classification species is defined through a relation by similarity, and this is an empirical definition referring to epistemology. Inclusion of both definitions in a single "formula" of species yields a certain problem caused by specific requirements imposed on the species concept at the levels of ontology and epistemology. Indeed, any theoretical definition, in order to be applied in practice, must be operationalized [Benjamin 1955; Magnus 1996; Hull 1997]. In the case of the generative concept, this means that for the allocation of particular plants to particular species to be substantiated, it is necessary to trace the entire chain of generations connecting them. Obviously, this condition is unfeasible in practice, and such a nonoperational character of generative concept serves for its criticisms [Sanson 1868, 1900; Bentham 1875; Gray 1876; Caruel 1883]. In contrast, phenetic definition (with some reservations) is operational [Sokal and Sneath 1963; Pavlinov 2018, 2021b], so it inevitably comes to the fore in the discussion of the practical approaches to species classifications.

This discrepancy results in different or similar treatments of species and higher categories. At a theoretical level, they are ascribed to different ontic status: the former is real, the latter are nominal (arbitrary), and this is one of manifestations of autonomization of species. This dichotomy will be subsequently supported by the Darwinian evolutionary concept, and it will become one of the tenets of biosystematics (see Section 5.3). However, at an empirical level, taxa of all ranks are defined uniformly as classification units through similarity: this argumentation line will become popular in the 20th century due to the influence of the positivist philosophy of science [Pavlinov 2018, 2021b].

4.3.5 "Ladderists" Against "Systemists"

In the second half of the 18th and the early 19th centuries, the old dilemma of realism *vs.* nominalism in the interpretation of species gets a new reading. The cause is that the species problematics becomes shaped not so much by scholastic philosophers as by naturalists relying on two the then most popular natural-philosophical ideas about Nature, "Systemic" and "Ladder". In this regard, the confrontation between their adherents comes to the fore: both these camps are equally committed to the idea of Divine creation, but the results of this creation, including the structure of Nature in general and the diversity of organisms in particular, are imagined by them significantly differently [Rieppel 2011a; Richards 2016; Pavlinov 2018].

As shown in one of the previous sections of this chapter, "systemists" are faithful to an idea of hierarchical System of Nature, which consists of discrete taxonomic units distributed over discrete ranks. With this, it is acknowledged (following *Old Testament* ontology) that the natural units of at least lower ranks are really discrete as a result of *special creation* (see Section 5.1.1 on this). This is the world picture of those naturalist systematicians, who see an ultimate end of cognitive activity in the development of *Natural system* as a taxonomic classification approaching an implied System of Nature.

The "ladderists" adopt an idea of a continuous linear Ladder of Nature (*Scala Naturae*), also known as the Great Chain of Being, in which there are no breaks between its individual links; it is the result of *general creation*. According to this concept, "all the orders of natural beings form but a single chain, in which the various classes, like so many rings, are so closely linked one to another that it is impossible for the senses or the imagination to determine precisely the point at which one ends and the next begins—all the species which, so to say, lie near to or upon the borderlands being equivocal, and endowed with characters which might equally well be assigned to either of the neighboring species" ([Lovejoy 1936: 149], quoting Leibniz). This idea, which has antique roots, at the time under consideration is scaffolded by the Leibnizean principles of *plenitude* and *continuity*; they are generalized by the abovementioned formula "Nature does not make leaps". Accordingly, for the "ladderists", the main task is to reveal this Ladder of Nature: it appears to them in the form of *Natural order*, in which there are no discrete taxonomic units separated by gaps, and any boundaries between both themselves and their ranks are arbitrary.

The general principles of the elaboration of the Natural system are considered in the previous sections, so below some significant details of the Natural order are exposed.

An important part of the "Ladder" ontological background is an idea that Nature as a whole is internally connected by an overall chain of affinities, conditioned by the unity of the creative principle of all things, viz., Divine plan of creation. This root principle is marked by the conception of *prototype* (matrix, mould), which largely agrees with the antique *arche*. From it, all bodies of inanimate and living nature arise by a successive "unfolding" (*evolutio*) to become ordered in a continuous ascending series from the simplest to the most complex and perfect. Thus, the "Ladder" natural philosophy, with its basic idea of "unfolding" prototype, implicitly contains something like a general idea of "stationary development" of Nature, according to which the latter is not only the world of being but also the world of becoming [Richards 1992; Ogurtsov 1993; Bowler 2003; Rieppel 2011a, 2016].

The ideas of our close concern, which are developed by the leading "ladderist" Charles Bonnet (1720–1793) in his *Contemplation of Nature*, may be summarized as follows. As far as "there are no gaps in Nature, it obviously follows that our classifications do not describe it. The classifications we create are completely nominal" and they are but "means corresponding to our needs and to the limitations of our knowledge" [Bonnet 1769: 28, 39]. This general statement holds true for the units (including species) distinguished by systematicians: they are arbitrary just because boundaries between them are so. Thus, Bonnet and his fellow "ladderists" follow Lock in their arguments against the reality of such units. It is noteworthy that this conclusion clearly contradicts the above-presented position of Leibniz about physical species (see Section 4.2), to whom "ladderists" repeatedly refer.

Among adherent proponents of the Ladder of Nature, one of the most influential is G.-L. Buffon [Kanaev 1966a; Stafleu 1971; Grene and Depew 2004; Hoque 2008]. He believes that all natural things represent a continuous realization of a single idea of Creator, and he asserts in the *Premier Discours* of his *Natural History* that "only individuals really exist in nature, while genera, orders, classes exist only in our imagination" [Buffon 1749: 54]. His main argument against these categories, borrowed from Lock and Leibniz [Sloan 1976, 1997], accentuates the absence of hiatuses in continuous Nature: "One can descend by almost insensible degrees from the most perfect creature to the most disorganized matter"; therefore, "it is impossible to give a general classification, a perfect systematic arrangement, not only for

Natural History as a whole, but even for a single one of its branches" [Buffon 1749: 12–13]. With this, however, Buffon cannot be qualified as an orthodox "ladderist": in his understanding, the real arrangement of Nature is not linear but multidimensional [Kanaev 1966; Gregory 2006]. The latter means that "the chain [of being] is not like a simple thread extending only in length; it is a broad structure or rather a bundle that emits side branches from link to link" [Buffon 1954: 417].

At the beginning of his career, the general attitude Buffon "the ladderist" toward species and other categories is quite nominalistic: as just said, this is because he does not believe in boundaries and hiatuses between them. In this regard, he considers "systemic" account of species "incompatible with the idea of Nature as a gradual continuum of forms in which there are no breaks" [Lovejoy 1959: 90]. Nevertheless, he does not reject these classification categories altogether but applies them as descriptive device to characterize diversity of animals. Thus, in the same *Premier Discours*, he indicates that "the history of an animal must not be the history of the individual, but that of the whole species" and promises to describe "every species of animal, plant or mineral" [Buffon 1749: 5]. And in the subsequent volumes of his *Natural History*, he actually applies certain elements of the genus-species scheme to describe animal diversity: for instance, in the *Natural History of Birds*, he groups them into genera and even families [Buffon 1770]. In addition, when describing the breeds of domestic sheep, Buffon writes that "physical genera actually consist of species modified in various ways by human hands, but which, nevertheless, have a common and uniform origin in nature" [Buffon 1769: 144]. By mentioning "physical genera", he probably follows Leibniz and refers to their persisting chains of generations. Anyhow, toward the end of his life, he abandons a purely nominalistic position (see Section 5.1.2).

A combined natural-philosophical and pragmatic attitude, similar to that of Buffon, is characteristic of Jean-Baptiste Pierre Antoine de Monet de Lamarck (1744–1829), who is the author of the first sufficiently formulated evolutionary concept. Being faithful "ladderist", he nevertheless does not think of the ordering of animals as linear: this is evident from the branching configuration of the evolutionary tree depicted in his *Zoological Philosophy*. There is a chapter in it devoted to species, where Lamarck provides its mixed phenetic+generative definition, quite common in this times: "any collection of like individuals which we see produces by other similar individuals" [Lamarck 1984: 35]. Further, he writes about the problems associated with the practical identification of species due to their variability, emphasizes the disappearance of hiatuses between them as evidence accumulates, and notes that "everything is more or less merged into everything else, that noticeable differences disappear", so the species "arranged in series according to their natural affinities, exhibit such slight differences from those next them as to coalesce with them. These species merge more or less into one another, so that there is no means of stating the small differences that distinguish them" [*op.cit.*: 37]. He adds to this that one of the reasons for the gaps between species may be the extinction of intermediate forms. Lamarck concludes his analysis by stating that "nature [...] definitely contains nothing but individuals which succeed one another by reproduction and spring from one another; but the species among them have only a relative constancy and are only invariable temporarily" [*op.cit.*: 44]. With this, however, he acknowledges a "semirealistic" account of species as classification unit, as far as it is defined as "any collection of like individuals perpetuated by reproduction without change, so long as their environment does not alter enough to cause variations in their habits, character and shape" [*loc.cit.*].

It is important to note that despite "ideological" confrontation of "systemists" and "ladderists", in practical applications of their natural-philosophical positions, the differences between them are not too much categorical. Indeed, as can be seen from the preceding, neither Buffon nor Lamarck considers the Ladder of Nature to be strictly linear: it "branches" reflect the complex structure of the natural affinities of organisms. This position is especially clear in A.L. de Jussieu's "systemic" classification method that "connects all forms of plants into an indissoluble whole and follows step by step from simple to complex [...] in an unbroken series, like a chain whose links represent countless species [...] or like a geographical map, in which the species are distributed over territories, provinces, and kingdoms" [de Jussieu 1789: xxxiv].

5 The Modern Times
Species as Biological Unit

The beginning of the Modern Times was marked by the birth of modern science in the 16th century [Whewell 1859; Gaydenko 2000, 2003]. This global scientific revolution, associated with the rise of Cartesian rationality, embraced mainly the section of natural science that F. Bacon dubbed as "natural philosophy" (physics, astronomy, chemistry, etc.). Unlike this, the nascent systematics, whose cognitive situation includes the species problem, being part of "natural history" (in the Bacon sense), developed at that time based on the genus-species classification scheme inhered from scholasticism (see Section 4.3). With this, certain elements of new rationality, assimilated by early systematics, appeared to be quite compatible with a scholastic methodology by mechanistic interpretation of organisms as a "sum" of diagnostic traits, and even strengthened the focus of systematics on the just-mentioned scheme, which made it scholastic in its foundations and intentions. For this reason, systematics of the 16th–17th centuries, figuratively speaking, partly remained in the Middle Ages, and "its own" Modern Times did not come yet then, delayed for two centuries. The whole pool of biological disciplines studying biodiversity was waiting for them until the mid-18th century, and their coming got marked by postscholastic scientific revolution in them [Pavlinov 2018, 2021b]. Due to this, systematics begins to turn from a universal (recall that Linnaeus systematized all three "Kingdoms of Nature" based on a unified genus-species scheme) to a biological one [Wilkins 2003a, 2009a; Pavlinov 2018, 2021b].

This revolution proceeds in the general context of the reformation of the entire complex of disciplines studying plants and animals. In its center is the biologization of understanding of organisms: its comprehension turns from a mechanical set of parts (diagnostic traits) to a living being connected by a multitude of relationships with its environment. This leads to an understanding of the fundamental difference between the living and inanimate nature, as evidenced by the appearance of the term "biology" at the beginning of the 19th century [Lunkevich 1960; Pozdnyakov 2015, 2018]. One of its important outcomes becomes a gradual *declassification* of wildlife research, meaning liberation of the rising biology from the dominance of classificatory program. An important understanding is reached that studying the living is not limited to the systematic catalogues with the lists of taxa and their characters [Sachs 1906; Zavadsky 1968].

* * *

The change in the dominant trend of the historical development of the species problematics at the turn of the 18th–19th centuries was prepared by the whole of its previous conceptual (pre)history. In the latter, the most important milestones look as follows.

DOI: 10.1201/9781003346296-6

In its beginning, one should place an initial distinction of living and inanimate matters made by antique thinkers, and especially by Aristotle, who endows the former with a generative power. This idea means that the living is specific by ensuring a potential eternity of *ousiai* and *eidê* embodied in mortal organisms by their perpetual chains of generations.

Another source is the *Old Testament* mythology describing the divine creation of animals and plants, which "disconnects" species as a category from an interaction of matter and form postulated by Aristotelian world picture. Although its first *Book of Genesis* historically precedes the Antiquity [Shifman 2007], European philosophical thought absorbs an idea of creation in the postantique time, and in a peculiar combination with certain elements of antique natural philosophy. The reason of considering it here is that the biblical model of world-making (according to one of its later interpretations) can be considered the first recognition of the particular nature of species of living organisms, different from the "species" of inert matter.

The movement of Renaissance thought leads to the undermining of antique and scholastic partonomic-taxonomic dualism, which results in taxonomization of the understanding of species. It ceases to be treated as parton, i.e., as an essence of organisms, and its principal and then the only account becomes taxonomic, meaning its treatment as an aggregate of organisms (herbalists and Cesalpino). In this new context, the generative idea is strengthened due to its incorporation, together with similarity, in the taxonomic interpretation of species (Ray).

The next becomes undermining of correlative conjugacy of genus and species and removal of genus-species uncertainty due to structuring of the genus-species scheme by making it ranked (Tournefort, Linnaeus). As a result, species becomes autonomized and assigned a fixed place in the hierarchy of Nature. Of prime importance becomes distinguishing between physical (natural) and logical (classification) species by Leibniz, which is supplemented by Linnaeus' separation of taxa of different ranks into "works of Nature" (species and partly genus) and "works of Nature and Art" (higher categories). This implies a realistic account of species and turns out to be one of the prerequisites for shifting an emphasis from the genus-species scheme to species as such.

A crucial role is played by the adoption of an inductive argumentation scheme, caused by the empirical methodology of all natural science (Adanson, Jussieu, etc.). According to it, the elaboration of classifications does not end with the species level, but begins with it, so the definition of species is to precede logically the definitions of genus and higher categories. All this aims researchers to focus on the criteria for recognition of natural species regardless of the solution of classification tasks, and the generative idea becomes the main "supplier" of these criteria at a theoretical level (Buffon, etc.). With this, reference to the continuous chains of generations in the species definition signifies an important worldview shift: for the developing ontology of species as a biological entity, its temporal dimension becomes critically important.

As a result of all these events in the considered conceptual history, by the end of the 18th century, species of living organisms acquires an independent substantive meaning, and of several its hypostases, its treatment as a taxonomic natural unit becomes dominating. Due to this, natural species shifts to the center of the cognitive situation being developed by systematics, and the genus-species problem is simplified to the species one. The postscholastic revolution at the end of the 18th and the early 19th centuries has a significant impact on shaping the species problem by strengthening its biologization, which expands this problem beyond the limits of systematic research [Pavlinov 2018].

* * *

In the second half of the 19th century, the structure of the cognitive situation, in which the species problem develops, begins to change significantly due to the extension of its "conceptual dimensionality". One of its new "dimensions" is shaped by the indication of a specific natural (instead of the former divine) mechanism of generation of species, viz., the process of biological evolution. The latter concept becomes very popular after Darwin's *Origin of Species* (1859): it is build on the generative idea and significantly contributes to the "temporarization" of species. Its essential novelty lies in the treatment of species as an *evolutionary unit*: the particular species arise, persist, and become extinct as a result of the evolutionary processes. Another important "dimension" is added to the new conceptualization of species by its account as an *ecological unit*: a certain type of relations with the environment within the framework of the general "economy of nature" becomes one of its principal characteristics. This conceptual change is a consequence of the rising of ecology as a fundamental biological discipline that begins at the turn of the 19th–20th centuries [McIntosh 1986].

As a result, the general trend of biologization of the species problem in the 20th century includes two partially interrelated components, evolutionary and ecological, so that the ontic frame of the cognitive situation takes the form of an *evolutionary-ecological continuum*. All this gives born to the general idea of species as a specific *biological phenomenon*, which is characteristic precisely and only for the living [Zavadsky 1961, 1968; Volkova and Filyukov 1966] and may be thought of as a special way of "realizing life" [Komarov 1940], so this unit can be rightfully called *biological species*, or *biospecies* in the general sense. Such consideration of species constitutes a part of the understanding of fundamental specifics of the natural phenomenon studied by biological science, which is the life [Mayr 1982, 1988].

One more dimension of the new conceptual framework of the species problem is provided by increased attention to the *intraspecific diversity*, which becomes treated as a biologically significant phenomenon deserving special investigation [de Gandolle 1855; Wallace 1858; Darwin 1859; Bateson 1894; Clements 1908; Filipchenko 1915; Zirkle 1959; Mayr 1988; Bowler 2005; Vinarskiy 2013]. This is resulted from focusing on the processes occurring in the local (microscale) areas of the evolutionary-ecological continuum. This leads to a shift of emphasis from the species itself, in its broad "Linnaean" understanding, to the intraspecific units, which are interpreted as real participants in the evolutionary and ecological processes in wildlife. In this shift, besides evolutionary and ecological considerations, a special role is played by emerging population genetics, which elaborates its own criteria for individuating the elementary natural units [Bateson 1899 (2009); De Vries 1904; Lotsy 1916]. A plethora of particular intraspecific units becomes recognized in the first half of the

20th century, and with time, *population* takes a central place among them to become a core of population biology. With the latter embracing a significant portion of our problematics, species loses its salient status and becomes but one of the population units of a certain level of generality [Mayr 1942, 1963; Simpson 1951; Dobzhansky 1970; Ereshefsky and Matthen 2005]. Accordingly, when discussing the structure of biota, a single *population-species level* is identified in its hierarchy [Zavadsky 1961, 1968; Kremyansky 1969; Bykhovsky 1972; Shkorbatov 1979].

Special attention to the variability is evaluated by the zoologist Ernst Mayr (1904–2005) as the "great conceptual revolution ever happened in biology", involving a shift from *typological* to *population* thinking [Mayr 1942, 1959a, 1982, 1988]. However, such drastic opposition is hardly quite correct: in fact, acknowledging reality of natural population presumes its endowment with certain emergent properties that constitute its "essence" and make it a whole irreducible to a set of individuals [Sober 1980, 2000; Stamos 2003; Pavlinov 2018] (the same is true for species, see Section 6.3).

This new conceptual development of the species problematics leads to a blurring of the general notion of species in its new evolutionary-ecological-variational account, which is declared a "species crisis" and even "destruction of species" [Mayr 1963, 1976, 1982; Skvortsov 1967, 1972; Zavadsky 1968; Shkorbatov 1976; Gilson 1984; Bonneuil 2002]. This "crisis" develops during the end of the 19th and the beginning of the 20 centuries to involve several aspects of species consideration. One of them is the impossibility to discriminate conceptually and distinguish practically species and various intraspecific categories, which gives rise to the *species question* [Johnson 1908; Greene 1910; Turrill 1925]. Another becomes recognition of several *kinds of species* and isolation of the corresponding particular conceptions instead a unified general species concept, which results in the rise of the *species problem* in its contemporary understanding [Romanes 1895; Turesson 1922; Robson 1928; Hawkins 1935; Camp and Gilly 1943; Cain 1954]. All these novelties are summarized by the emergence of a *conceptual species uncertainty* as an important part of our cognitive situation [Hey et al. 2003; Pavlinov 2017, 2021a, 2022a].

To conclude this introduction, it is to be noted that a complete declassification of the species problem, as it is seen from the contemporary, turns out to be impossible, since any group of organisms in any of its conceptualizations, when individualized, is regarded basically a classification unit [Gross 1988; Hey 2001a,b; Richards 2010]. Therefore, the contemporary researchers, when discussing the issues concerning biospecies, often consider it in a classification context (e.g., [Huxley 1940; Mayr 1957a; Starobogatov 1985, 1996; Leroux 1993; Dupré 1999; Hull 1999; Wilkins 2003a, 2009a; Richards 2010; Ellis 2011; Kunz 2012; Sloan 2013]).[1] With this, they (as Leibniz did in his time) almost do not explore relations between classification and natural accounts of species (e.g., [Cain 1954; Grant 1981; Tsvelev 1995; Mahner and Bunge 1997]), and even if they do, they still declare that the species problem is the main one of systematics, and not of natural history [Heptner 1947; Zavadsky 1961].

[1] Is it not strange that in one of the most recently published books [Mishler 2021], the species problem is discussed in the context of taxonomic nomenclature?

Hence, a continuing juxtaposition of considering species either a classification or an evolutionary-ecological unit: the first is an element of a universal meta-language, in which representatives of various sections of biology communicate, and the second is thought of as an element and participant of the processes occurring in Nature [Dupré 1993, 1999].

* * *

This chapter first considers the formation of early conceptual prerequisites for the development of ideas about species as a specific biological unit. This includes its objectification in the context of the biblical world picture and maturation of generative idea providing a specific biological mechanism of species integrity. This consideration, facing the past, may seem out of its proper position, but placing it here allows highlighting a long-term continuity in the development of contemporary ideas about biospecies. Thereafter, the principal manifestations of the biologization of the species problematics during the 19th–20th centuries are characterized, including the formation of evolutionary and ecological ideas about species, the development of the species question by the proponents of biosystematics, and the fate of the species problem caused by the multiplicity of particular species conceptions.

5.1 THE FORERUNNERS: INITIAL ROUTS OF OBJECTIVATION OF BIOLOGICAL SPECIES

Throughout the conceptual history of the species problematics, the question of reality *vs.* nominality of species of living organisms was and remains one of the basic hotspots. Substantively, this question can be formulated as follows: does biospecies exist as a natural phenomenon, i.e., as a certain unit of the structural and functional organization of biota, occupying a certain fixed level in its hierarchy? Within the framework of natural-philosophical consideration of this issue, its key point is shaped by the search for possible mechanisms that distinguish ontologically species of the living from "species" of inanimate nature [Pavlinov 2013a, 2022a]. Initially, these mechanisms are supposed to be separate divine creation of the biospecies "after their kinds" and their subsequent self-reproduction in the form of continuous chains of generations. They come to the fore in the species problematics with the development of scholastic systematics, but reflections on them began much earlier. Their history, which predates the contemporary biological account of species, is outlined in the following sections.

5.1.1 Biblical Motifs

According to the biblical world picture, the fundamental cause of the entire universe, including both material and supra-material worlds, is its divine "creation out of nothing" (*creatio ex nihilo*). It constitutes the basis of the biblical deistic ontology, characteristic of the entire postantique natural philosophy. It is set forth in

Chapter I of the *Book of Genesis*, with which the *Old Testament* (Tanakh) begins; its first written versions date back to the 13th–12th centuries BC [Shifman 2007]. Of our concern here is everything that relates to the living and constitutes the so-called *biblical biology*, which is rightfully regarded as one of the branches of ethnobiology [Berndt 2000].

The biblical natural philosophy frames the ontological basis of the cognitive situation of early systematics that deals with the elaboration of a new understanding of species in the 16th–18th centuries. Accordingly, the reference to divine creation and to subsequent immutability of the particular species serves as the main argument in favor of their reality and discreteness. In this regard, it is to be mentioned that the popular treatises are then published, in which biological issues are considered strictly in the biblical context, such as *Biblia Naturae* by J. Swammerdam [Lindeboom 1982]. Recall that for C. Linnaeus, "Nature is the behest of God" (see Section 4.3.3); his confidence in this and his great authority make some modern authors attributing to him a "responsibility" for the prevalence of the idea of the creation and immutability of species among naturalists of that time [Quinn 2020]. Indeed, the mathematician William Hopkins (1793–1866), in his criticism of the just published Darwin's *Origin of Species*, is convinced that "every natural species must by definition have had a separate and independent origin, so that all theories [...] which assert the derivation of all classes of animals from one origin, do, in fact, deny the existence of natural species at all" [Hopkins 1860: 747].

* * *

According to the canonical biblical legend, the creation of all things took six days, and the living organisms were created on the third, fifth, and sixth days [*Gen.* 1: 11–12, 20–22, 24–28]. With this, God seems to have "entrusted" His creative power to "the earth" and "the waters" and let them to "bring forth vegetation" and "living creatures of every kind" (https://www.vatican.va/archive/bible/genesis/documents/bible_genesis_en.html). This scenario of divine creation partly resembles an antique concept of hylemorphism (see Section 2.1.2): species of living beings have arisen as a result of the interaction of Divine archetypes (an analog of form, έίδοσ) with the earth and waters (an analog of matter, ὕλη). The basic notion of our concern here is *creation unit*, which denotes an entity that appears due to its divine creation. The issues concerning them constitute a significant part of the theoretical section of biblical biology; these are as follows: what were the particular creation units, how exactly they were created, what is their status in the overall hierarchy of Nature, etc.

> It is interesting to note that, when placing the species problem in the context of biblical mythology and considering it with regard to the multiplicity of species conceptions, D. Hull formulates the following curious question: "God created numerous different species, but did He create a single sort of species or many different sorts of species?" [Hull 1999: 23].

Medieval natural theologians distinguish two basic scenarios of divine creation of the living [Locy 1915; Marsh 1944]. The version of *special creation* suggests that its immediate result was every single creation unit containing certain organisms according to their "kinds", and these units remain unchanged after their creation. The version of *general creation* implies that the creation units were the prototypical (primary) forms, which then brought forth all the diversity of the living, following the laws of nature predestined by the Deity. There are different details of general creation, supposing different extent and nature of the Creator's intervention in the fate of the world created by Him. Examples are the theories of multiple spontaneous generation of living beings from inanimate matter (goes back to Antiquity), Cain of Being (Bonnet), repeated catastrophes (G. Cuvier), "theistic evolution" (L. Agassiz), etc.

One of the reasons for the endless discussions about creation units, which the Christian natural theologians and the biblical biologists were and are still conducting, is lexical semantic. It lies in the ambiguity of terminology in the original text of the *Genesis* written in the Biblical Hebrew. In it, the term *min* (מִין) is used for the designation of any creation unit, which etymologically is rooted in a general idea of grouping and dividing [Marsh 1941, 1944; Payne 1958; Jones 1972a; Schafer 2003]. Thus, based on the latter, one may suppose that the ancient authors of the *Genesis* imply initially a taxonomic interpretation of *min*. With this, it seems evident that *min*, as biblical folk taxon, has a clear naturalistic meaning, unencumbered by any subsequent theological (ontological, logical, etc.) interpretations. The basis of the integrity of *min* is its creation "according to (or after) its *kind*": it is this specific "kind unity" that is responsible for the constancy and discreteness of each *min*, so its progeny retains likeness to its initial *kind* and can change only within certain narrow limits [Marsh 1941, 1944, 1960; Jones 1972a]. *Mins* and their progeny are endowed with an ability to "be fruitful and multiply", they can be counted, and their representatives can be gathered in one place—for example, in Noah's Ark.

Here special attention should be paid to a possible meaning of "kind", "according to (after)" which organisms were created: it seems reasonable to suppose that this term refers to what early Christian theologians call Divine archetypes interpreted partonomically (see below). What is of paramount importance to us is that, as the contemporary interpreters believe, biblical *min* refers specifically and exclusively to biological organisms and is not associated with any forms of inert matter [Payne 1958; Jones 1972a; Seely 1997]. Thus, if it is correct, an implicit boundary between living and inanimate nature is drawn from the very beginning by the biblical world picture to become an important prerequisite for the subsequent biologization of the entire species problem.

* * *

When discussing the impact of biblical mythology on the development of the species problematics, it is to be taken into consideration that the *Old Testament* is introduced into the Christian world in the form of Greek and Latin translations. In them and in the early commentaries on them, the basic ancient terms of the genus-species division are used as equivalents of *min*: these are γένοσ in the Greek *Septuagint* (the second century BC) and *genus*/*species* in the Latin

Vulgate (the fourth century AD). The subsequent translations into modern European languages are mainly based on this classical terminology: for example, the Slavic translation of the *Septuagint* (the 9th century) employs "*род*" (genus in its common sense), in the English translation of the *Vulgate* (the 14th century) the neutral term *kind* is used throughout, and in its German translation (the 16th century) the term *Art* is used (corresponds to species).

As an example, an excerpt from Verse 12 of Chapter 1 of *Genesis* of three main translated editions, Greek, Latin, and English, can be cited:

> Καί εζηνεγκεν η γη βοτάνην χόρτου, σπεῖρον σπέρμα κατά γένος και καθ' ομοιότητα, και ξυλον κάρπιμον ποιούν καρπόν, ου το σπέρμα αυτοί εν αυτω κατά γένος έπι της γης.
>
> Et protulit terra herbam virentem, et facientem semen juxta *genus* suum, lignumque faciens fructum, et habens unumquodque sementem secundum *speciem* suam.
>
> And the earth brought forth the herb of grass bearing seed according to its *kind* and according to its likeness, and the fruit tree bearing fruit whose seed is in it, according to its *kind* on the earth.
>
> *([Hetzenauer 1914; Brenton 1980], italics in original)*

Thus, whatever the original semantics of the Biblical Hebrew terminology, the linguistic, and with it also substantive, basis of the medieval species problematics is provided by the Latin-language thesaurus of the *Vulgate* with its central notions of *species* and *genus*. Both these notions and their meaningful interpretations are inherited from Neoplatonism and retain a certain continuity with antique natural philosophy [Gilson 1955, 1991]. This means that the medieval philosophical interpreters of the *Vulgate* regard the creation units not so much as specific groups of organisms (taxa), but as the universals, rooted in the Divine archetypes (partons) of Divine Mind as *genê* (biblical "kinds"): this is one of the central tenets of the Christian Platonism (see Section 3.1). Of paramount importance is a fundamental idea of the realist theologians (Augustine, Calcidius, Anselm of Canterbury, Abelard, etc.) that these universals are as real as the Divine archetypes (ideas, forms) [van Winden 1965; Charlesworth 1982; Gilson 2010; Panchuk 2016]. It seems to be the root of an initial substantiation of the Christian idea of the reality of species to be subsequently picked up and developed by early systematics.

According to Christian Platonic eidology, the main biblical creation units are *genera*, which are then divided into *speciei*: this is, for example, the doctrine of Eriugena [Marenbon 1998]. This becomes biblical ontic substantiation of the reality of hierarchical genus-species relations (Pierre Abelard, Nicolaus of Cusa, etc.). If elements of Aristotelian ousiology are added to the Christian world picture, *forma* is considered the basic creation unit. It is most often referred to as *species* and usually has a partonomic meaning of essence: in this account, Divine archetypes are primary forms (*speciei archetypae* of Calcidius), and their incarnations in the created world are secondary forms (*speciei secundae* of Calcidius, *formae nativae* of Bernard of Chartres) [van Winden 1965; Gilson 2010; Albertson 2014].

An apparent hint of distinguishing between partonomic *forma* and taxonomic *species* in the accounting of divine creation can be found in a short essay *On Creation* by Aurelius Augustine. According to him (and as noted above), God had not created living nature himself, but "delegated" to the earth and waters His creative power as an array of *formae*, and the latter, as creation units, appeared those "kinds", according to which plants and animals became actually created [Augustine 2020]. So, the particular species belonging of individual animals and plants is accounted for by their mutual similitude in their features inherited from the respective original "kinds" [Maurer 1982; Simonetta 1995]. Echoes of such vision of divine creation are present much later in Linnaeus' *Philosophy of Botany* (see below).

* * *

Having passed through various interpretations of the medieval and renaissance philosophers, biblical ideas of the divine origin of genera and species become an important part of the ontic component of the cognitive situation of early systematics of the 16th–18th centuries, in which context the contemporary species problematics becomes shaped. It should be particularly noted that the first naturalist systematicians follow principally the tradition of realistic scholasticism in the interpretation of genera and species. Here, we briefly analyze the views of two key figures, J. Ray and C. Linnaeus, who most definitely say about the divine origin of the taxonomic categories.

John Ray is not only a naturalist but also an amateur theologian, and it is his biblical enthusiasm that concerns us hear. As a systematician, he professes a scholastic genus-species scheme, subordinating species to genus and considering them as classification units. However, as a natural theologian, he puts the species in the first place, being sure that these were the species that were created in accordance with the ideas in Divine Mind [Ray 1696, Praefatio: (viii)]. Indeed, in the introductory section of the *History of Plant Species*, Ray writes that "the number of species in nature is strictly determined: God interrupted on the sixth day his great work, the creation of new species" [Ray 1686: 40]. This argument evidently corresponds to the special creation model, and Ray repeats it in his *The Wisdom of God* and *Discourse on the Species differences* to substantiate, with reference to divine creation, the constancy of both species themselves and their number [Ray 1714, 1928a]. When outlining this standpoint, he does not mention genera: it seems that in Ray's world picture, God did not create genera, limiting himself to the creation of the species—contrary to the fact that the canonical Greek edition of the Bible says about genera only, and its Latin edition says about both genera and species (see above). At any rate, J. Ray seems to be the first among systematicians to substantiate the reality of species as a natural unit with a direct reference to divine creation. With this, however, it is to be reminded that he, on such account, does not discriminate between species of animate and inanimate natures: they all are divine creation units of the same kind (see Section 4.3.2).

The position of Carl Linnaeus regarding the divine origin of plants changes significantly throughout his lifespan. At the beginning of his scientific career, he follows Ray to assert that "all the species of botanists, existing now or in the future in their number, were created by the wave of the hand of the almighty Creator" [Linnaeus 1737a: § 310]; therefore, "there are as many *species* as there are different *forms* created by an initially Infinite Being" ([Linnaeus 1737b, *Ratio operis*: § 5; Linnaeus 1751: § 157], italics added), and he clarifies this thesis by stating that "the Creator has entrusted the generation of species to Nature" [Linnaeus 2003: § 259]. At this point, Linnaeus adopts the special creation model; with this, of interest is that he seems to agree with the Augustinian standpoint. First, with distinguishing species and forms terminologically, these are the latter that are regarded by Linnaeus as the initially creation units, and second, he believes that the species themselves were created not by Creator himself, but by Nature that had been "entrusted" by Him to give born to those units. Regarding genus, Linnaeus' position is not so definite: in § 159 of the *Philosophy*, he considers it both secondary relative to species and of the same ontic status, being "made in the first place such as it is".

However, one and a half decade later, Linnaeus interprets divine creation in a markedly different manner that agrees with the general creation model. This new viewpoint is at first stated in the dissertation *Foundations of fructification* "officially" authorized by his student Johannes Mart. Gråberg [Ramsbottom 1938; Hofsten 1958; Larson 1967, 1968; Wilkins 2009a; Barsanti 2011]. However, Linnaeus prepares his students' dissertations mostly by himself, so the opinions expressed in them are almost certainly Linnaean [Usov 1888; Yuzepchuk 1957]. Beginning with a doubt, he writes at first that

> it will not be so easy to prove whether [all] the species were created with a wave of the hand of Creator's immediately from the very beginning or they multiplied in number over time due to the way Nature fulfilled the will of Creator".
>
> *[Linnaeus 1762: 18]*

And yet, a few pages later, he quite confidently asserts:

> I. That in the beginning the Creator created only one plant of each *natural Order*, differing from others by habitus and fructification.
>
> II. That He (1) inseminated them with each other, so that from their offspring, which slightly changed fructification, there turned out as many natural *Genera* as there were different variants of parents; and since this could hardly continue further, even by His imperious hand, we place it at the very beginning. Thus, all original genera with a single permanent species [each].
>
> III. That there arose as many genera as there were individuals at the beginning (2), and after that these plants (perhaps over time?) were fertilized by others of different genera, and thus other Species are produced until so many [thereof appeared], how many are now in existence, who preserved unchanged fructifications of Mother and changed their habituses by Father. [...] So that some genera produced many species, while others did not.
>
> *([Linnaeus 1762: 20–21], italics in original)*[2]

---

[2] Another version of this passage is provided by James Larson with reference to the Linnaeus' *Lectures* [Larson 1971: 69].

Thus, according to an updated world picture of Linnaeus, God first created prototypical plants, one for each order, which were probably still the same "forms". They then, by mutual crossing, gave birth to the genera, with one species each, and these prototypical "generic species" produced by mutual crossing the overall existing species diversity. On this basis, it seems reasonable to conclude that at the end of his life, Linnaeus considers the higher categories as natural as species and genera, since their prototypical forms were the created units given forth by Creator. As is seen, this interpretation of the arising of plant diversity combines two general ideas, divine creation and partly generative.

This creation model is reproduced in clearer versions in a number of posthumous editions of Linnaeus, including his *Lectures on the Natural Orders of Plants* [Linnaeus 1792: 16, 18] and in the writings of some of his disciples. Among the latter, Johan Andreas Murray (1740–1791) outlines most succinctly Linnaeus' viewpoint in the *Plant System* attributed by him to Linnaeus. He writes that in the beginning God created

> as many different plants as there are *Natural orders*. Then the plants of these orders were mixed together so that in subsequent generations as many [different] plants arose as there are *Genera* today. Then the Divine nature mixed these generic plants and produced plethora of *Species*.
>
> *([Murray 1784: 8], italics in original)*

* * *

In the 19th century, an idea of divine creation of species becomes significantly pushed by the evolutionary model of the gradual diversification of life due to speciation. Nevertheless, the biblical model continues to develop new and partly revives old versions of the world making. One of these versions tends to Platonic eidology; it can be regarded as a continuation of Christian Platonism that developed in the first centuries AD. During the period under consideration, it is stimulated by typological ideas [Amundson 1998; Pavlinov 2018]. The striking example of this trend is provided by the doctrine of the zoologist Richard Owen (1804–1892) about the hierarchy of archetypes, understood in the sense of Platonic *eidê* and treated as attributes of Divine Mind [Rupke 1993; Camardi 2001].

The ideas of Christian Platonism, which most closely concern the species problematics, are developed by the paleozoologist Jean Louis Rodolphe Agassiz (1807–1873). His world picture, outlined in *An essay on classification*, is a kind of combination of natural science and biblical religion [Agassiz 1859]; preferring the latter, he actively opposes the evolutionary idea at the end of his life [Lurie 1960; Winsor 1979; Rieppel 1988; Stamos 2005; Irmscher 2013]. Being a disciple of G. Cuvier, he proceeds from an idea that the general structure of the living is basically a hierarchy of body plans. For Agassiz, the latter are a result of the embodiment of *prophetic types* of different levels of generality being "the categories of [Deity] mode of thinking" [Agassiz 1859: 8, 176]. They are elements of Divine plan of creation and are realized step by step, starting from the highest and ending with the lowest, and this partonomic hierarchy yields a taxonomic one with species at its lowest step.

As is seen, the divine creation in this version is quite similar to Plato's sequential emanation of *eidê* and differs significantly from the original biblical scheme. Moreover, the process of genesis "after Agassiz" is significantly extended over time: an influence of his deep knowledge in paleontology and geochronology is quite evident. The most essential part of his doctrine, in which ontology and epistemology meet, constitutes the belief that "the human mind is only translating into human language the Divine thoughts expressed in nature in living realities" [Agassiz 1859: 204]. According to his Platonic standpoint, he is sure, contrary to Linnaeus and natural systematicians, that "genera, families, orders, classes and types have the same foundation in nature as species" [*op.cit.*: 8]. And he specifically emphasizes "the existence in nature of distinct species, persisting with all their peculiarities, for a time at least" [*op.cit.*: 3]; with this, he insists "that species did not originate in single pairs, but were created in large numbers" [*op.cit.*: 253]; see also [Agassiz 1869].

The contemporary anticreationists, due to their presentist attitude (see Section 1.2), perceive Agassiz's opinion purely negatively; with this, their interpretations of Agassiz's statements (e.g., [Mayr 1959b; Lurie 1960; Wilkins 2003a, 2009a]) are sometimes not adequate to his ideas. For instance, J. Wilkins, when discussing Agassiz's ontology, refers to his passage in the *Methods of Study in Natural History* that "individuals are the transient representatives of all these organic principles" [Agassiz 1886: 136], and based on this, he ascribes to Agassiz an idea that "species are not composed of organisms, in other words, organisms at best 'represent' species. They are not things in the physical sense" [Wilkins 2003a: 78]. Accordingly, Wilkins concludes that for Agassiz, species are not real—they "exist as ideas, which represent the relations actual individuals bear to the world" [Wilkins 2009a: 113]. However, this particular criticism is perverse on two points. First, this Agassiz's book has a didactic character: it deals with how to "represent" different categories of the animal kingdom in lectures, and not about what the species and other categories "represent" themselves ontologically. Second, for Agassiz, being a Platonic biblical scholar, Divine ideas and their embodiments are as real as the "things in the physical sense".

* * *

In the contemporary, biblical worldview finds itself in the very background of the cognitive situation because of the domination of the global evolutionism; this holds true for the species problematics. However, there is a section of modern biblical biology, relating to this problematics, known as *baraminology*: the latter term is derived from the notion of *baramin* (בָּרָא + מִין), which is translated as *created kind* [Marsh 1941, 1944] (the above creation unit). At the level of ontology, baraminology develops a conception of theistic evolution as one of the versions of the doctrine of general creation [Marsh 1941, 1944; Wise 1990; Frair 2000; Wood 2002, 2008]. In it, the emergence of *baramins* is attributed to divine creation, but, unlike in the orthodox biblical special creation, subsequent evolutionary transformations of species and higher categories are presumed, including the appearance of the new ones [Morris and Morris 1997; Glover 2007; Lamoureux 2008].

Among the issues considered by baraminology, one concerns the natural status of *baramin*, including its rank, i.e., its position in the hierarchical structure of Nature [Jones 1972a; Seely 1997; Frair 2000; Wood et al. 2003; Gishlick 2006]. The contemporary studies by baraminologists show that this question has no simple and unambiguous answer that would fit a predominant view, dating back to the early translations of the *Old Testament* [Marsh 1941, 1944; Awbrey 1981; Seely 1997; Turner 2009]. According to one of their leaders, Frank Lewis Marsh (1899–1992), *min* does not "present but one idea in the mind of the reader; not the broad Linnaean species, nor his narrow one, nor the modern 'species', but only the Genesis kind" [Marsh 1944: 48]. F. Marsh thinks that the only extant *baramin* of undoubtedly species rank is *Homo sapiens*, while the status of other *Genesis kinds* cannot be strictly and definitely determined [Marsh 1941, 1944]. Such fuzzy understanding of the status of *baramins* yields a diversity of opinions regarding their ranking: according to various authors, they may correspond to species, genus, or groups of higher ranks [Marsh 1941, 1944, 1960; Jones 1972b, 1973; Gish 1993; Wood 2008, 2016; Lightner 2013]. One of the notable aspects of consideration of this issue is associated with the legend of the Flood and Noah's Ark, set forth in Chapters 6–8 of the *Genesis* [Wilkins 2009a; Minelli 2022].

5.1.2 Generative Idea

The generative idea is an extension of what J. Wilkins calls the generative species concept [Wilkins 2003a, 2009a, 2010]. In its interpretation by antique and scholastic natural-philosophers, organisms possess a certain *generative power* that underlies their striving to copulate and produce progenies looking like progenitors. This yields chains of generations, and it is their continuity that provides a long-term existence and self-identity of each species. In a broadened interpretation, it is permissible to consider in such milieu phyletic lines consisting of continuous chains of populations rather than of generations [de Queiroz 1998, 1999, 2005a].

The special significance of generative idea (concept) is in that its awareness represents the first and most important step towards the biologization of the species problematics, for it substantively separates the living from inanimate matter. Indeed, all living organisms are mortal, so their reproduction is an indispensable condition for their species to persist, while the elementary units of the inert matter are eternal in a sense, so their temporal duration does not need to be reproduced after they arises some way.

This idea probably origins with Epicurean natural philosophy [Wilkins 2009a]; however, it is Aristotle who explicitly expresses it in his treatise *On the Soul* as a particular Law of Nature, according to which "like begets like" [Smith 2009]; it will be called "law of propagation" by I. Kant (see below). Aristotle interprets natural species as ὁμοειδῆ, which ensures the transmission of *ousiai* from progenitors to their progenies in the chains of generations and, thereby, the potential eternity of both primary and secondary *ousiai*, i.e., the species as *eidê* [Lones 1912; Sloan 1970; Lennox 1985; Stamos 2003; Smith 2009; Berti 2016]. On this basis, Aristotle is often called a herald of "generative" understanding of natural species of the living beings.

Another important source of the generative idea is the above-considered *Genesis*, according to which God has precepted for the living beings, including man, to "be fruitful and multiply" (*Gen.* 1: 22, 28). This testament is actively discussed by medieval theologians; in particular, Augustine draws species attention to it in his book *On the City of God*, when emphasizing that all human tribes, no matter how different they may be, "if they are human, they are descended from Adam" [Augustine 2000: 713]. Islamic calamists, whose natural philosophy tends to Aristotelianism (Avicenna, Averroes), are also convinced that "all biological species are eternal, so you have an infinite series of past ancestors" [McGrade 2003: 151]. At last, let us recall that Albert the Great writes in the 13th century that "species [...] participate in an eternal existence by reproducing one individual from another" [Albertus Magnus 1999, Vol. I: 72].

In the initial period of arising of systematics, the generative idea begins to be pronounced as a key element of the essential definition of natural species. Thus, A. Cesalpino in his *16 Books on Plants* writes that "the proper work of the plant soul is to generate its own kind, which ensures the eternity of species (*specierum aeternitatem*)" [Cesalpino 1571:1]. For this reason, when classifying plants, Cesalpino focuses on the organs of reproduction, since they are involved in the production of new generations and thereby provide the potential infinity of the existence of species.

John Ray is the first among systematicians to explicitly express a germ of generative idea: in the *History of Plant Species*, he specifically indicates "special origin from seed" as a criterion of species belonging: "one species never springs from the seed of another and vice versa" [Ray 1686: 40]. He estimates this criterion higher than similarity: "no matter what variations occur in the individuals or the species, if they spring from the seed of one and the same plant, they are accidental variations and not such as to distinguish a species" [*loc.cit.*]. In a small essay *A Discourse on the Seeds of Plants* (1674), Ray extends this criterion to animals, based on the likeness of a plant seed and an animal egg [Ray 1928b: 76], and further argues this thesis in the treatise *The Wisdom of God* [Ray 1714]. For this reason, the formulation of J. Ray is traditionally counted as a forerunner of the modern reproductive (biological) species conception [Cain 1954; Glass 1959; Zavadsky 1961, 1968; Mayr 1982; Wilkins 2003a, 2009a; Zachos 2016]. This is partly true, but it should not be forgotten that Ray understands species as a universal category, which is uniform by its divine origin, so the "species" of living and inanimate nature are equivalent for him in this regard (see Section 4.3.2).

To complete this fragment of the conceptual historical picture, the embryologist William Harvey (1578–1657) should be mentioned. He is a follower of Cartesian mechanism in the interpretation of organisms [French 1994b], but his view of species will influence the shaping of Buffon's naturalistic attitude [Sloan 1970, 1986, 2009]. In one of the sections of his *Experiences on the birth of animals*, when considering the relation between individual and species, Harvey writes that the generation "cycle makes the fowl genus eternal:[3] first a hen, then an egg, this series continues as an endless lineage. From weak and mortal individuals to an immortal species"

[3] In this case, genus is understood as a single genealogical chain. This term will sometimes be used in the same sense by Buffon when describing breeds of domestic animals [Grene and Depew 2004].

[Garveo 1651: 86]. P. Sloan believes that in this context, Harvey's "species" is almost the same as Aristotle's "substantial form" [Sloan 1970, 1986], that is, in our terminology, he presumes partonomic rather than the taxonomic account of species. At the same time, declaring his philosophical position in the introductory section of the *Experiences*, Harvey regards species as a context-dependent universal, just like Abelard. Indeed, he writes:

> although an object is singular in the external perception, [...] nevertheless, when an impression of it should become an abstraction, which internal perception judges and understands, it becomes universal. At the same time, it happens that several people abstract *several species* and imagine different notions, considering *the same object* at the same time.
>
> *([Garveo 1651, Prefatio: (4)], italics added)*

* * *

In the second half of the 18th century, the generative idea becomes nearly completely shaped and acknowledged by both philosophers and naturalists [Look 2009; Smith 2009, 2011; Mensch 2013]. So, G. Leibniz considers this idea to be key for comprehending a real cause of the integrity of both plant and animal species, and he writes in the aforementioned treatise *New Essays on Human Understanding* that

> in the case of organic bodies—i.e. the species of plants and animals—we define species by generation, so that two similar individuals belong to the same species if they did or could have come from the same origin or seed.
>
> *[Leibniz 1996: 310]*

Furthermore, none other than I. Kant in his lecture *On Different Races of Man* says that

> the natural division into species and [genera] in the animal kingdom is grounded on the common law of propagation, and the unity of the species is nothing other than the unity of the generative power that is universally valid for a certain manifoldness of animals. For this reason, Buffon's rule, that animals which produce fertile young with one another (whatever difference in shape there may be) still belong to one and the same physical species, must properly be regarded only as the definition of a natural species of animals in general in contrast to all [scholastic] species.
>
> *[Kant 2007: 85]*

It is to be emphasized that the Kantian "law of propagation", quite particular in this context, will become one of cornerstones of his global generative principle embracing the entire universe [Sloan 2006, 2009; Demarest 2017]. In the *Critique of Judgment*, he calls this principle "theory of evolution" [Kant 2004: 379] and makes it almost the first substantiation of a natural-philosophical idea now called *global evolutionism* [Mensch 2013].

It is noteworthy that some natural-philosophers of this time (e.g., Jan van Helmont, F. Bacon) attribute an ability to such ensuring of the "species eternity" only to higher forms of life, whereas its lower forms, in their opinion, persist due to spontaneous self-generation [Smith 2009]. This opinion also goes back to antique natural philosophy: it implicitly contains an idea that spontaneously generated (from soil, water, or whatever) organisms acquire certain species belonging due to their embodying particular species essences [Lunkevich 1960; Smith 2009].

Turning to the early biologists' contribution to the development of generative idea and making it a concept, it is to be reminded that they all are postscholastics in their worldviews and cognitive intentions. Some of them are naturalists of a "Ladder" kind (like Buffon); others are natural systematicians developing their inductive natural method (Jussieu, de Candolle, Lindley, etc.). As shown above, they develop a combinatorial account of species, which theoretical part implies generative idea, while empirical part refers to phenetic idea (see Section 4.3.4). This means that the antique generative idea of reproduction of forms (*ousiai*), after its biblical interpretation by Augustine and placed in the systematic context by Ray, becomes transforming in a concrete *generative conception* of taxonomically treated species. The abovementioned declassification of the species problem becomes an important precondition for this by aiming naturalists at searching for an appropriate "biological" definition of species, as a natural unit, regardless of a context set by deductive taxonomic hierarchy.

The first person to be mentioned in this regard is certainly G.-L. Buffon [Sloan 1970, 1976, 1979, 1986, 1995; Farber 1972; Bowler 1973; Gayon 1996; Wilkins 2003a; Grene and Depew 2004; Talpsepp 2013]. Having softened his orthodox "Ladder" natural philosophy (see Section 4.3.5), he writes in one of the volumes of his *Natural History* that "in nature there are individuals and sequences of individuals, i.e., species" [Buffon 1769: 144]. Elsewhere in the same encyclopedia, he includes a short opus "On Nature" and begins it with the statement that "an individual [...] is nothing in Nature; a hundred and a thousand individuals are still nothing in Nature. Species are the only beings of Nature, eternal and unchangeable, like herself" [Buffon 1843: 52].[4] Arguing for species reality and essential immutability, Buffon appeals to the generative idea, according to which "this chain of successive existences of individuals [...] constitutes the real existence of the species" (cited after [Sloan, 1976: 370; Sloan, 1995: 131]). Buffon believes that each species has at its beginning an "internal model" (*moule intérieur*) created by God, on which basis all organisms of this species, when reproducing, are replicated as its copies [Sloan 1970, 1986, 2009; Bowler 1973; Ibrahim 1987; Grene and Depew 2004).[5] Buffon maintains therefore that it is the ability to produce fertile offspring that "constitutes reality and unity of what should be called species, both in animals and in plants" [Buffon 1755a: 64]. As a result, Buffon is sure that species "not to be conceived of as a *class* of organisms, but rather as the *lineal* series of parent generating offspring" ([Sloan 1970: 336], italics

---

4 With this, Buffon understands species significantly differently than the systematicians: for instance, he believes that "all quadrupeds can be reduced to a small number of real species and genera" [Sloan 1976: 375].

5 Ph. Sloan believes that the "moule intérieur" is equivalent to a real essence, which allows considering Buffon's position a "*crypto-essentialism*" [Sloan 1970].

in original). As noted above, the natural systematicians fundamentally distinguish by this feature natural species from higher categories as classification (logical) units.

This "generative" account of species is most consistently summarized by Buffon in his article "Species" in the fifth volume of the D. Diderot's *Encyclopedia*, which begins as follows:

> All similar individuals that exist on the surface of the Earth are regarded as constituting an indivisible species; however, neither the number nor the similarity of individuals make their totality a species, but only a constant and continuous sequence of renewal of these individuals that constitutes it.
>
> *[Buffon 1755b: 956]*

Buffon adds subsequently to this "formula" an important refinement that could be called *two-way reproductive criterion*: it is not just that "like begets like", but that parents produce fertile offspring, which are able to interbreed freely only among themselves but not with the members of other species, and it is this breeding pattern that allows species to preserve its discreteness [Lovejoy 1959; Sloan 1970, 1976, 1986, 2009, 2019; Bowler 1973; Grene and Depew 2004]. Thus, the reproductive criterion of Buffon implies (a) fertility of progeny within a generative chain and (b) its sterility between the chains, so "all animal species are separated from each other by a gap that Nature cannot overcome" [Buffon 1755a: 59].[6] It seems that it is this formulation that can be considered the first precise "biological" definition of species as a natural unit [Ghiselin 1966, 1969, 1974]. The Buffon's clarification that individuals "exist on the surface of the Earth" is intended to emphasize that he understands species as a "physical" unit (in the sense of Leibniz), actually existing in the physical space-time, and not a logical universal [Lovejoy 1959; Ghiselin 1969, 1974; Sloan 1986, 1995, 1997, 2009; Gayon 1996; Stamos 2003]. The novelty of Buffon's version of the "generative" account of species is so significant and fateful that Ph. Sloan rightfully calls it *Buffonian revolution* [Sloan 1995].

As noted above, an emphasis on generative idea becomes one of the important components of postscholastic revolution in systematics, and with it in the whole of biology. Among its "founding fathers", M. Adanson should be mentioned here for his writing in the *Families of Plants*, with reference to Buffon, that "species consists of a constant and continuous sequence of individuals generated by crossing two sexes" [Adanson 1763: clx]. Similarly, G. Cuvier comments more than definitely on this point in a letter to his student Christopher Pfaff:

> We imagine that a species is the total descendence of the first couple created by God [...]. I maintain that this is the sole certain and even infallible character for the recognition of a species. All other proofs are merely presumptions.
>
> *[Coleman 1964: 145]*

[6] Something similar is expressed later by abovementioned I. Illiger [Illiger 1800], who is also listed for this among the forerunners of the modern reproductive (biological) species conception [Kanaev 1966; Mayr 1968].

One more implicit idea of Adanson should be particularly highlighted in this regard: he says that in different groups of organisms with different reproduction systems (i.e., differently realizing their "generative power"), species are distinguished on different grounds: in unisexual organisms, the main criterion should be the chain of generations, whereas in asexual and bisexual ones, it is the similarity of individuals [Adanson 1763: clxviij]. Perhaps this is the first indication in the considered conceptual history of an important point, which will lead to the recognition of different kinds of species (see Section 5.4). It is significant for a more complete biologization of the general notion of species due to the linking ontic status of natural species with certain aspects of species biology (see Section 6.3).

The acceptence of generative idea leads to a new and very important conceptualization of the species notion. The zoologist and historian of biology Oliver Rieppel in his book *Phylogenetic systematics* draws attention to the fact that the idea of species as a sequence of generations leads to its idea as a dynamic system [Rieppel 2016]. As an illustration of the transition of these two ideas into one another in the 19th century, Rieppel cites an aphoristic phrase of the botanist Anton Friedrich Spring (1814–1872), who states that due to constant self-reproduciton, "species exists not in a state of being, but of continuous becoming" [Spring 1838: 49].

From this understanding of "species as a process" [Rieppel 2009; Dupré 2020, 2022] follows a corollary, important for a new understanding of its ontology, viz., an attribution of (quasi-)individual status to it (see Section 6.2). In the 19th century, it is generated by organismal natural philosophy (F. Schelling, L. Oken, and their followers) that likens Nature to a living superorganism [Meyer-Abich 1949; Raikov 1969; Gottlieb 1992; Richards 1992; Sloan 2009; Rieppel 2011b, 2016]. Thus, Gottfried Reinhold Treviranus (1776–1837) in his fundamental *Biology* writes that "the entire animal kingdom has its evolutionary periods [...] similar to the periods which are expressed in individual organisms. Those animal species and genera that have undergone an evolution [...] can be compared with the organs that have vanished in the course of the evolution of each animal" (cited after [Richards 1992: 45–46]). Carl von Nägeli (1817–1891) regards this fundamental issue in a special book *The Origin and concept of the natural-historical species*, and he argues that each species is a whole that is born, develops, and dies like an organism and compares speciation with a born of child from its mother [Nägeli 1865]. Being an evolutionist, he equates in this regard all taxonomic categories: "each systematic category is conceived as a natural unit, which represents a transit point in the major development-historical movement" [*loc.cit.*: 32]. The same idea is expressed in more detail by Ernst Heinrich Philipp August Haeckel (1834–1919), who devotes to the species issues a chapter "Evolutionary History of Species" in his *General morphology*, in which he outlines his phylogenetic theory [Haeckel 1866, Vol. 2]. He considers species a *genealogical individual* of the second order (with monophyletic group, or *phylon*, being genealogical individual of the third order) and defines it as "the unity of all generative circles, which have the same forms under the same conditions of existence" [*op.cit.*: 353]. According to Haeckel, this "genealogical individual" passes in his historical development through three stages, which he designates as *epacme*, *acme*, and *paracme specierum* [*op.cit.*: 361–363].

5.2 THE 19TH CENTURY AND BEYOND: BIOLOGIZATION OF THE SPECIES PROBLEM

One of the most important breakthroughs in the conceptual history of the species problem, occurring at the turn of the 18th–19th centuries, is set by the shift from a predominantly classification to a predominantly biological interpretation of species [Wilkins 2009a; Pavlinov and Lyubarsky 2011; Zachos 2016; Pavlinov 2018]. This transition is truly revolutionary: it is one of the manifestations of a fundamental scientific revolution in understanding what living nature is, how it functions, how it is structured and develops, etc., causing arise of contemporary biology [Pozdnyakov 2015, 2018].

Previously, species was considered in the context of the classification genus-species scheme, so it was understood as a universal genus-dependent device for describing the qualitative structure of being. It was defined by a certain commonality of characters of things, and there was no difference between "species" of living and inanimate nature. Realistically treated, this view corresponded to a stationary world picture, in which species appeared as an unchanging entity in the structure of Nature, at first without a certain place in its hierarchy and afterwards being ascribed such place.

A new comprehending of species of the living is set by the context of arising evolutionary-ecological paradigm, so it becomes considered an element of the developing and functioning biota. In such world picture, species appears as a separate dynamic unit, not only maintaining its relative stability but also capable of change. In both cases, this essential characteristic of species is determined by its certain intrinsic and extrinsic interactions inherent only in living nature. As stated above, this brings a fundamental demarcation between the latter and inert matter and makes biological (in the general sense) species a particular phenomenon of life [Zavadsky 1961, 1968; Volkova and Filyukov 1966; Mayr 1982, 1988].

This new conceptualization of the species problem involves two largely interrelated basic aspects of species as a natural unit, evolutionary and ecological, each framed by a corresponding idea. They are complemented by one more, genetic aspect, which, from a holistic standpoint, is secondary to them in the sense that it is a specific means of their implementation.

5.2.1 Evolutionary Idea: Species as an Actor of Biological Evolution

The modern idea of the universe as an infinite dynamic self-developing hypersystem goes back to an antique idea of active Nature as the cause of itself (*causa sui* of scholastics). The *Old Testament* mythology supersedes this idea with its stationary world picture, according to which the universe arose as a result of a single act of divine creation and has not changed since then. This orthodox biblical version of the origin of life becomes more complicated over time due to acknowledgement of a certain dynamics of the structure of the living resulted from the multiple emergences of new species for various reasons (multiple creations, transmutations, cross-breeding, etc.). These conceptual complications gradually prepare the ground for a general *idea of transformism*, or *evolutionary idea*, also known under more familiar names of *theory of evolution* or *theory of development* [Chambers 1844;

Miller 1849; Spencer 1864; Fausek 1904]. It is designed to explain materialistically, i.e., without reference to the supernatural forces, multiple appearances of species of living organisms as a result of natural causation. This idea (or theory), with changing the ontic component of our cognitive situation due to comprehension of natural species as an evolving unit, affects its subjective component by replenishing it with *evolutionary thinking* and aiming the subject to clarify historical causes of the species diversity [Mayr 1982; Gould 2002].

With some coarsening, it can be argued that the evolutionary idea, with regard to the biological forms, grows as an "add-on" over the generative one. The latter presumes that each species is a chain of generations of organisms; according to the abovementioned organismal natural philosophy, this makes it an individual-like developing whole. To this is added an idea of the Ladder of Nature, according to which the particular species are the links in a single Chain of Being, which appears as a result of a sequential "unfolding" (*evolutio*) of the original prototype. Finally, the time dimension is added to this "evolutionary" model—and an idea of the historical process of the origin of species thus emerges. As a result, the sequence of generations of organisms within each species, with the time scale expanding to embrace all their chains, turns into a multispecies sequence, in which different species originate from one another in the long-lasting process of biological evolution.

In the 19th century, this evolutionary idea, fueled by different conceptual origins, takes the form of several particular interpretations by Lamarck, Baer, Darwin, Haeckel, Cope [Bowler 1975, 2003; Burkhardt 1987; Richards 1992]. They all focus on the origin of different groups of organisms, including species, from their common ancestors and from each other; in this regard, the transformist concept, as applied to the living, is termed *theory of origin* by its first creators [Darwin 1859; Haeckel 1866; Cope 1887]. E. Haeckel designates the historical process of origin of species as *phylogeny*; the now well-known term *evolution*, in its historical interpretation by philosophers (Kant and Schelling, see [Rieppel 2016]), acquires its specific biological meaning after the publication of *The Principles of Biology* by the philosopher Herbert Spencer (1820–1903) [Spencer 1864].

> It should be noted, however, that many naturalists of the 19th century believe that the constancy of species as a natural unit is an indispensable condition for acknowledging its reality [Mayr 1957a]. In this regard, for example, indicative is the standpoint of the geologist and naturalist Charles Lyell (1797–1875), whose geochronological findings substantiate significantly a transformist world picture. He devotes a considerable part of the 2nd volume of his *Principles of Geology* to the question of the reality of animal and plant species and inclines to an idea of their constancy [Coleman 1962; Sloan 2009].

One of the most important hallmarks of the evolutionary idea, which largely determines the general trend of biologization of the species problem in the 19th century and beyond, is the adaptational explanation of biological evolution. It presumes that the transformations of some biological forms into others are caused by external conditions and are resulted from specific interactions of organisms with their environment. According to this, species becomes regarded as a natural unit in the structure of the overall evolutionary-ecological continuum combining evolutionary and ecological aspects [Zavadsky 1961, 1968; Schwartz 1980; Boero 2015; Pásztor et al. 2016]. The first aspect refers to an understanding of species as an actor in the evolutionary process (considered herewith); the second aspect refers to its understanding as a participant in the ecological processes in natural ecosystems (see the next section on it). Due to this, the naturalist theoreticians of the early 20th century place the species problem, depending on their natural-philosophical preferences, in two contextual frameworks, either evolutionary [Dobzhansky 1937] or ecological [Turesson 1922] or occasionally in both [Hall and Clements 1923].

Most evolutionary concepts presume a *gradual* transformation of biological forms (species, etc.) due to progressive accumulation of distinctive features. This is supported by a reference to the natural-philosophical principle *Natura non facit saltus*, repeatedly mentioned above. In some evolutionary models, going back to the classical concept of Natural System, the accent is placed on the origins of higher categories (families, orders, etc.), explained by the changes in their structural plans (archetypes). Other models mostly consider the transformations of intraspecific categories (races, subspecies, etc.) into species due to turning their minute differences into larger interspecific ones. The models of the first group will be called *macro*evolutionary, the second group—*micro*evolutionary [Leavitt 1909; Filipchenko 1915]. In the latter, intraspecific variation becomes treated as "one of the principal properties of species, the form of its existence" [Kuzin 1962: 146].

An alternative to gradualism is the concept of *saltational* evolution, according to which some forms (species, etc.) turn into others due to abrupt changes. Such a model is initially developed based on an assumption of drastic genetic transformations, including those caused by interspecific hybridization [De Vries 1904; Lotsy 1916; Goldschmidt 1940; Lysenko 1952; Stegny 1993]; its possibility is now discussed by evolutionary developmental biology (*evo–devo*) [Nazarov 1991; Minelli 2003; Hall and Olson 2006; Laubichler and Maienschein 2007; Minelli and Fusco 2008]. The conception of *punctuated equilibrium* takes an intermediate position, according to which the process of speciation occurs gradually, but in a relatively short period, after which a new species passes into a relatively steady long-lasting state [Gould and Eldredge 1977; Mayr 1988; Eldredge 1993; von Vaupel 1994; Sterelny and Griffiths 1999; Pagel et al. 2006).

* * *

The first fully shaped evolutionary conception (theory) is proposed by J.-B. Lamarck in his *Zoological Philosophy*, based on the natural-philosophical idea of the Ladder of Nature. Simultaneously with him, an evolutionary idea is expressed by the German naturalists committed to organismic natural philosophy (K. von Baer, G. Treviranus, F. Tiedemann, etc.), who conjecture it based on embryological data [Lunkevich 1960; Raikov 1969; Bowler 2003; Rieppel 2016]. Neither Lamarckian nor other evolutionary views of that time receive wide recognition, partly due to the influence of typological ideas, representing the living matter stationary, and partly due to their certain prematurity: the scientific community appears not yet ripe for adoption of the historically interpreted evolutionism [Lunkevich 1960; Coleman 1964; Burkhardt 1977; Shatalkin 2009; Sloan 2019]. The Lamarckian evolutionary model involves a nominalistic treatment of species [Oxenham 2015] by the assumption that "what we call species were imperceptibly fashioned among them one after another and have only a relative constancy, and are not as old as nature", that "the species [...] have only a relative constancy and are only invariable temporarily", and that, finally, "it is useful to give the name of species to any collection of like individuals perpetuated by reproduction without change, so long as their environment does not alter enough to cause variations in their habits, character and shape" [Lamarck 1984: 40, 44].

By the mid-19th century, the evolutionary idea is already, that's to say, "in the air". In its initial "materialization", a significant role belongs to the popular scientific treatise *Vestiges of the natural history of creation* by the naturalist and book publisher Robert Chambers (1802–1871) [Chambers 1844]; it rapidly gains a great popularity and has withstood more than a dozen reprints in several European languages [Secord 2000; Bowler 2003; Khramov 2021]. From the modern viewpoint, his sketch should be considered a version of "theistic evolution", since Chambers repeatedly refers to "the Divine Author of all things" [Chambers 1844: 189, 198, etc.].[7] But apart from these references, he presents a very detailed evolutionary model of the historical development of earth and the living, quite advanced for his time, with direct references to the data of geology and paleontology. For Chambers, species are real groups of organisms that exist at different times, arising and extincting, gradually increasing in number and diversity. In particular, he writes "that the earliest species comparatively soon gave place to others, and that they are not represented even in the next higher group of rocks" [Chambers 1844: 60].

* * *

By its fundamental consequences for the conceptual framework of the considered cognitive situation and the entire biology, the beginning of the "evolutionary revolution" is undoubtedly marked by the book *On the Origin of Species by Means of Natural Selection* by Charles Robert Darwin (1809–1882) [Darwin 1859]. Its title clearly indicates the main task of its author: to explain the mechanism of origin of some species from others in the process of biological evolution [Mayr 1959c, 1991]. The method of solving this task proposed by Darwin largely sets an incentive for the development of all species issues in a substantially new conceptual content [Stamos 2007, 2013].

In his book, Darwin repeatedly notes that he uses the notion of species in a rather "naturalistic" sense without loading it with a theoretical content [Mayr 1959c; Sulloway 1979; Beatty 1983, 1985; McOuat 1996; Mina 2010]. The reason is his confidence that, although "every naturalist knows vaguely what he means when he speaks of a species" [Darwin 1859: 37], there is currently no satisfactory definition

[7] The same, though, can be said about Lamarck, who writes that "nothing exists but by the will of the Sublime Author of all things" [Lamarck 1984: 36].

of species. He himself provides a very traditional definition elsewhere in his work: "species [are] those collections of individuals, which have been so designated by naturalists" (cited after [McOuat 2001: 4]). Therefore, "in determining whether a form should be ranked as a species or a variety, the opinion of naturalists having sound judgment and wide experience seems the only guide to follow" [Darwin 1859: 47].

However, in fact, Darwin's understanding of species is realistic and hardly pure "naturalistic" [Stamos 1996, 2007, 2013; Mallet 2010a, 2013a]. In general, it fully corresponds to the "combinatorial" definition of early postscholastic systematicians that becomes widespread in the first half of the 19th century (see Section 4.3.4). His understanding is based on the Buffonian version of generative idea: in one of his notebooks preceding the *Origin of Species*, Darwin writes that every "species is real with regard to contemporaries—fertility must settle it" (cited after [Sulloway 1979: 29]). On the other hand, in the same notebook, Darwin adds another "definition of species: one that remains at large with constant characters" [*loc.cit.*]: this is a clear manifestation of phenetic idea.

Darwin's most important contribution to the development of an evolutionary idea is a conception of *natural selection* as a specific acting cause of evolution (*causa actualis* of Aristotle) that transforms one species into others. Regarding the problem under consideration, of special importance is Darwin's great attention to intraspecific variability. Before him, variations of species, including their possible transformations (transmutations), were discussed mainly from a typological perspective: how broad they are and whether they affect the structural plans of genera, families, etc. (e.g., [Besnard 1864; Usov 1888]). Darwin considers them in an evolutionary aspect in a quite particular manner: variations serve as a raw material for selection, so if there is no variability, there is no selection, and thus there is no "Darwinian" evolution [Mayr 1988; Barberousse and Samadi 2010].

His microevolutionary model implies that the species arise as a result of gradual divergence of the intraspecific varieties by accumulating their differences and subsequently losing an ability to interbreed with each other. This process is illustrated by Darwin's scheme of a gradual differentiation of phyletic lineages corresponding to particular varieties [Darwin 1859: diagrams between pages 117 and 118]. Based on this model and the scheme illustrating it, Darwin argues that "varieties have the same general characters as species, for they cannot be distinguished from species" [*op.cit.*: 58] and that "varieties are species in the process of formation, or [...] incipient species" [*op.cit.*: 111]. And Darwin concludes his analysis of the "origin of species" with hoping that over time, "we shall at least be freed from the vain search for the undiscovered and undiscoverable essence of the term species" [*op.cit.*: 485]. This evolutionary model initiates *Darwinian revolution* in the conceptual history of the species problem: it means that natural species (a) is an actor in evolutionary process and (b) is not a naturally salient unit, but just a final phase of gradual changes in local varieties.

One of the consequences of the Darwinian evolutionary model to the practical systematics is that it explains in its own fashion a distinction of "bad" and "good" species, which has long been known to practicing taxonomists. It asserts that the former are incipient, while the latter are fully formed species units [Darwin 1859; Kerner 1866].

Contemporary theorists continue to discuss this distinction [Mallet 1995; Shaw 1996; Amitani 2010, 2015, 2017, 2022; Wilkins 2022].

Darwinian model yields a specific dualism in the understanding of species, which can be regarded in two hypostases—with or without referring to the time dimension, which corresponds to either diachronic or synchronic aspects of species consideration [Simpson 1951; Mayr 1988; Otte and Endler 1989; Stamos 1996, 1998, 2002, 2007; Stamos 1998, 2002; Lee and Wolsan 2002; Haffer 2006; Richards 2010; Rieppel 2010a], and they are allegorically called "vertical" and "horizontal". The components of this *diachronic-synchronic dualism* are complementary to each other and linked by an uncertainty relation: fixing one aspect of species as a studied object does not allow examining it in the same detail in another aspect [Skarlato and Starobogatov 1974; Starobogatov 1980; Pavlinov 2018, 2021b].

The "vertical" aspect of species consideration presumes its representation as an array of crossing phyletic lineages: in this regard, it corresponds to a fragment of the temporal chain of varieties replacing one another in time. This aspect, as noted above, is built over the classical generative idea, in which the chains of generations are substituted by the chains of varieties; in this, it is a forerunner of the contemporary general lineage conception [de Queiroz 1998, 1999]. When considering these chains of gradual transformations, the paleontologist L. Wurtemberger, in his work of 1880, clearly expresses a skepticism about species: "since it is impossible to indicate where a species begins and where it ceases to exist, then the concept of species loses all scientific ground" (cited after [Zavadsky 1961: 41]). The synchronic aspect of species is set by its considering in a local time and space ("nondimensional" conception of Mayr); if addressing to Nature, simultaneously coexisting forms as elements of ecosystems are meant. These forms are discrete species due to reproductive barriers between them, for which reason "species come to be tolerably well-defined objects, and do not at any one period present an inextricable chaos of varying and intermediate links" [Darwin 1859: 177].

Thus, the Darwinian evolutionary model fills the species problem with a rather specific dual content. On the one hand, it presumes the reality of coexisting "nondimensional" species as a natural biological unit, which both integrity and discreteness are ensured by their being (almost) closed reproductive systems, and this approves reality of the species category. On the other hand, the interpretation of "diachronic" species as the sequential stages of divergence of varieties without clear boundaries deprives the species category of its salient status attributed to it by natural systematics. As a result, as the botanist Vladimir Leontievich Komarov (1869–1945) notices, "after the revolution accomplished by Darwin's doctrine, the concept of species became much less definite than it has been in Cuvier's time" [Komarov 1940: 38]; E. Mayr will accuse Ch. Darwin of his "eliminating the species as a concrete natural unit" [Mayr 1963: 14]. The latter, of course, is incorrect: Darwin denies not the reality of species as a natural unit, but a salient position of the species category in the hierarchy of both Nature and classifications [Ghiselin 1969, 1997; Stamos 1996, 2007; Ereshefsky 2009a, 2011; Sloan 2009; Pavlinov 2017, 2018]. So, one can say that he is a realist regarding species unit and a nominalist regarding species category.

* * *

Focusing on the intraspecific local forms (subspecies, races, varieties, etc.), stimulated by the microevolutionary model of Darwinism, as well as of Neolamarckism [Regel 1917; Yuzepchuk 1939, 1958; Zavadsky 1961, 1968], yields an acknowledging the evolutionary and taxonomic significance of the minute intraspecific variations. As a result, many systematicians committed to this model, when classifying organisms, begin to focus not on the species, but on the intraspecific forms as the participants in the real microevolutionary processes. This gives rise to *classification Darwinism* as a particular taxonomic theory [Pavlinov 2018, 2021b] and causes a new wave of species nominalism [Zavadsky 1968; Stamos 2007; Pavlinov 2018]: since species are constantly changing, they are "a human contrivance [so] the making of species is an expediency" [Bailey 1896: 457, 458]. In an extreme form, it is asserted that "species and variety are only abstract concepts: they do not exist in nature" [Timiryazev 1904: 81].

According to a more moderate position, which is designated as *bionominalism* [Mahner and Bunge 1997], the intraspecific forms are real, whereas the "Linnaean" species are not. Thus, the botanist Sergey Ivanovich Korzhinsky (1861–1900) states in his *Flora of the East of European Russia* that "races are true systematic and geographical units. They are subject to research as something that really exists. Meanwhile, species and subspecies represent something conditional" [Korzhinsky 1893: 76]. V. Komarov adheres to a similar position and emphasizes in the introductory section of his *Flora of Manchuria* (1901) that "the notion of 'species' [...] is an ideal view of a common type, completely coinciding from the logical standpoint with the concept of 'genus'" [Komarov 1949: 74].

As said above, this viewpoint yields a "species crisis" that involves negation of its broad interpretation inherited from Linnaeus and early postscholastic systematicians. However, an underlying influence of the classical classification tradition persists, according to which the species is still given a key importance (as evidenced by the title of the Darwin's book). Accordingly, the critics of "Linnaean" species, committed to its narrow interpretation, apply this term to the local geographical forms, designating them by the "Linnaean" binomials. As a result, in the second half of the 19th century, an informal school of species systematics arises, called *analytical* [Stevens 1994; Bonneuil 2002]. Its proponents become known as *species splitters*, or "species-makers" [de Candolle and Cogniaux 1876; Clements 1908], and their aspiration to describe more and more local forms as species represents a kind of "nomenclatural itch" [Evenhuis 2008; Pavlinov 2022b]. The botanist Edward Lee Greene (1843–1915), discussing this issue, points out a possible curious "bibliopole" aspect of such species-making: the more species are delineated, described, and illustrated in monographs, the thicker and hence more profitable they become [Greene 1910].

The botanist Anton Kerner von Marilaun (1831–1898), considering "good" and "bad" species in his *Natural History of Plants*, calls geographical races *small species* and believes that their "combination in one ideal form [...] is unacceptable, [it] is the result of speculation" (cited after [Kamelin 2004: 18–19]). Some botanists arrange such local forms into a hierarchy of natural units of an "around-species" complex (see Section 5.3 on this term). For example, V. Komarov writes about species of the *first* and *second* order: the former are local geographical races, and the latter are broadly interpreted "Linnaean" species [Komarov 1940, 1949]. Refining this treatment, he calls *tribe* any diagnosable form limited to a certain territory, which in traditional terminology is *elementary species*, while a group of similar tribes ("union of tribes") is a *composite species* [Komarov 1927].[8]

In the Darwinian microevolutionary model, species multiplication is explained mostly by geographical isolation, according to which a new understanding of species acquires spatial dimension. Therefore, its adherents highlight that the natural species, unlike the logical "Linnaean" one, is characterized not only by morphological features but also by a specific distribution range, and therefore, geographical criterion must be included in its definition [Korzhinsky 1893; von Wettstein 1898; Komarov 1927, 1940, 1949; Sinskaya 1961]. V. Komarov expresses this idea as follows: "a species is a morphological system multiplied by geographical certainty" [Komarov 1927: 39]; this formulation will be subsequently called a *Kamchatka aphorism* [Pavlov 1951; Skvortsov 1967].

One of the important sources of an increasing species splitting become experimental approaches, developed at first in the 18th century by the botanist Joseph Gottlieb Kölreuter (1733–1806) and actively implemented in the 19th century by his colleagues [Clausen et al. 1939; Komarov 1940; Turrill 1940; Rozanova 1946; Zavadsky 1968; Hagen 1982; Bonneuil 2002]. Their proponents propose to reveal the stable traits and boundaries of the particular species by experimenting with hybridization and cultivation of plants under different conditions. One of their pioneers, the experimental botanist Claude Thomas Alexis Jordan (1814–1897), calls elementary *monotypic* species those morphologically uniform varieties which reproduce themselves steadily in a series of generations under controlled conditions [Jordan 1847, 1873; Planchon 1874; Boas 1898]. His disciple Charles Victor Naudin (1815–1899) identifies up to five "steps" of species, from "bad" to "good", based on experimental verification of the degree of stability of their traits and the degree of sterility of their crosses [Naudin 1852, 1862; Marza and Cerchez 1967].

However, there many *species lumpers* remain as adherents of the Linnaean broad treatment of classification species, who constitute *synthetic* school of species systematics [Stevens 1996; Bonneuil 2002]. They protest against an unrestrained description of new elementary species, in their opinion threatening the established order in systematics; for obvious reasons, there are a lot of nonevolutionists among them. One of their leaders, the abovementioned botanist George Bentham, announces a "crusade" against the species splitters: he and his coauthor Joseph Dalton Hooker (1817–1911) accuse them of "killing botany as a science". Guided by an imperial mentality and taking advantage of their position as leaders of the principal Kew botanical center [Endersby 2008], they in their floral summaries (like *Index Kewensis*) adopt

8 The notion of *composite species* (= *conspecies*) is introduced in the mid-19th century to denote "Linnaean" species, consisting of several subspecies [Schlegel 1844]; the notion of *elementary species* is proposed later [De Vries 1904].

and consistently apply something like an "international standard" for a broad understanding of plant species [Bonneuil 2002]. Thus, announcing a program for preparing a new worldwide botanical checklist, Bentham promises that it will be based on "the species in the Linnean sense, which, though not susceptible of a strict definition, is pretty generally understood amongst botanists" [Bentham 1875: 45].

* * *

In the 20th century, the evolutionary idea is given a special emphasis in conceptualizing the species notion, based on the confidence that (a) the process of evolution is a basic universal mechanism of emergence and development of the living and (b) species is an actor of this process. Proceeding from this idea, G. Simpson emphasizes in his *Principles of Animal Taxonomy* that "a concept that omits this consideration is incomplete if not quite inappropriate" [Simpson 1961: 154]. This standpoint is brightly expressed by an aphorism of the evolutionary geneticist Theodosius Dobzhansky (1900–1975), according to whom "nothing in biology makes sense except in the light of evolution" [Dobzhansky 1973: 125]. On this basis, for example, the zoologist Joel Cracraft believes that "species concepts serve two masters, evolutionary theory on the one hand and taxonomy on the other. [...] Ontological and biological difficulties can be alleviated if species are defined in terms of evolutionary units" [Cracraft 1987: 329]. However, every thesis always finds its antithesis in the scientific realm: for example, Bradley Wilson, paraphrasing Th. Dobzhansky, argues that "while nothing in biology may make sense except in the light of evolution, evolutionary biology can make sense without species" [Wilson B. 1996: 419].

Since the evolutionary idea develops in many aspects as an "extension" of the generative one, particular species conceptions, arising on its basis throughout the 20th century, realize in one form or another an understanding of species as a temporary lineage of populations (see Section 5.3). The abovementioned classification Darwinism takes the shape of biosystematics that considers species as a population unit [Hall and Clements 1923; Huxley 1940; Camp and Gilly 1943]. Another is based on a simplified (cladistic) version of evolution and identifies species with metapopulation lineage [de Queiroz and Donoghue 1988; de Queiroz 1998, 1999]. At last, a quite complex configuration of species, regarded in this vain, is suggested by its evolutionary conception [Simpson 1951, 1961].

5.2.2 Ecological Idea: Species as a Component of Ecosystem

An ecological idea, which composes a basis of *ecological thinking*, implies in its most general and simplified sense that, when recognizing the natural aggregates of organisms, including species, it is necessary to take into account the peculiarities of their interactions (a) with each other and (b) with their environment. Accordingly, these interactions are divided into two principal categories: *autecological* corresponds to intraspecific relations and *synecological* corresponds to interspecific relations [Odum 1953; McIntosh 1986; Mirkin and Naumova 2005]. In such general understanding, this idea, minded in a natural-philosophical framing, goes back to the antique understanding of the universe as an active *Physis*, which components, including organisms, are in the permanent mutual interactions.

In the earliest systematics of the 16th–17th centuries, which inherited some important ideas about characterizing genera and species from scholasticism of the realistic kind, the principal attention is paid to the intrinsic properties of organisms related somehow to their essences. Accordingly, any extrinsic characteristics, before all ecological ones, appear excluded from the conceptual framework of the respective cognitive situation. However, naturalists of the 18th century show a growing interest in the generative idea "implemented" by the reproduction system, which is autecological by its principal content, and emerging postscholastic systematics of the late 18th and the early 19th centuries fully assimilates this idea (see Section 5.1.2).

Based on this foundational premise, mycologist Heinrich Klebahn (1859–1942) introduces in the 1890s a notion of *biological species* as the one that is characterized more by biological than morphological features (see [Klebahn 1904]). At the same time, a narrower notion of *physiological species* is coined, which main characteristics are borrowed from physiology [Cockerell 1897]. This new trend in the considered conceptual history becomes appreciable in the early 20th century [Klebahn 1904; Clements 1908; Kholodkovsky 1910] and marks the beginning of *ecologization* of the species problem: according to the botanical ecologist Göte Wilhelm Turesson (1892–1970), the latter "is thus seen to be in a large measure an *ecological problem*" ([Turesson 1922: 101], italics added), which is epitomized by a notion of *ecological species* [*op.cit.*]. This new emphasis in the interpretation of species marks a significant shift in our problematics deserving being designated *Turessonian revolution.*

A notable part of this new trend becomes *geneticization* of the species problem based on the approaches developed by ecological (population) genetics that begins to take shape at that times [Vorontsov 2004]. Its emergence is promoted by the autecological context that presumes an analysis of genetic processes within species: they are considered an important part of intraspecific interactions that ensure both species stability and variability [Ghiselin 1987]. William Bateson (1861–1926), one of the leaders of nascent genetics, believes that "what we call 'species' is a mixture of different phenomena confounded under one name" [Bateson 1899 (2009): 169], and that only those phenomena, revealed experimentally, are truly natural species; the botanist geneticist Hugo De Vries (1848–1935) agrees with him [De Vries 1904]. Early geneticists emphasize the role of crossing and genetic isolation in the origin of new races and species [De Vries 1904; Lotsy 1916], and W. Bateson devotes the first chapter of his book *Problems of genetics* to the species issues, beginning it with a note that "nowhere is our new knowledge about heredity and variability more directly applicable than to the problem of what is a species and what are variations" [Bateson 1913: 3]. Thereafter, G. Turesson combines its ecological and genetic aspects of species consideration to provide its understanding as a complex *genecological* unit. An emphasis on its ecological meaning is highlighted by the term *coenospecies*, whereas its genetic aspect is highlighted by the term *genospecies* [Turesson 1922, 1930].

An important step in developing an idea of biological species is made by the zoologist Edward Bagnall Poulton (1856–1943) in the article symptomatically entitled

What is a species? He distinguishes between classification and natural species in a notable way by defining the first as *syndiagnostic* (unity of diagnostic traits), and the second as *syngamous* (unity of cross-breeding system) [Poulton 1904]. A specific accentuation on the latter's autecological content is provided by the zoologist L. Plate, who writes in his *Principles of Systematics* that "the members of a species are tied together by the fact that they recognize each other as belonging together and reproduce only with each other" [Plate 1914: 160]. This treatment gets fixed by the notion of *syngameon* [Lotsy 1931; Dobzhansky 1935; Boecklen 2017][9] to denote species as a (quasi)closed genetic system. It gets enriched soon by the ethological and other mechanisms involved in maintaining species discreteness, which results in a *biological* conception in the sense of Dobzhansky–Mayr, one of the most popular in the middle and partly the second half of the 20th century [Dobzhansky 1937; Mayr 1942, 1963, 1969, 1982, 1988, 2000; Lehman 1967; Giray 1976; Rubtsov 1996; Rieseberg and Burke 2001; Friedmann 2007, 2013]. Since it refers to the mechanisms of reproductive isolation that ensure the stability of the species gene pool, the names *genetic* or *reproductive* or *isolation* seem more appropriate for this conception [Dobzhansky 1935; Darlington 1940; Simpson 1951, 1961; Coyne et al. 1988; Brooks and McLennan 1999; Mallet 2001; Pavlinov 2018, 2022a]. The Plate's idea results in a *recognition* conception, focusing on the selective behavioral responses of sexual partners to each other, which leads to their assortative crossing [Paterson 1985; Masters et al. 1987; Friedmann 2007, 2013; Masters and Génin 2018].

This conception is based on Buffon's two-way reproductive criterion, which refers to the "internal fertility and external sterility" [Cuénot 1936: 251]. The significance of this criterion for the development of the modern understanding of species is that it can be determined *absolutely*, i.e., not through relative differences, but through its own biological nature, and this serves as an argument in favor of the reality of species category [Plate 1914; Hatch 1941; Simpson 1951, 1961; Lehman 1967; Mayr 1968, 1969; Collier 1985; Lee and Wolsan 2002; Lee 2003; Bock 2004; Claridge 2010]. This essentially differs species from both intraspecific forms (not reproductively isolated from each other) and from supraspecific taxa (not internally connected by crossing system), an idea that has been considered since the early 19th century (see Section 4.3.4).

* * *

A synecological idea, filled mostly with a "naturalistic" connotation, becomes actively argued by G.-L. Buffon in the mid-18th century. He is convinced that the comprehension of the living is unfeasible without taking into account an inseparable connection of organisms with their habitat, i.e., their lifestyles with all ecological, physiological, behavioral, etc., features [Kanaev 1966; Stafleu 1971; Grene and Depew 2004; Hoque 2008]. At the early 19th century, this idea is being worked out in more detail by Friedrich Wilhelm Heinrich Alexander

[9] It should be recalled that an idea of different definitions of the "kinds of species" is expressed for the first time by M. Adanson (see Section 5.1.2). J. Mallet thinks that E. Poulton solves the species problem by his approach [Mallet 2004], but in fact he initiates it [Wilkins 2009a; Pavlinov 2022a].

Freiherr von Humboldt (1769–1859). He, following Buffon, objects to counting any classification of organisms natural if it is based on their "formal" diagnostic traits and does not reflect their basic life forms [Humboldt 1806]. The latter, though, coincide with the higher taxa of folk systematics and have no direct concern to the species diversity at that time [Atran 1990; Berlin 1992; Pavlinov 2010].

Regarding species, the above-emphasized adaptational explanation of evolution appears of paramount importance for its "ecological" account [Zavadsky 1961, 1968; Schwartz 1980]. Indeed, the main result of adaptive evolution is that each species turns out to be fitted by its features to a certain environment, which frames species-specific ecological niche. This yields an important corollary that a profound conceptualization of species as a natural biological unit should involve its interaction with the environment. Accordingly, *ecological certainty* becomes one of the most significant characteristics of biospecies thus understood [Zavadsky 1961, 1968]. An ecological emphasis on the understanding of species appears to be one of the culminating events in the fundamental trend of biologization of all the species issues, so this shift may deserve being designated as *Turessonian revolution*.

5.3 BIOSYSTEMATICS AND OTHERS: FALL AND RISE OF SPECIES

The philosophical background of the cognitive situation embracing the species problem is developing during the 20th century following certain movements of "historical pendulum" [Pavlinov 2018, 2021b]. In its first half, the positivist philosophy of science dominates in biology: according to its basic ideas, dating back to philosophical empiricism of the 16th century (F. Bacon, J. Locke, etc.) and taking shape in the second half of the 19th century (O. Comte, E. Mach, etc.), everything complex must be decomposed into elementary "bricks", and the relations between the latter serve as the basis for explaining complex phenomena. In its second half, postpositivist philosophy of science (K. Popper, W. Quine, etc.) comes to the fore to revive interests in the complex natural phenomena regarded from a conceptualistic standpoint.

Each move of this "pendulum" results in a significant change in the onto-epistemic framework of the considered cognitive situation, which inevitably involves the respective change in the view of the structure of biodiversity. The positivist philosophy stimulates research interest in the latter's lowest levels; a microevoluitonary model meets these conditions to a large extent by its focusing on populations and genes as the just-mentioned "bricks". The postpositivist philosophy revives attention to the macroevolutionary process of structuring biota, in which species formation is treated as an important "nodal point" in switching microevoluiton to macroevolution.

* * *

The shift of cognitive emphasis from species to intraspecific units, stimulated by the popularity of microevolutionary models at the beginning of the 20th century, generates a specific research program in systematics, which primary task is declared to be the study of processes and mechanisms of microevolution [Clements 1908; Hall and Clements 1923; Huxley 1940; Mayr 1942; Heslop-Harrison 1960; Dean 1979, 1980]. The biologist Julian Huxley (1887–1975) suggests to "look at some of the aspects of the species problem and of taxonomy in its function of representing evolution in action" [Huxley 1940: 4]. Its predecessor was the abovementioned informal analytical school of species systematics; however, to emphasize its novelty as compared to the old "orthodox" systematics based on an analysis of museum materials, this new program is called at first the *new systematics* [Hubbs 1934; Turrill 1938; Huxley 1940; Mayr 1942], and then *biosystematics* [Camp and Gilly 1943; Camp 1951; Myers, 1952]. It is also known as *population* systematics to emphasize its attention to the natural populations and biological processes in them [Mayr 1969; Pavlinov and Lyubarsky 2011; Pavlinov 2018], and also *Darwinian* systematics to reflect its commitment to Darwinism [Mayr and Bock 2002]. Biosystematics arises and most actively functions as a branch of botanical systematics [Camp and Gilly 1943; Camp 1951; Heslop-Harrison 1960; Takhtajyan 1970; Solbrig 1970; Dean 1979, 1980; Stace 1989; Hoch and Stephenson 1995]. In zoology, it (as a "new systematics") is at first hotly popularized by E. Mayr [Mayr 1942], but then he loses interest in it, and therefore this term practically disappears from zoological publications.

Guided by the microevolutionary idea, biosystematics inherits from the early Darwinism and Neolamarckism a conviction that not "Linnaean" species, but natural local grouping of organisms, viz., *population*, is real. This term, borrowed partly from genetics, is suggested in this meaning by the botanist Gustaf Einar Du Rietz (1895–1967) to be applied "to all sorts of taxonomical and plant-sociological units" [Du Rietz 1930: 337]. This new accentuation is most clearly formulated by E. Mayr in his monograph *Systematics and the origin of species*:

> The *old systematics* is characterized by the central position of the species. No work, or very little, is done on infraspecific categories (subspecies). A purely morphological species definition is employed. […] The major problems are those of a cataloguer or bibliographer, rather than those of a biologist. The *new systematics* may be characterized as follows: The importance of the species as such is reduced […]. The population […] become the basic taxonomic unit. The purely morphological species definition has been replaced by a biological one, which takes ecological, geographical, genetic, and other factors into consideration.
>
> *([Mayr 1942: 6–7], italics in original)*

Thus, in the cognitive situation of the "new systematics" of the first half of the 20th century, the natural-classificatory dualism of species decisively shifts toward its understanding as a natural unit. Accordingly, an increase of "natural" emphasis laid by the first postscholastic systematicians becomes more evident: for any species classification to be natural, the species-as-taxon should correspond as much as possible to the natural species-as-population [Turrill 1938; Huxley 1940].

Due to the preserved significant influence of classificatory cognitive program, studies in the evolutionary and genetic-ecological processes at the population level tend to be reduced to the classification tasks, including delineation and ranking of biosystematic units. A movement in this direction includes the elaboration of a specific language for describing the structure of biodiversity and generates a lot of new terms: biosystematics actively experiments with its intraspecies units, defining them on different grounds. One of the first terminologically fixed is above-mentioned elementary species, equated with geographical race [De Vries 1904; Johnson 1908; Komarov 1927]; it is designated later as *microspecies* [Grant 1981]. Certain units are distinguished based on an experimental genetic approach: stable local variations are called *jordanons*, broadly interpreted traditional species are *linneons* [Lotsy 1916, 1931]. The higher intraspecies categories distinguished in biosystematics include fairly stable morpho-geographical, ecological, and/or genetic units, which are called *subspecies*, *race*, and *natio*; the lower ones include units that do not have such certainty, these are *morph*, *aberration*, and *phase* [Semenov-Tyan-Shansky 1910; Bianchi 1916]. With an emphasis on the ecological criterion, the main lower intraspecific unit is defined as *ecotype* [Turesson 1922; Valentine 1949]; according to the genetic criterion, two lower-level hierarchies are proposed: *komparium*, *kommiskuum*, and *konvivium* [Danser 1929]; *biotype*, *pure line*, and *clone* [Du Rietz 1930]. Combining phenetic and genetic criteria yields recognition of *homogenon*, *alloploidon*, *mycton*, *phenogen*, *phenon*, etc. [Camp and Gilly 1943]. Fragmentation of intraspecific forms yields at first recognition of up to a dozen local units, then their number reaches several dozen [Sylvester-Bradley 1952; Baranov 1958], and by the end of the 20th century, about a hundred biosystematic units will be identified on various grounds and termed by various authors [Stace 1989].

This general trend of rank fragmentation also involves traditional species rank, which becomes divided into several subranks corresponding to various degrees of evolutionary differentiation. To begin with, it is proposed to insert *circle of forms* (Formenkreis) or *circle of races* (Rassenkreis) between "Linnaean" species and subspecies (geographical race) [Kleinschmidt 1900, 1926; Rensch 1929, 1932; Stresemann 1936; Mayr 1942, 1963; Williams 2007]. Over time, the following linear hierarchy of ranks becomes popular (listed in an ascending order): *subspecies* > *semispecies* = *allospecies* > *conspecies* > *superspecies* = *ex-conspecies* [Mayr 1942, 1963; Amadon 1966; Stepanyan 2011; Mallet 2013b].[10] This sequence reflects the phases of microevolution and speciation [Huxley 1940], and it corresponds to how C. Darwin explains gradual transformations of geographical forms first into "bad" and then into "good" species. These phases are embedded in an idea of the "life cycle" of species: its "age stages" were first considered in the mid-19th century (Nägeli, Haeckel, see Section 5.2.1) and get again "re-opened" in the mid-20th century to be called *neospecies*, *mesospecies*, *euspecies*, and *telospecies* [Dillon 1966].

10 We can also mention the *circle of species* (Artenkreis), equivalent to the superspecies [Rensch 1934]; Vernon Grant combines semi- and microspecies under a rather strange category of *nonspecies* [Grant 1981].

As a result of all these multidimensional fragmentations of the species-intraspecies hierarchy, caused by microevolutionary interpretation of differentiation of populations, the already mentioned species rank uncertainty arises [Hey et al. 2003; Coyne and Orr 2004; Rasnitsyn 2007; Pavlinov 2013a, 2018, 2022a]. It means that both the very ranked systems of natural and classification units, which tend in their meaning to the general notion of species, and the criteria for distinguishing their ranks turn out to be significantly blurred. The latter, as said above, yields an introduction of an intuitively obvious, but not strictly formalized notion of "*around-species*" unit [Mikhailov 2003; Pavlinov 2017, 2022a; Pfander 2018]. This view of taxonomic hierarchy receives methodological justification from the fuzzy logic [Pavlinov 2018, 2022a] (see Section 6.1 on the latter).

The species rank uncertainty was first recognized by the postscholastic systematicians of the early 19th century. In its new capacity, it is caused by an adoption of microevolutionary model that yields the so-called *species question* to become one of the most actively discussed by biosystematicians of the early 20th century. In this new sense, it first appears, probably, in the *Origin of Species* by C. Darwin, who writes about "the evidence advanced by our best botanists on the *question* whether certain doubtful forms should be ranked as species or varieties" ([Darwin 1859: 248], italics added).

Ch. Darwin raises one more question about species, "namely whether all the individuals of the same species have descended from a single pair, or single hermaphrodite, or [...] from many individuals simultaneously created" [*op.cit.*: 355]. This question is of a far more general meaning than previous, as it involves the key dilemma of creation *vs.* origin of species, which Ch. Lyell discusses *in absentia* with Darwin [Wilson 1971]. For this reason, some authors count this dilemma as the principal content of the Darwin's "species question" [Wilkins 2009a; Mishler and Wilkins 2018]. Perhaps, in the 19th century, this was so [Herber 1963; Coleman 1964; Winsor 1979]; however, at the beginning of the 20th centuries, it loses (for various reasons) its sharpness, and the first Darwin's "species question" comes to the fore, viz., the question of relative ranking species and intraspecific forms [Johnson 1908; Greene 1910].

The species question in its classificatory account becomes "the main theoretical problem of systematics" of that time [Heptner 1947: 7]. It is discussed in 1908 at a special symposium organized by the Botanical Society of America, in which taxonomists, ecologists, and physiologists participate [Johnson 1908]. It is noteworthy that the participants of this discussion constantly mention the notion of *conception of species*, thereby introducing it into a full-fledged scientific circulation. Systematicians (C. Bessey, N. Britton) insist on the preference of "Linnaean" species: "may have been too broad, but it had the merit of being understood. [...] It is still well fitted to our mental needs and mental capacity" [Johnson 1908: 220]; it is admittable to divide it into races, but not into smaller forms. Ecologist H. Cowles, on the contrary, insists that the broad "Linnaean" species should be abandoned and replaced with the elementary species identified experimentally. Physiologists and ecologists (J. Arthur, F. Clements, Cowles) believe that in higher plants, "Linnaean" species should be identified by morphological traits, while natural intraspecies units by physiological ones based on experiments; J. Arthur discusses in this vein "physiological species" of prokaryotes. It should be particularly noted that H. Cowles writes about the "nonhomology" of species in groups of organisms with different biology, contributing, along with E. Poulton, to the initial development of the species problem in its modern interpretation (see Section 5.4).

* * *

The skeptical attitude toward species, inherent in early biosystematics, is brought to its logical end by those philosophical biologists who are committed to the positivist philosophy of science. For them, there do not exist in nature not only species in any meaning but also lower population units: in this respect, they do not differ from nominalistic scholastics like Ockham, logicians like Mill, and "ladderists" like early Buffon and Lamarck. As a result, in their eyes, the species issues lose their actuality, since any and all classification universals are considered fictitious and having no objective relevancy. For example, the botanist Charles Edwin Bessey (1845–1915) expresses this idea more than definitely: "Nature produces individuals, and nothing more. [...] So species have no actual existence in nature. They are mental concepts, and nothing more" [Bessey 1908: 218]. Later, his colleague John Scott Lennox Gilmour (1906–1986), in his article *Taxonomy and philosophy*, with direct reference to the ideas of positivism, exposes a "new philosophy" of systematics. He asserts in the spirit of Lockean sensualism "that the individual is a concept, a rational construction from sense-data, and that the latter are the real objective material of classification" [Gilmour 1940: 466]. Gilmour believes that any traditional reasoning about the "naturalness" of species gives it "a false appearance of metaphysical objectivity", which in fact does not and cannot be [*op.cit.*: 467]. In this regard, the same question is raised that scholastics had vividly discussed in the Middle Ages: "whether the species is a purely artificial device employed for making the bewildering diversity of living beings intelligible, or corresponds to something tangible in the outside world" [Dobzhansky 1935: 345]. With this reasoning in mind, the notion of species is compared with the notion of phlogiston in chemistry or ether in physics and is proposed to be removed from the conceptual apparatus of biology as having no correspondence with "biological reality" [Burma 1949, 1954; Davidson 1954; Beatty 1983]. This movement of positivist thought, reviving the understanding of species in the traditions of nominalistic scholasticism, continues to nowadays [Simonetta 1982, 1992; Kober 2008; Casetta 2010, 2012; Crane and Sandler 2011].

In the middle and second half of the 20th century, such a positivistic view of the problem in question yields an increased demand for treating species as formal classification units of the nominalistic kind, i.e., without reference to any biologically meaningful theoretical contexts. For this reason and contrary to the dominating general trend of the biologization of the species problem, the natural-classificatory dualism becomes resolved in favor of classification interpretation of species and intraspecies units. According to an accent made on the procedural operationality of biological notions, any classification units are recognized as *operational taxonomic* ones, which are recognized by similarity and arranged in a rankless hierarchy [Gilmour 1940; Mason 1950; Michener 1962; Sokal and Sneath 1963; Colless 1967, 2006; Sneath and Sokal 1973; Cronquist 1978; Levin 1979]. Those allocated to the lowest level are called *least-inclusive taxonomic* units [Pleijel and Rouse 2000], whereas those "tied" somehow to the traditional species rank are qualified as *para-species*, sometimes also *surrogate species* [Nowak 1992; Oliver and Beattie 1996; Krell 2004; Ward and Stanley 2004; Minelli 2019]. Interpreted purely pragmatically [Brasier 1997; Claridge et al. 1997; Seifert 2014, 2020], they are "biologically empty" [Oren 2004; Ward and Stanley 2004; Pavlinov 2018] and may hardly be counted as biologically meaningful entities; nevertheless, distinguishing them is sometimes considered the core of the nowadays species problem [Miller 2016].

Another support for a "nihilist" attitude to species in the second half of the 20th century gains from a revived interest in phylogenetics in its now most popular guise of *cladistics*. It includes certain elements of positivist rationalism, so its ontic background is shaped by a simplified version of phylogeny as *cladogenesis* (Hennig 1966; Eldredge and Cracraft 1980; Wiley 1981; de Queiroz and Gauthier 1992). Accordingly, the hierarchical structure of biodiversity is interpreted as *phylogenetic pattern*; its definition is based entirely on the time dimension of evolution and discards all other aspects of the latter. This entails an appeal for the elimination of the "Linnaean" ranked hierarchy [Ereshefsky 2001], which implicitly revives an old scholastic idea of rankless hierarchy in a new format [Pavlinov 2005, 2018, 2021b].

Consideration of this hierarchy at its lower levels yields a *phylogenetic* (or rather genealogical) conception of species [de Queiroz and Donoghue 1988; de Queiroz 1998, 1999], which identifies it with a certain fragment of a phyletic (generative) chain designated as *phylospecies*. It is proposed "that existence as a separately evolving metapopulation lineage be considered the only necessary property of species" [de Queiroz 2005b: 1263]. This author is sure that "all of [species conceptions] either explicitly or implicitly equate species with separately evolving (segments of) metapopulation lineages" [*loc.cit.*], so a *general lineage* conception is suggested as having a universal biological application [de Queiroz 1998, 1999, 2005a, 2007; Miller 2006, 2016; Staley 2006, 2009a,b, 2013]. Thus, this conception, just like the one developed by biosystematics, presumes a "population" account of species compatible with the Darwinian microevolutionary model, so it may be thought of as a version of biosystematic multilevel hierarchy of population units regarded in a rankless guise. The latter, in an extreme manifestation, yields rejection of the species notion together with the species rank uncertainty [Zachos 2016; Mishler 2021], which agrees with the positivists' declarations. However, if a traditional notion of species is preserved, particular treatments of phylospecies result in recognizing its various kinds, depending on the methods of "cutting" metapopulation lineages (see Section 5.4).

At the end of the 20th century, a contemporary "microphylogenetics", better known as *phylogeography* [Avise et al. 1987; Avise 2000, 2009; Abramson 2007], also contributes significantly to an erosion of the "around-species" hierarchical system. It presumes a rankless hierarchy of population units of different levels of generality that are uniformly identified as *phylogroups*. A "by-product" of this approach provides a conception of an *evolutionarily significant* population unit [Moritz 1994;

Mayden and Wood 1995; Riddle and Hafner 1999; Casacci et al. 2014]; its weakness is in its lacking the clear criteria of "evolutionary significance".

* * *

The understanding of species is developed in a considerably different way by certain approaches that partly continue the tradition of "Linnaean" systematics, but in a new conceptual framework, which is now evolutionary-genetic-ecological. One of them is advocated by a branch of biosystematics, whose proponents assert that "if systematics is to exist as a science it should at least define the species, perhaps the most fundamental of the concepts with which it deals" [Camp and Gilly 1943: 380]. Being conceptualists, these two botanists are sure that "the species is a unit, but it achieves this state only by a definition of the concept upon which its biological unity is based", and this unit is defined rather aphoristically as a "dynamic population within a genetic system" [*loc.cit.*]. In a more traditional manner appealing to generative idea, the biospecies "in most cases can be defined as separate self-reproducing units that objectively exist in nature" [Huxley 1940: 4]. This unit is suggested to call *biosystematic* species [Grant 1981]; to demarcate it from a "classical" one, the latter is suggested to designate as "*binom*" to highlight its basically nomenclatural meaning [Camp 1951].

A powerful support of a salient status of natural species in its new account is provided by arising synthetic theory of evolution (STE), according to which the formation of a new species is a critical ("nodal") final stage of microevolutionary processes; this new understanding is highlighted by designating the latter as *speciation* [Dobzhansky 1935, 1970; Huxley 1942; Simpson 1953; Vorontsov 1980; Mayr 1982, 1988; Cracraft 1983, 1987, 1989b; Stebbings 1987; Ghiselin 1997; Coyne and Orr 2004; Shaw 2012].[11] At this stage, the innovations acquired by populations during preceding microevolution get fixed: this determines the characteristic of species as a qualitatively distinct unit, from which anagenetic macroevolution begins [Futuyma 1987]. Th. Dobzhansky, one of the founders of STE, writes that "the fundamental importance of this stage is due to the fact that it is only the development of the isolating mechanisms that makes possible the coexistence in the same geographic area. [...] This, in turn, opens the possibility for the organisms dwelling together to become adapted to different places in the general economy of nature" [Dobzhansky 1935: 354]. It is noteworthy that the chapter of Dobzhansky's *Genetics and the Origin of Species*, which is dedicated to species, is called emphatically "Species as Natural Units" [Dobzhansky 1937]. In it, he asserts an important idea that as far as evolution is a real natural process, such understanding of species means its treatment as "a real and important phenomenon in nature" [*op.cit.*: 313]. This understanding of species becomes really "synthetic" and culminates the overall trend of biologization of the considered cognitive situation and complements its subjective component with *species thinking* [Richards 2010].

[11] However, this term is coined much earlier [Cook 1906].

The already mentioned conception of punctuated equilibrium, complementing the classical version of STE [Gould and Eldredge 1977; Pagel et al. 2006], allows resolving in a specific way a contradiction between the continuous nature of microevolution and the discrete nature of species, i.e., between dynamic and static aspects of its consideration [McCabe 2011].

Recall that, according to the Darwinian model, there are no qualitative differences between geographical race, incipient mesospecies, and euspecies that reach complete isolation. As shown above, this yields consideration of species as a fragment of a continuous and monotone genealogical lineage and is ultimately used as an argument against its reality (in the Lockean sense) [Darwin 1859; Bailey 1896; Starobogatov 1985; Casetta 2012]. J. Cracraft designates this contradiction as "the paradox of discrete entities", which is caused by a "microscale" view of microevolution: if you trace "in thought" a continuous lineage generation by generation, any gap in it can hardly be detected [Cracraft 1987, 1989a].

A solution proposed by the conception of punctuated equilibrium is based on a "macro-scale" view of Darwinian evolution. Accordingly, it is hypothesized that a transition from one species to another within a single phyletic lineage is "quantized": it proceeds in a much shorter period (suppose, several to tens of thousands of years) than the time during which euspecies exists in a quasi-stationary state (suppose, several million years). This means a possibility of drawing "macro-scale" boundaries between the consecutive "adjoining" species,[12] though they are not sharp. The occasionally hybridizing close allopatric species can be regarded similarly: their rare hybridization does not deny their isolation but indicates a fuzziness of their boundaries [Mayr 1968; O'Hara 1994; Abbott et al. 2008; Huang 2010]. In botany, a vague delineation of species depends partly on different modes of speciation [Rieseberg et al. 2006], and sometimes this serves as a basis for the denial of species units [Levin 1979; Iwatsuki et al. 1986]. In general, according to this viewpoint, biospecies should be interpreted as a quasi-discrete unit with blurred vertical and horizontal boundaries [Van Regenmortel 1997; Hey 2001a; Sites and Marshall 2004; Hanage et al. 2005; González-Forero 2008; Hanage 2013; Pavlinov 2013a, 2017, 2018, 2022a; Minelli 2015; Zachos 2016, 2018]. It is acknowledged consequently "that completely nonarbitrary species delimitation is impossible" [Zachos 2018: 811].

Such interpretation contradicts one of the basic conditions of binary logic that underlies the classical systematics inherited from scholasticism, which presumes discreteness of classification units; it is sometimes called *xenotaxonomy* (Greek ξένος, alien) [McCabe 2008]. But this interpretation is fully compatible with the modern ideas that the natural complexly structured objects most usually have no discrete boundaries [Kosko 1993]. The description of such "vazzily" viewed world is formalized by the aforementioned fuzzy logic [Zadeh 1992; Kosko 1993].

A rather complex configuration of species is developed within the framework of STE in the form of *evolutionary species* conception [Simpson 1951, 1961; Wiley 1978, 1980; Szalay 1993; Wiley and Mayden 2000]. The latter's author, the zoologist George Gaylord Simpson (1902–1984) defines species as "a phyletic lineage

[12] They are commonly referred to as *chronospecies* [Simpson 1961; Stanley 1978; Dzik 2005].

(ancestral-descendent sequence of interbreeding populations) evolving independently of others, with its own separate and unitary evolutionary role and tendencies" [Simpson 1951: 289]. His colleague Leigh Van Valen (1935–2010) reinforces the ecological aspect of this conception by defining specifics of the evolutionary role of species with reference to specifics of its ecological niche [Van Valen 1976]. Due to this, this conception turns out to be one of the most "multidimensional" by combining evolutionary, ecological, and partly genetic ideas [Wiley and Mayden 2000].

Acknowledgement of the evolutionary significance of species inevitably entails recognition of its fundamentality and salient status as a classification unit in the taxonomic theories that are conceptually based on STE. However, the same holds true for those theories that continue early postscholastic tradition and reject a close connection between evolution and systematics (e.g., [Blackwelder and Boyden 1951; Borgmeier 1957]). Therefore, almost all taxonomic manuals of the second half of the 20th century discuss the issues related to species. In this regard, E. Mayr's "conceptual evolution" is indicative: at the beginning of his scientific career, he rejects the significance of species (see above), but then changes his viewpoint to an opposite [Beurton 2002; Chung 2003; de Queiroz 2005c]. This is clearly reflected in his later definition of systematics as "a science of species" [Mayr 1969, 1982], and his active defense of reproductive (biological) species conception becomes one of the main topics of his publications [Mayr 1982, 1988, 1996, 1997, 2000].

Such development of the species problematics in the first half and mid-20th century results in the acknowledgement of biospecies as a complexly structured system, divided into population units of different levels of generality under the influence of various factors. It becomes very significant to recognize that the differentiation of species is ordered, so it is interpretable as a specifically organized natural polymorphic system [Vavilov 1931; Huxley 1940; Mayr 1942, 1957a, 1959a, 1963; Grant 1957; Dobzhansky 1970]. On this basis, a conception of broad *polytypic* species is adopted as an alternative to a conception of narrow *monotypic* microspecies: according to the former, the particular intraspecies units (races, subspecies, etc.) are regarded as the "forms of existence" of the widely distributed species in different regions [Schwartz 1959, 1980; Zavadsky 1961, 1968; Mayr 1963, 1976, 1988]. Such consideration of species and its structure is supported in synecology by a conception of fundamental niche subdivided into local realized ones [Odum 1953; Chase and Leibold 2003; Mirkin and Naumova 2005]; it is also characteristic of some theoretical models of biodiversity (e.g., [Barraclough 2019]).

Accentuation of population differentiation of the widespread species stimulates detailed explorations of its geographical variation, which reveal a phenomenon of continuous spatial variation called *clinal* [Arnoldi 1939; Huxley 1939, 1942]. This yields a contradiction between the continuous nature of this kind of intraspecies diversity and a traditional means of its description with discrete classification units (subspecies, etc.). This contradiction gives rise to a *subspecies problem* [Huxley 1940, 1942; Starrett 1958], which in fact should be called *subspecies question*, in many respects analogous to the above-considered species question [Clausen 1941; Weatherby 1942; Meikle 1957]. Its radical theoretical solution, which is opposite to the basics of biosystematics, removes species rank uncertainty by the rejection of any intraspecies classification units and acknowledging "Linnaean" species as a finite (indivisible) taxon [Wilson and Brown 1953; Terentyev 1968; Malikov and Golenishchev 2016; Burbrink et al. 2022].

5.4 "MULTIPLICATION OF ESSENCES": THE KINDS OF SPECIES

The cognitive situation, with the general species notion as its focal element, has always been and remains problematic. The latter means that a kind of "envelop" is built around this notion, designated as the *species problem* [Stamos 2003; Amitani 2010; Richards 2010; Chambers 2012; Pavlinov 2013a, 2017, 2018, 2021a, 2022a; Bartlett 2015; Ereshefsky 2017; Wilkins et al. 2022]. As noted in Chapter 1, this problem is generally caused by the difficulties associated with an elaboration of species definition that would be uniform for all biological disciplines studying the diversity of organisms.

For the sake of fairness, it should be noted that the acknowledgement of relevance of the species problem is not unanimous. For example, D. Hull is sure that

> reference to the species problem today might sound quaint and vaguely anachronistic. Perhaps the species problem was of some importance ages ago in the philosophical dispute between nominalists and essentialists, or a century ago in biology when Darwin introduced his theory of organic evolution, but it certainly is of no contemporary interest.
>
> *[Hull 1977: 91; 1989: 79]*

Arising of the species problem in its more narrow contemporary understanding begins at the end of the 19th and early 20th centuries and results from a fundamental reformatting of the entire species problematics, which is caused by the shift from predominate classification to predominately natural account of species. The probably most significant hallmark of the latter is linking a general understanding of species with its natural and evolutionary history: this leads to an awareness that the species diversity of the living is structured by an array of biological mechanisms, which ratio varies among different groups of organisms. This new conceptualization yields an increasing heterogeneity of the ontic component of the cognitive situation caused by splitting the general species notion into particular conceptions, which means a "multiplication of essences" in the general understanding of species.

David Stamos, when evaluating the species problem of the second half of the 19th century, lists the following principal interpretations of species ([Stamos 2003], with some changes): it is (a) a product of the cognizing mind (nominalism), (b) a result of divine creation (creationist realism), (c) characterized by essences (essentialist realism), (d) a reproductive community (biological realism), and (e) a final stage of the divergence of races (evolutionary realism). In the first half of the 20th century, recognition of different *kinds* (sorts) *of species* described by different conceptions becomes a significant part of the overall problematics [Du Rietz 1930; Dobzhansky 1937; Faegri 1937; Turrill 1938; Huxley 1942; Cain 1954]. A multiplicity of interpretations of species ontology subsequently contributes to this "multiplication of essences" (see Section 6.2).

An awareness of the meaningfulness of distinguishing different kinds of species originates in the mid-18th century, when it is suggested to define differently the species with sexual and asexual reproduction (M. Adanson). An idea of recognition of those kinds is expressed explicitly for the first time by the evolutionary biologist George-John Romanes (1848–1894) to highlight the diversity of the ways of evolutionary origins of the species [Romanes 1895]. Echoing Romanes, Henry Cowles writes about "nonhomology" of species in the groups of organisms with different biology [Cowles 1908], which are characterized (in our terminology) by different species integrators. Accentuating this point, William Turrill asserts that "no single absolute test for a species is yet known, and it is debatable if such is ever likely to be found" [Turrill 1925: 365]. E. Mayr, somewhat exaggerating this situation, notes "that there are probably as many species concepts as there are thinking systematists and students of speciation" [Mayr 1942: 115]. This multiplicity of particular accounts of species is posed as a principal content of the contemporary species problem in a book with the same name [Robson 1928; Hawkins 1935].[13]

One of the important aspects of the species problem thus understood is highlighted by Lancelot Hogben, who notes that "the word 'species' has no single meaning. Hence, there is no one problem of the origin of species. There are many problems of the *origins* of species" ([Hogben 1940: 269], italics added). Summarizing this idea, J. Huxley writes that "we must not expect a single or a simple basis for definition of [species], since species arise in many different ways" [Huxley 1942: 167]. As a result, an understanding of a certain association between the conceptions of species and speciation arises [Huxley 1942; Polyansky 1959; Scudder 1974; Cracraft 1983, 1989a; Starobogatov 1985; Chandler and Gromko 1989; Mallet 1995, 2008; Magnus 1996; Ghiselin 1997; Coyne and Orr 1998, 2010; Harrison 1998; Sluys and Hazevoet 1999; Levin 2000; Miller 2001; Noor 2002; Claridge 2010; Aldhebiani 2018; Nathan and Cracraft 2020].

> In the context of this "bundle", a *stochastic* species conception can be mentioned, in which the origin of species units is considered a stochastic process providing for the local "splash-outs" of the fluctuating intraspecific diversity on the species level [Pavlinov 1992, 2013a]. This results in emergence of *virtual* species that spontaneously arise and exist for but a short time. It can be assumed that the quasispecies of microorganisms (see below on them), as well as many cryptic species that do not have obvious ecological and physiological specificity, belong to this category. If this is true, then the virtual species are quite widespread in the living nature.

The main reason for the fragmentation of species conceptions is an emphasis on different hypostases attributed to the natural species, which are evolutionary, ecological, genetic, behavioral, etc. It turns out that the biological characteristics implied by these hypostases are so different and in a sense mutually "orthogonal" that the species conceptions based on them appear irreducible to each other. Therefore,

[13] Regarding the history of the notion of "species problem", it appears as early as 1922 in the article of G. Turesson [Turesson 1922]. Paleontologist Warren Allmon dates its appearance to 1924 in an article by Arthur Trueman [Allmon 2016]. However, the latter writes about "species question and the problem of palaeontological nomenclature" [Trueman 1924: 355], i.e., he does not mean the actual "species problem" in its theoretical account.

recognition of the kinds of species can be thought of in a more general sense as their consideration within different "species paradigms" [Morgun 2002, 2005]. In general, this means recognition of a certain *nonequivalence*, or *heterogeneity* of the species units in different groups of organisms [Kuprevich 1949, 1950; Polyansky 1956; Zavadsky 1961, 1968; Nikolskiy 1962; Wilkins 2003a; Nathan and Cracraft 2020]. The abovementioned idea of H. Cowles of "homologization" of the species based on the similarity of their reproduction systems receives an important refinement: their similarity in this regard is called *homotypy*, and dissimilarity is called *heterotypy* [Manceau and Lambert 2019]. Indeed, the major groups (prokaryotes and eukaryotes, plants and animals, etc.) differ in biological features that structure their species diversity; therefore, they are characterized, according to the figurative expression of John Wilkins, by "various ways of being a species (species *modes*)" ([Wilkins 2003b: 621], italics added). Perhaps it can be argued that this idea, which is crucial for the understanding of the nature of biological species, becomes by the middle of the 20th century a kind of "accepted view". So subsequently, it is not the species heterogeneity that is discussed, upon which almost all researchers agree, but how and why it manifests itself in different groups of organisms (see Section 6.3).

The development of the species problem in this new understanding yields a multiplication of criteria, by which different kinds of species can be distinguished. An example is the following one-level system of grouping species conceptions on different grounds: taxospecies, biospecies, genospecies, agamospecies, paleospecies, and morphospecies [Cain 1954; Grant 1981]. Obviously, this segregation is logically inconsistent, as it is framed by an incompatible bases of division [Zavadsky 1961, 1968; Mayr 1969]. E. Mayr, in the works just cited, suggests other basic "concepts", namely, typological, nondimensional, and biological; as can be seen, his species "kinding" system bears the same logical error as Cain's.

One of the important consequences of the recognition of different kinds of species based on their breeding systems is a suggestion to regard "real species" only those population units that correspond to Buffon's two-way reproductive criterion, whereas a special terminology should be applied to other units that are not such "species" [Dobzhansky 1937; Faegri 1937; Turrill 1938; Huxley 1940, 1942; Burma 1954; Polyansky 1959; Doolittle and Papke 2006; Doolittle 2018]. These "other units" are especially relevant in the case of prokaryotes, parasitic organisms, lower fungi, etc., with their very particular natural histories. At first, specifics of their "around-species" diversity does not receive due attention [Khokhryakov 1955; Golovin 1958; Parmasto 1986b; Cohan 2002; Domingo 2002; de Meeûs et al. 2003], but later even philosophers get interested in this issue [Wilkins 2006a; Ereshefsky 2010b].

In an extreme case, the entire diversity of organisms with asexual reproduction is counted possible to treat as "*nonspecies*" ([Khokhlov 1946], cited after [Polyansky 1959]), while the diversity of viruses is treated as "*pre-species*" [Kuprevich 1949, 1950].[14] With the development of this idea, the units with the amphimictic

[14] This extreme standpoint differs significantly from the position dating back to antique natural philosophy, according to which every (any) organism necessarily belongs to a certain species [Zavadsky 1961, 1968; Volkova and Filyukov 1966; Wiley 1981].

reproduction system are suggested to be designated *euspecies*, while other "around-species" units become *quasi-* or *pseudospecies* [Dobzhansky 1970; Eigen 1983; Van Regenmortel 1997; Domingo 2002; Stamos 2003; Wilkins 2006a; Stich et al. 2007; Ereshefsky 2010b; Andino and Domingo 2015; Pavlinov 2018, 2022a]. The other terminological variants are proposals to divide such units into *population* and *clonal* [Takhtajyan 1955] or *Mendelian* and *non-Mendelian* categories [Echelle 1990]. However, an opposite opinion is that the traditional species terminology is quite applicable in all these groups (e.g., [Kursanov 1945; Polyansky 1956, 1958; Bobay and Ochman 2017, 2018]).

* * *

The development of the species problem in this vein in the second half of the 20th century yields a significant increase in the number of species conceptions. This gives born to an impression that "there is no more popular pastime in the literature on the philosophy of biology than analyzing the concept of species" [Caplan 1980: 71], and that they "established a minor industry devoted to the production of new definitions for the term species" [de Queiroz 1998: 57]. This multiplicity of particular species conceptions becomes known as *species pluralism*, and it follows from acknowledging that the structure of biota is so complex and multifaceted that it is not amendable to a single biologically meaningful species "superconcept" [Mishler and Donoghue 1982; Reig 1982; Kitcher 1984a; Otte and Endler 1989; de Queiroz and Donoghue 1990; Ereshefsky 1992b, 1998, 2001, 2017; Stanford 1995; Dupré 1993, 1999; Hull 1999; Reydon 2004, 2005a; Pavlinov 2007, 2009, 2013a, 2021a; Amitani 2010, 2017; Slater 2013; Bzovy 2016; Zachos 2016; Nathan 2017; Talpsepp 2013; Villuendas 2019). Such a situation with the current problem, in which general understanding of species is distorted in many particular conceptions, fully deserves to be assessed as another "species crisis".

As noted in Chapter 1, one of the main reasons for this plurality is the consideration of different issues when exploring species diversity: some of them concern the essential properties and ontology of species as a natural phenomenon, and others involve mechanisms of species arising and sustainable persistence, more others concern the ways of delineating and ranking the particular species. Currently, the total number of such conceptions reaches almost three dozen; they differ in the levels of theoretical load and generality, methods of defining species units, criteria for their distinguishing, etc.; each of them has certain specific advantages and disadvantages, and all are more or less restricted in their applications. Their comparative analysis is conducted in a number of studies [Häuser 1987; Williams 1992; Mayden 1997, 1999; Sterelny and Griffiths 1999; Mallet 2001; Ereshefsky 2001, 2017; Wilkins 2002, 2003a, 2006b, 2011; Pavlinov and Lyubarsky 2011; Kunz 2012; Zachos 2016; Pušić et al. 2017; Pavlinov 2018], they are briefly overviewed below.

It is important to distinguish between two main forms of species uncertainty, *ontic* (metaphysical) and *epistemic* (cognitive) [Richards 2008, 2010; Slater 2013; Bzovy 2016, Pavlinov 2018, 2022a]; their meaning is evident from the structure of the cognitive situation described in Chapter 1. In the first case, we imply essentially different aspects of the diversity of organisms, to which particular species conceptions correspond; this aspect-dependent consideration of biodiversity is supported by conceptualism.[15] This also includes the multiplicity of interpretations of the ontic status of species units; this is a *status* pluralism [Slater 2013]. In the second case, it means that, whatever the ontology of the studied biological units, the resolving capabilities of any single species conception, as a heuristics, are too limited to embrace its entire multidimensional structure. It is also reasonable to distinguish between *conceptual* and *empirical* forms of species uncertainty: the first concerns different ways of defining species and the second refers to different ways of elaborating species classifications; Matthew Slater designates them as *category* and *taxic* pluralism, respectively [Slater 2013].

According to a certain "accepted view", a pluralistic position presumes that "anything goes", so delineation of species under such circumstance is purely conventional, and this means that there are no species in Nature at all [Casetta 2010, 2012]; however, this interpretation is untenable [Kitcher 1984a,b; Boyd 1999; Pavlinov 2018, 2021b]. The classical dichotomy "realism *vs.* nominalism", rooted in antique and scholastic natural philosophy, proceeds from acknowledging that, provided that the objective world is the only one, every natural entity exists and can be individuated only in a single way. Accordingly, if this cannot be done, such an entity cannot be thought of as objectively existing (real, natural), i.e., it is subjective (nominal). But modern conceptualism implies that in a cognitive situation, the species units do not appear "by themselves" as given in a certain unique way, but are "grasped" from the overall structure of an objectively existing and complexly organized multidimensional diversity by means of particular conceptions [Sloan 2002; Brigandt 2003a; Pavlinov 2009, 2013a, 2021a; Bartlett 2015] (see Section 1.1.1). This standpoint is based on an assumption "that the theoretical perspectives we adopt correspond to real levels in nature but that nothing in the structure of nature picks out one level as privileged" [Kitcher 1987: 190] (I would say here about "aspects" instead of "levels"). Some philosophers designate such standpoint as a *promiscuous* or *pluralistic realism* [Dupré 1981; Kitcher 1984a,b; Wilson R. 1996].

Thus, it is crucial to emphasize here that this cognitive position rejects neither the real existence of essentially different "around-species" units nor the possibility of their generalization by some single notion [Polyansky 1956, 1958; Sluys 1991; Brigandt 2003a; Barker 2019b; Pavlinov 2022a]. In this, a conceptualist pluralistic account of species differs fundamentally from a nominalistic idea of "biology without species" advocated by positivists [Burma 1954; Davidson 1954; Kober 2008]. From a cognitive standpoint, species pluralism is indispensable for its aiming the researchers at identifying and careful studying those particular species integrators that are responsible for their different "modes of existence" [Grantham 1993; Pavlinov 2022a]. This standpoint focuses on an intentional nature of explorations

[15] D. Stamos regards conceptualistic justification of species pluralism far-fetched: he reduces the whole species problem to a confrontation between realism and nominalism and treats pluralism as one of the latter's manifestations [Stamos 2003].

of the real multidimensional diversity of organisms: different species conceptions are introduced to capture particular manifestations of this diversity, significant from certain cognitive positions.[16]

As noted above, one of the extreme variants of a pluralistic view of the species problem is a suggestion to refuse the species notion in favor of a new terminology. As J. Huxley believes, "the existence of different kinds of species [...] makes us sometimes wonder whether the term itself should not be abandoned in favor of several new terms, each with a more precise connotation" [Huxley 1942: 156]. M. Ereshefsky embodies this idea in the so-called *eliminative pluralism* ([Ereshefsky 1992b, 1998, 1999, 2001]; also [Reydon, 2005b; Nathan 2017; Barker 2019b]). In this connection, the famous parable about blind people, groping an elephant and describing it in different ways, is mentioned: "much of the so-called problem results from the fact that there isn't just *one* elephant and that they aren't all even *elephants*" ([Winston 1999: 43–44], italics in origin; also [Brooks and McLennan 1999]). Agreeing upon this viewpoint, several philosophical biologists and biological philosophers note that the use of a single notion for the fundamentally (ontologically) different natural entities masks their biologically significant differences and should be abolished [Szalay and Bock 1991; Szalay 1993; Ghiselin 1997, 2002; McOuat 2001; Rapini 2004; Reydon 2005a,b]. Realizing this idea, Thomas Reydon introduces the following terms corresponding to different ontologies of different hypostases of "as if species" units: *evolveron*, *phylon*, *organism-kind*, and *evolveron-kind* [Reydon 2005a: 45–46).[17] In a more moderate guise, some "neutral" species-based terms are invented that are not loaded with a traditional unified understanding: *species aggregate* (Heywood, 1963), *eidological unit* [Kamelin 2004], and *specion* [Dubois 2011]; the notion of "around-species" unit can also be mentioned in this regard.

> It is pertinent to note that species uncertainty is not unique: in that overall cognitive situation, in which the diversity of organisms is explored, another case of a pronounced conceptual pluralism is the notion of homology that cannot be defined univocally [Brigandt 2003b; Jamniczky 2005; Kleisner 2007; Pavlinov 2011, 2018; Minelli and Fusco 2013; Minelli 2016]. Their "parallelism" is explained formally by the fact that delineation of species and homologues is based on an equivalent classification procedure: the difference is that the former belongs to taxonomy, and the latter to partonomy [Jardine 1967; Nelson 1994; Brigandt and Griffiths 2007; Pavlinov 2012, 2018].

The pluralistic position is opposed by *species monism*, which in its most general form implies "acknowledging the fundamental unity of contemporary ideas about species" [de Queiroz 1999: 74]. As noted above, this implied unity serves as something like an informal disciplinary matrix (in the sense of [Kuhn 1962; Boyd 1999], see Section 1.1.1), which is shaped by a species thinking style (in the sense of [Richards 2010]). For it, it is possible to specify the same aspects as for pluralism, of which the most fundamental are ontic and epistemic. In this case, it is

16 It is to be reminded that intentionality is initially inherent in the conceptualistic cognitive standpoint laid down in the Middle Ages (Abelard etc., see Chapter 3).

17 The term "phylon" was proposed by E. Haeckel to designate the monophyletic group [Haeckel 1874].

acknowledged that the diversity of the living is unified in its basic causes and structure (ontology) and therefore is susceptible to being represented by a unified species concept (epistemology).

This monistic account is a legacy of those previous stages of the conceptual history of the species problematics, during which the natural-philosophical idea of species as a fundamental unit of the System of Nature was persisting in one form or another. Accordingly to it, the question whether natural species exists or not should have strictly one, preferably affirmative undoubted answer [Dupré 1999]. This corresponds to a monistic ideal of classical science, which expresses the striving for a certain final knowledge expressed by a single "formula" [Weinberg 1992]. It is rooted in a primordial faith in a single cause of the being, be it Platonic the One, Arisotelean Prime Mover, biblical Divine Mind, or else. An underlying striving for monism manifests itself in a suggestion that although "an absolute definition of the term 'species' satisfying the condition of logic is impossible, its definition should be applicable equally to organisms with sexual and asexual reproduction" [Paramonov 1951: 147]. As Samuel Bartlett expresses this standpoint, the "species problem [is] in need of an actual solution that is at once definitive, provable, and conclusive" [Bartlett 2015: 4], and this solution should provide a kind of *ultimate* species concept [Miller 2016]. The search for the latter is sometimes compared to the search for a unified field theory in physics [Smith 1990] or (in a somewhat humorous form) for The Holy Grail, and such "holy grailing" (as Mark Twain would have called it) looks as fascinating as it is fruitless [Wilson 1992; Cracraft 1997; Hull 1997].

The *monism-pluralism dilemma*, taking the form of a moderate monistic account, is expressed by K. Zavadsky as follows: "species as a phenomenon of nature [...] represents a universal form of organization of the living, but depending on various causes it manifests itself very differently" [Zavadsky 1961: 101]. Similarly, J. Wilkins believes that "there is *one* species concept. There are *two explanations* of why real species are species [...]: ecological adaptation and reproductive reach. There are *seven* distinct definitions of 'species', and *27* current variations and mixtures" ([Wilkins 2011: 59], italics in original). The pluralist philosopher P. Kitcher believes that we can think of the species concept as being a union of overlapping more particular species conceptions [Kitcher 1984a]. M. Ereshefsky paradoxically counts it possible to regard "eliminative pluralism as just a complicated form of monism" [Ereshefsky 1992b: 688]: this is because he excludes from consideration the pluralistically interpreted general species notion and replaces it with several monistically treated particular ones.

An antirealistic kind of monism suggests its own radical solution to the species problem by abandoning the general species notion: indeed, if there is no species, there is no species problem altogether [Kober 2008; Mishler 2010, 2021]. However, this is not a solution to the problem, but an evasion of it, which is similar to a curious childish game: if you just close your eyes, the world around you disappears.

* * *

The recent development of the species problem, facing the monism-pluralism dilemma and a seeming impossibility of elaboration of a unified "ultimate" conception, concentrates on the search for possible ways to decrease conceptual uncertainty. Accordingly, the cognitive situation is customized by a pragmatic idea of desirability to declare some conception preferable according to certain criteria (scientific consistency, biological meaningfulness, operationality, etc.), and to treat all others as secondary and auxiliary [Ereshefsky 1992b, 1998, 2001, 2017; Dupré 1993, 1999; Szalay 1993; Stanford 1995; Hull 1997, 1999; Amitani 2010, 2017]. Such dropout yields exclusion of the insignificant conceptions from consideration and thereby leads to a "reasonably controlled" species pluralism [Hull 1999; Minelli 2020]. P. Sloan highlights this standpoint by asserting that many particular species conceptions are "artifacts of a complex conceptual history" [Sloan 2002: 247]; therefore, from the evolutionary epistemology viewpoint, a "culling" of such subsidiary conceptions is a normal course of the conceptual history of the species problem [Hull 1999; Ereshefsky 2001]. Thus, the balance of species pluralism *vs.* monism becomes in many ways similar to what is known in systematics as a *moderate pluralism*: it is possible to develop a variety of classifications, but only one of them is a priority to serve as a *general reference system* [Gilmour 1940, 1961; Sokal and Sneath 1963; Hennig 1966]. However, different background theories give priority to different "scales of priorities" of species conceptions, so the very basic conceptual uncertainty is preserved, which is evident from the following.

In the context of ontology, the significance of species conceptions is associated with the structure of biodiversity: it is estimated according to an adequacy of particular conceptions to certain fixed aspects of this diversity [Pavlinov 2013a, 2022a]. If an emphasis is on species integrators, a priority can be given to the cohesion conception, which considers them complexly [Templeton 1989; Wu, 2001; Achtman and Wagner 2008; Barker and Wilson 2010; Hausdorf, 2011; Bobay and Ochmana, 2018; Barker 2019a,b]. D. Stamos believes that the "horizontal dimension has priority (pragmatic, epistemological, logical, and ontological) over the vertical dimension" [Stamos 2002: 171]. According to an opposite view, supported by phylogenetic theory, this is the vertical dimension that has a priority [Løvtrup 1979; Nixon and Wheeler 1990; Sterelny 1994; de Queiroz 1998, 1999, 2005a, 2007; Miller 2006, 2016; Staley 2006, 2009a,b, 2013].

The phylogenetic conception ("vertical" species) is considered a priority because (a) phylogeny is a universal process producing the entire diversity of the living, and therefore (b) all other conceptions presumably imply somehow interpretation of species as a metapopulation lineage [de Queiroz 2005a]. Based on this, Queiroz believes that "the species problem has, for the most part, already been solved [since] all are encompassed by a single, general concept that equates species with segments of population-level lineage" [de Queiroz 1999: 49]. However, an opinion that ecological and reproductive conceptions are the parts of phylogenetic one [Ridley 1989] seems erroneous: they all belong to the same level of generality and only partially overlap (see below). Therefore, a purely "vertical" account of species does not allow for description of the "horizontal" structure of ecosystems, for which ecological species conception is designed [Van Valen 1976; Schwartz 1980; Andersson 1990; Vrba 1995; Khlebosolov 2003; Frankham et al. 2012]. So the Queiroz' argument is true to the extent that the biological diversity is reduced to a phylogenetic pattern [Eldredge and Cracraft 1980; Eldredge and Salthe 1984; Pavlinov 2005; Velasco 2008], which is certainly an oversimplification.

In the context of epistemology, a segregation of conceptions can be justified by reference to their fitting certain criteria of scientific knowledge [Ereshefsky 1992b, 1999, 2001; Hull 1997, 1999]; accordingly, "in deciding which species concepts to take seriously, we seem inextricably caught up in the issue of what counts as genuine science and what not" [Hull 1999: 43]. However, a fundamental question arises, what the "genuine science" is: it is answered differently by particular scientific philosophies, so this issue does not have an obvious unequivocal solution. For example, within the framework of the positivist philosophy of science, a priority conception should be neutral with respect to any ontological considerations: the better it is, (a) the more effectively it solves particular cognitive and applied tasks, and (b) the wider the range of such tasks [Gilmour 1940, 1961; Sonneborn 1957; Sokal and Sneath, 1963; Rojas 1992; Stanford 1995; Bachmann 1998; Sarkar and Margules 2001; Mace 2004; Sarkar 2005]. From this viewpoint, phenetic conception is preferable, which operates with the similarity as such [Sneath and Sokal 1973; Pigliucci 2003]; however, this is not true from a conceptualistic standpoint, according to which similarity is "substantively empty" if regarded out of a certain metaphysical context [Tversky 1977; Sober 1984; Dupré 1993; Murphy 2002; Rieppel and Kearney, 2002; Bartlett 2015; Pavlinov 2018, 2021b]. R. Richards connects an assessment of the scientific significance of species conceptions with the "division of conceptual labor solution": a certain priority conception is to be fixed, while others are linked to it by performing a function of "correspondence rules" (in the sense of R. Carnap), and according to the *principle of proliferation*, the more of such "rules" can be associated with a priority conception, the more it preferable [Richards 2008, 2010].

Seeing in a pragmatic context, the main problem is that different conceptions imply different criteria for the practical delineation of species and, accordingly, generate different species classifications for the same groups of organisms. This means that a certain organism can be allocated to different species according to different characteristics presumed by particular conceptions.[18] As a result, *empirical species uncertainty* arises [Hey et al. 2003; Isaac and Purvis 2004; Pavlinov 2017, 2018; Maxwell et al. 2020], which is a particular case of *taxonomic uncertainty* [Sneath and Sokal 1973; Zarenkov 1988; Pavlinov 2018] or *taxic pluralism* [Slater 2013]. Meanwhile, classical systematics, based on binary logic, demands discreteness of classification units, which means that each organism may belong to only one "true" species. Expressing this monistic position, R. Richards writes that "efforts to preserve biodiversity require that we get our understanding of biodiversity *right.* That requires we get species groupings and counts *right*" ([Richards 2010: 206], italics added).

[18] It is worth noting that this form of pluralism is considered quite acceptable when classifying the life forms [Du Rietz 1931; Remane 1943].

Within the framework of positivistically oriented approaches, the main way to eliminate any species uncertainty from the cognitive situation is to reduce the species problematics to the practical tasks of distinguishing species units on a unified empirical basis [Peterson and Navarro-Siguenza 1999; Sterelny 1999; Agapow et al. 2004; Ward and Stanley 2004; Balakrishnan 2005; de Queiroz 2005c; Yang and Zhou 2010; Wilkins 2018, 2022; Padial and De la Riva 2020]. Such units, called *parataxonomic*, are delineated to describe the structure of species diversity in the first approximation without its conceptual justification [Krell 2004; Ward and Stanley 2004]. One of the possible options is based on an idea of *conventionalism* developed by scholasticism, justifying recognition of the "species by agreement" [Mina 2007, 2010]. Such idea can be implemented by a kind of "species convention" based on an adoption of a unified protocol for describing species diversity, including the selection of appropriate diagnostic traits [Nazarenko 2001; Kunz 2002; Barker and Velasco 2013]; an example is the *species rank standard* approved by the British Ornithological Union [Helbig et al. 2002]. The respective pragmatically substantiated priority conception can perform an "umbrella" function [Reydon 2005a,b; Kendig 2014; Amitani 2015, 2017] to make species classifications, elaborated on different conceptual bases, phenomenologically comparable [Sterelny 1999; Pavlinov 2022a]. This proposal is supported by a computer simulation of species classifications based on different evolutionary models [Kampis et al. 2007].

However, the just above reference to the "appropriate traits" means that any such decisions could be valid only for particular groups of organisms. Thus, this approach does not eliminate uncertainty altogether, but only within certain major taxa. With this, limits of each of the latter seem to be specifiable only empirically based on a trial and error approach; this is clearly illustrated by the recent search for the standard DNA markers known as *DNA barcode* [Hebert et al. 2003; Blaxter et al. 2005; Savolainen et al. 2005; Puillandre et al. 2012; Seifert 2014, 2020; Sigwart and Garbett 2018].

The preceding consideration of monism-pluralism dilemma concerns the species unit; it is equally relevant to the species rank [Ereshefsky 1999; Slater 2013]. A "moderate" monism presumes the latter's realistic account: there is a certain level of generality in the overall structure of biodiversity, which is traditionally associated with the species rank, though it is fuzzily defined and is manifested differently in different groups of organisms [Pavlinov 2009a, 2017, 2022a; Huang 2010; Wilkins 2011; Lyubarsky 2018]. An "absolute" pluralism implies the nominalistic interpretation of species rank: there is no such universal level of generality in the structure of biodiversity, so it is evidently subjective and conventional [Stanford 1995]. This nominalistic standpoint is characteristic of the abovementioned phenetic and phylogenetic treatments of species, with their abandoning any traditional ranking system.

A pragmatic standpoint presumes the possibility to allocate all the species units, regardless of their particular conceptualizations, to the same universal species category to make them comparable [Richards 2010; Bzovy 2016]. The latter can be thought of as corresponding to a certain "ideal" (even if imaginary) species conception, which is considered scientifically and biologically consistent as regards certain onto-epistemic criteria [Pavlinov 2017, 2022a] (see Section 6.3). It can also be a kind of omnispective conception, such as "combined" or the abovementioned "umbrella" one [Faegri 1937; Collier 1998; Amitani 2017].[19]

* * *

The conceptual splitting of the general species notion, characteristic of the most recent development of the species problem, impels the scientists to arranging the "flock of species conceptions" [Sterelny and Griffiths 1999; Sigwart 2018] into some systems to navigate their diversity. Since such arranging can be carried out according to different principles and grounds, it turns out to be multidimensional, i.e., pluralistic in itself [Nathan 2017; Holmes 2000; Bzovy 2016; Pavlinov 2018].

According to the various onto-epistemic grounds, species conceptions are divided into the following general groups (kinds), which overlap to some extent: aspect-based and object-based ([Pavlinov 2011, 2018, 2021b], see Section 1.1.1), ontic and epistemic [Reydon 2004, 2005a,b, 2022; Bzovy 2016; Reydon and Kunz, 2019], theoretical (natural, causal, functional) and descriptive (formal, classification, operational) [Hull 1997; Rheins 2011; Kunz 2012; Slater 2013; Pavlinov 2017, 2018; Barker 2022], static and dynamic [Dobzhansky 1935; Levin 1979; Cracraft 1983, 1987; Mallet 2001], synchronic (ecological, "horizontal") and diachronic (historical, "vertical") [Stamos 1998, 2002; Sluys and Hazevoet 1999; Lee and Wolsan 2002; Zachos 2016], procedural (including historical) and structural ("mechanistic") [Kitcher 1984a; Luckow 1995; Levin 2000; Stamos 2003]. To this could be added segregation of the conceptions of species according to the latter's implied ontic status as natural kind, quasi-individual, etc. (see Section 6.2).

Of the possible ways to "kinding" species conceptions from more specific, biologically sensible standpoints, the most significant are those, in which they are grouped according to the causes responsible for the very existence of the species units [Ereshefsky 1992b; Hull 1997, 1999; Mayden 1997, 1999; Bartlett 2015; Pavlinov 2017, 2018]. Of course, there are also several possible options here. For instance, K. Zavadsky follows "classical" recognition of two principal kinds of species based on their reproduction systems, apomictic and amphimictic, which provides a "rational typology of species" [Zavadsky 1961, 1968]. M. Ereshefsky identifies three such kinds of species, viz., interbreeding, ecological, and phylogenetic [Ereshefsky 1992b]. D. Hull divides the plethora of species conceptions into other three groups, based on similarity, evolutionary interpretation, and (for an unclear reason) that of W. Hennig [Hull 1997, 1999].

Below summarized is a variant of grouping species conceptions based on the latter's onto-epistemic consideration and accentuates on the presumed causes of commonality of "around-species" units [Mayden 1997, 1999; Coyne and Orr 2010; Pavlinov and Lyubarsky 2011; Zachos, 2016; Pavlinov 2017, 2018, 2022a]. These causes are the species integrators, which are given the greatest importance in the respective conceptions, and they constitute a *fundamentum divisionis* of this arrangement.

[19] This "umbrella" conception should not be confused with the notion of "umbrella species" used in biodiversity conservation projects [Roberge and Angelstam 2004; Ozaki et al. 2006].

They are of two principal types: (a) ontic integrators refer to the natural mechanisms structuring the diversity of organisms and (b) epistemic integrators refer to the cognitive means of describing this diversity. The first type includes conceptions based on historical, reproductive, ecological, and typological commonality; the second type includes conceptions based on similarity as such; these are phenetic, phenomenological, nominalistic, pragmatic, and classification conceptions.

Treating species as a **historical commonality by kinship relations** is rooted in the antique generative idea, which was accentuated by naturalists in the 17th–18th centuries and transformed into the evolutionary one in the19th century; it is currently being developed by phylogenetics as a *genealogical* or *phylogenetic* conception (*phylospecies*) [Mishler and Brandon 1987; Ridley 1989; de Queiroz and Donoghue 1990; Nixon and Wheeler 1990; Baum and Donoghue 1995; Baum and Shaw 1995; Wilson 1995; Horvath 1997a; de Queiroz 1999, 2005a, 2007; Wheeler 1999; de Pinna 2000; Meier and Willmann 2000; Wheeler and Platnick 2000; Staley 2009a; Shanker et al. 2017; Padial and De la Riva 2020; Mishler 2021]. Phylospecies is generally identified with a fragment of phyletic lineage, and its particular interpretations depend on the ways of "cutting" the latter: it is defined as a segment of the lineage between two cladistic events called *metaspecies* [Donoghue 1985; de Queiroz and Donoghue 1988, 1990; Kluge 1990; Kornet 1993; Frost and Kluge 1994; Lee 1995; Samadi and Barberousse 2006], as a strictly monophyletic crown group called *cladospecies* or *apospecies*, whereas a paraphyletic group is *plesiospecies* or *paraspecies* or *composite* species [Eldredge and Cracraft 1980; Olmstead 1995; Crisp and Chandler 1996; de Pinna 2000; Mishler and Theriot 2000; Kornet and McAllister 2005]. An understanding of phylospecies as a *genetic continuity* [Rollins 1965] yields a *phylogenomic* conception [Staley 2006, 2009a,b, 2013].

The phylogenetic conception in its strictly cladistic version presumes the rankless hierarchy of "around-species" units and nullifies traditional rank-based terminology [de Queiroz 1999, 2005a, 2007; Mishler 1999, 2021; Pleijel 1999; Hendry et al. 2000; Mishler and Wilkins 2018]. However, supporters of this version continue to use species notion when discussing the structure of biodiversity, with the only reservation that it be interpreted in terms of phylogenetic theory [Crowe 1999]. This notion is preserved in the *PhyloCode* to provide a certain continuity with "Linnaean" systematics [Cantino and de Queiroz 2020].

As noted above, the proponents of phylogenetic conception claim that it is universal and applicable to any organism to the extent that their species diversity is a consequence of divergent evolution [de Queiroz 1999, 2005a, 2007; Mishler 2021]. However, its cladistic formulation is based on a very narrow interpretation of phylogeny and is therefore too reductionist to work effectively as a heuristic outside the cladistic research program [Pavlinov 2022a]. For example, the conception of cladospecies is hardly applicable in the studies of ecological aspects of species diversity, since the traditionally delineated "around-species" units are mostly paraphyletic (e.g., [Crisp and Chandler 1996; Funk and Omland 2003; Ross 2014; Freudenstein et al. 2017].

A more biologically meaningful implementation of the idea of historical commonality is provided by *evolutionary species* conception [Simpson 1951, 1961; Wiley 1978, 1980; Mlikovsky and Zemek 1983; Brothers 1985; Szalay 1993; Frost and Kluge 1994; Wiley and Mayden 2000]. As said above, it unites phylogenetic, partly ecological, and partly genetic treatments of species, so it is considered "synthetic" [Ellis 2011].

Understanding of species as an **isolated reproductive commonality** is also rooted in the generative idea, but it does not refer to any historical fate; instead, it is based on the Buffon's two-way reproductive criterion. This includes primarily *reproductive* (biological in the narrow sense) conception of Dobzhansky–Mayr, whereas *recognition* conception, going back to L. Plate, focuses on the selective behavior of sexual partners [Paterson 1985; Masters et al. 1987; Coyne et al. 1988; Eldredge 1995; Friedmann 2007, 2013; Masters and Génin 2018]. An advantage of this conception is that it allows, at least potentially, for an experimental operationalization, which makes the inferences of reproductive isolation testable hypotheses [Dobzhansky 1935; Mayr 1988, 2000; Magnus 1996]. However, in reality, conducting needed experiments faces significant difficulties and is possible with a very limited number of species. With this, this conception, in its strict understanding, is applicable only to amphimictic organisms, and for this, its pretension for the "all-biological" meaning is strongly criticized [Donoghue 1985; Häuser 1987; Zink and Mckitrick 1995; Wang et al. 2020].

Focusing on the species gene pool makes it an *integrated gene pool* conception applicable to the organisms with any system of reproduction [Giray 1976; Bobay and Ochmana 2017]. This aspect of commonality is most comprehensively treated by *cohesion* conception, which tends to regard all biologically sound species integrators [Reig 1982; Brooks and Wiley 1986; Templeton 1989; Pigliucci 2003; Achtman and Wagner 2008; Bond and Stockman 2008; Barker and Wilson 2010; Neto 2016; Bobay and Ochmana 2018; Barker 2019a,b, 2022], though it is sometimes reduced to the ethological and/or genetic mechanisms [Templeton 1989; Barker and Wilson 2010; Grant 2011]. It is considered reasonable to combine reproductive and phylogenetic species accounts to describe different aspects of "around-species" diversity [McKitrick and Zink 1988].

Interpretation of species as an **ecological commonality** yields conceptions of *ecospecies* and *cenospecies*, also *functional* conception ([Khlebosolov 2003]; not in the sense of [Barker 2022]). The ecospecies appears in the studies on niche structure of ecosystems (e.g., [Odum 1953; MacArthur I965; Whittaker 1972; He and Legendre 2002]), which justifies considering mainly local "around-species" units in this regard [Mikhailov 2003]. A combined ecological and morphological (phenetic) interpretation of species results in its *ecomorphological* conception, which brings it close to biomorph (ecomorph), a unit of low level of generality recognized in synecology and ecomorphology [Pavlinov 2010, 2018]. It is regarded as one of the important structural elements of ecosystems [Levushkin 1976; Chernov 1991; Protasov 2021], and the identification of species with it is characteristic for a "naturalistic" view of wildlife [Panov 1993; Mikhailov 2003; Mina 2007, 2010]. It is most relevant in those groups of organisms whose biological organization and evolutionary modes are too complex to apply reproductive or phylogenetic conceptions in their "pure form" [Parmasto 1986a; Cohan 2002; Kunz 2002; de Meeûs et al. 2003; Pavlinov 2017].

To distinguish ecologically defined species from ecomorph, its general definition includes also a reference to its genetic commonality [Blackman 1995; Rakauskas 2003], i.e., it is a *genecological* species of Turesson (see Section 5.2.2), defined by *ecogenetic* conception [Levin 2000]. This group of conceptions marginally embraces an understanding of species through coadaptive fitness of its gene pool [Dobzhansky 1937, 1970; Mayr 1963, 1988, 1998, 2000; Mallet 2010b], and species thus understood is called *syngen* [Sonneborn 1957]; this is formalized by *genic* or *differential fitness* conception [Wu 2001; Hausdorf 2011].

Interpretation of species as a **typological commonality** includes essentialist (goes back to antique natural philosophy) and typological (Cuvier, etc.) conceptions; they are based on the commonality of organisms in their vital properties, be they functional essences or structural body plans. *Epigenetic* conception belongs to this group, which focuses on the commonality of specific epigenetic (developmental) programs [Mikhailov 2003; Kendig 2010; McCabe 2011];[20] it is particularly relevant in the organisms with complex life cycles [Kursanov 1945; Polyansky 1956; Golovin 1958]. The epigenetic conception generally corresponds to the tenets of ontogenetic systematics, though the latter pays more attention to supraspecific taxa [Pavlinov 2021b]. The conceptions of this group are generally proposed to designate as *intrinsic* [Crane 2004], and the integrators implied by them are considered by *intrinsic biological essentialism* [Devitt 2008, 2018a].

An understanding of species as a **commonality by similarity** appears in the works of scholastics (especially nominalists); it is grounded on a new natural-philosophical basis by early postscholastic systematics as a *phenetic idea* (Adanson, Jussieu, etc., see Section 4.3.4). Nowadays, the latter is justified by the reference to the positivist philosophy of science. Unlike typological understanding, this one is based on overall similarity of phenotypes and polythetic interpretation of classification units [Sokal and Sneath 1963; Panchen 1992; Pavlinov 2018]; it is the same as *family resemblance* of Wittgenstein [Beardsmore 1992], sometimes mentioned in systematics [Hull 1965; Sneath and Sokal 1973; Andersen 2002; Pigliucci 2003; Pigliucci and Kaplan 2006; Zachos 2016]. Conceptions of this group are considered epistemic-based because from a conceptualistic standpoint, the similarity as such is not an attribute of the objective reality but rather a mental construct, viz., an assessment judgment resulted from comparisons of things [Tversky 1977; Sober 1984; Rieppel and Kearney 2002; Pavlinov 2018, 2021b], though it is not rarely regarded as objective [Gilmour 1940; Simpson 1961; Sokal and Sneath 1963; Colless 1967; Meyen 1977; Lyubarsky 1996; Rheins 2011].

This group is "typified" by the *phenetic* conception based on the similarity of phenotypes; its application results in *phenospecies*, or *morphospecies*, or *phenon*, which is a metaphysically neutral rankless classification unit [Cain and Harrison 1958; Sokal and Sneath 1963; Sokal and Crovello 1970; Levin 1979; Boyd 1999; Krell 2004]. Of nearly the same methodological meaning is *genetic* conception as an overall similarity of genotypes (*genospecies*, *genotypic cluster*) [Eigen 1983; Masters and Spencer 1989; Mallet 1995; Baker and Bradley 2006; Hanage et al. 2005; Seifert 2014, 2020]. Recognition of such units constitutes the basis of various *operational* conceptions that have mostly a pragmatic meaning [Doyen and Slobodchikoff 1974; Sokal 1974; Pleijel and Rouse 2000; Sarkar and Margules 2001; Sites and Marshall 2004]. Phenetic idea is sometimes united with generative (phylogenetic) one to provide some "hybrid" conceptions, such as *minimally diagnosed monophyletic* unit [Cronquist 1978; Cracraft 1989b, 1997; Rosselló-Mora and Amann 2001] or *phylophenetic* unit (Rosselló-Mora and Amann 2001; Kämpfer and Rosselló-Mora 2004].

The simplest variant of commonality by similarity is a *phenomenological* understanding of species; in its contemporary application, it is based not on scholastic

[20] Ph. Kitcher deduces epigenetic concept from antique identification of *eidos* with *forma* [Kitcher 1984b].

(philosophical), but on its purely empirical account as a *salient phenomenal* object [Wilkins 2018, 2022]. This is a "species" in its commonplace sense, which is characteristic of folk systematics [Atran 1990, 1998] and is present in most of the practicing research; Ch. Darwin wrote about it that although "no one definition has satisfied all naturalists; yet every naturalist knows exactly what he means when he speaks of a species" [Darwin 1859: 44]. Contemporary proponents of this viewpoint agree with Darwin by stating that "no 'definitions' of species or subspecies can bring uniformity and clarity to the understanding of these taxa" [Tsvelev 1995: 66], and that all classification tasks are best to solve guided by intuition, common sense, and established tradition [Darlington 1971; Heywood 1989; Timonin 1998; Stekolnikov 2003]. However, this does not rule out an interest in a philosophical justification of the phenomenological view of species [Sterelny 1999; Oskolsky 2001; Wilkins 2018].

In practical research, the analysis of similarity is usually considered primary with respect to any substantive theorizing; therefore, definitions of species based on the phenetic idea are suggested as an "empirical resolution" of the species problem [Drooger 1954; Mayr 1957c; Nitecki 1957; Reydon 2004, 2005a; Silcox 2014]. Indeed, "the species defined in this way will represent different kinds of evolutionary situations and will be equivalent only by designation, and not by virtue of the nature or extent of their evolutionary differentiation" [Heywood 1963: 27]. However, particular similarity-based conceptions are quite "local" to the extent that they imply different interpretations and measures of similarity relations. Because of the above-emphasized subjective nature of similarity, the biological content of the units delineated on its basis without any unifying metaphysical consideration remains substantively uncertain and may refer to any of both "around-species" and "nonspecies" units.

A biologically meaningful approach, combining phenetic idea with a realistic standpoint, is provided by *integrative* species conception. It distinguishes the "around-species" units based on as many biological properties of organisms as possible [Winston 1999; Dayrat 2005; Bond and Stockman 2008; Goulding and Dayrat 2016; Liu 2016]. However, it differs from pure phenetic conceptions in that these properties are included in the species classifications not formally, but after their proper estimations ("weighing") in accord to certain biologically meaningful criteria, and by taking into account the biological specificity of particular groups of organisms.

In the epistemic block, a **pragmatic** idea, going back to scholastic nominalism (Ockham, etc.), takes a noticeable place. It tends to the similarity-based group of conceptions and substantiates an expediency of elaborating different special purpose species classifications for solving particular research and applied tasks [Gilmour 1940; Sonneborn 1957; Kitcher 1984a; Ereshefsky 1992b; Stanford 1995; Bachmann 1998]. Delineation of such "species [...] is primarily designed for the convenience of biologists and is not intended to represent a biological reality" [Sonneborn 1957: 314]. P. Kitcher calls "*cynical*" the species conceptions presumed by this idea, but he believes that this is true for all particular interpretations to the extent that they adopt the general notion of species to particular cognitive tasks. From a conceptualism standpoint, Kitcher's thesis is righteous in many respects, but its "cynicism" is hazardous by ignoring metaphysical substantiation of particular cognitive (and eventually applied) tasks [McDade 1995; Pavlinov 2018].

A general **nominalistic** interpretation of species is to be mentioned, which goes back to nominalistic scholasticism and "Ladder" natural philosophy; it is currently declared by supporters of positivist philosophy with emphasis on pragmatic implication [Gilmour 1940, 1961; Sonneborn 1957; Michener 1962; Sokal and Sneath 1963; Sneath and Sokal 1973; Simonetta 1982, 1992, 2014; Casetta 2010, 2012]. In a biologically meaningful context, it is supported by the microevolutionary theory of the late 19th and the early 20th centuries. S. Løvtrup refers the species notion thus understood as the "third world" in the sense of K. Popper [Løvtrup 1987]. It presumes a "formal" account of species as a classification unit having no ontic correspondence; such formalization yields attempts at a "logical" (symbolic) representation of species [Kornet et al. 1995; Mahner and Bunge 1997; Hudson and Coyne 2002; Dress et al. 2010; Alexander 2013; Manceau and Lambert 2019]. In a certain sense, an "*organizational*" species is close to it [McKelvey 1982], as well as an *axiological* species, "described by the totality of Aristotelian categories" [Crane and Sandler 2011: 300].

From the viewpoint of systematics, a formal interpretation of species as a **classification** unit can be regarded as a "common denominator" of all the above conceptions to perform an "umbrella" function for them [Heywood 1963; Ereshefsky 1992b, 1998; Mahner 1993; Mahner and Bunge 1997; Dupré 1999, 2001; Sluys and Hazevoet 1999; Bock 2004; Rapini 2004; Baum 2009; Reydon and Kunz 2019]. It goes back to antique natural philosophy, according to which Nature is structured uniformly in all its manifestations due to the unity of its causes; therefore, its structure can be described with a certain unified cognitive tool, viz., a system of universals including species. This interpretation is reinforced by nominalistic philosophies (scholastic, positivist, etc.): they see its advantage in its freedom from any particular metaphysical (including biological) contexts, which makes it cognitively universal.

* * *

The development of the species problem within the monism-pluralism dilemma in the 20th century and beyond is noticeably ordered by an *operationalization* trend, which is typical of empirical science [Benjamin 1955]. According to the latter, any scientific notion/concept is "good" to the extent it can be operationalized, so this trend presumes elaboration of *operational* conceptions of species, which serve as research tools for uncovering the structure of species diversity [Levin 1979, 2000; Hull 1997; Sarkar and Margules 2001; Richards 2008; Donegan 2018; Kunz 2012; Conix 2019a,b]. It is noteworthy that almost all existing species conceptions are sometimes considered operational, and only treating species as natural kind (see Section 6.2) can be regarded as theoretical [Leroux 1993]. However, several problematic issues arise in this regard, both formal and substantive.

First, this trend results in the elaboration of a plethora of methods, which are used in practical research of species diversity; they are not reducible to each other at a methodological level to the degree they are based on different axiomatic systems [Williams and Dale 1965; Sneath and Sokal 1973; Shatalkin 1983b; Pavlinov 2018]. Due to this, the multiplicity of operational conceptions and the methods they apply inevitably contributes to the growth of species uncertainty at the empirical level by producing a multiplicity of particular species classifications. Besides, if they are developed without theoretical substantive background (an ideal of operationalism), this yields a widening gap between theoretical and practical research [McDade 1995; Mayden 2002; Pavlinov 2018].

Second, according to the *principle of fallibilism* [Popper 1935; Lakatos 1978], every research method provides a biased estimate of a real diversity to be studied by it. Due to this, the results of the analyses of the structure of species diversity obtained by formalized methods are likely to be "defect" due to the errors of two kinds: the type I errors lead to an overestimation of the studied diversity, while the type II errors lead to its underestimation [Frost and Hillis 1990; Maxwell et al. 2020]. The probability of such errors remains uncertain in the absence of a corrective function of biologically meaningful models.

Third, the attempts to resolve these particular methodological problems on certain formal grounds is impeded by the problem of *instrumentalism*, which means a certain "closure" of the cognitive activity onto a method as such [Rieppel 2007; Pavlinov 2018]. Accordingly, a consistency of species classifications is estimated not meaningfully through compliance with the structure of diversity, implied by a certain biologically meaningful theoretical model, but through a formally (axiomatically) justified quality of the method itself. Due to this, the final results obtained with it are perceived (or passed off) as what "really" is, without their critical substantive examination. This counter-biologization "solution" of the species problem means a return to the scholastic stage of its development, when the formal genus-species scheme was regarded as a universal methodological device for describing the diversity of all things (see Chapter 3).

Lastly, a "formal" elaboration of species classifications yields an uncertainty of their biologically meaningful interpretation: it makes unclear to what extent they reflect the "nature of things", and not the certain properties of classifying methods [Pavlinov 2018]. This kind of uncertainty is minimized by a closer link between natural and classification accounts of species, according to which the latter is treated as a representation (hypothesis) of the former. This standpoint is designated as *classificatory realism* [Slater 2013], and it is summoned to solve the "long-lasting controversy between nominalists and realists" [Reig 1982: 497]. Some researchers see no problem in this; they keep these two accounts separately by treating natural and classification species on different conceptual grounds (e.g., [Shatalkin 1983a; Lidén and Oxelman 1989; Mahner and Bunge 1997; Dupré 1993; Pozdnyakov 2007; Bzovy 2016; Zachos 2016; Nathan 2017; Nathan and Cracraft 2020]). However, this standpoint drops out of an overall biologization trend inherent in the entire postscholastic development of the species problem (see Section 1.3), which is directed against the treatment of species as a formal classification unit.

* * *

From a conceptualistic perspective, the modern development of the species problem toward conceptual pluralism fits into the ideas about the structure and dynamics of theoretical scientific knowledge. This trend implies that species pluralism is a consequence of an aspiration to elaborate more adequate ideas about the structure of biota, including mechanisms and external manifestations of its both dynamics and statics. However, such pluralism aggravates the contemporary species problem by an issue of *incommensurability* of conceptions based on different theoretical premises [Kuhn 1962; Körner 1966; Wang 2002].

The reason is that the different semantic constructs cannot be directly correlated with each other, which prevents their mutual "linear" interpretations. In our case, figuratively speaking (see Section 1.1.1), different species conceptions shape particular "conceptual realities" [Otte and Endler 1989; Sankey 1998; Sterelny and Griffiths 1999; McOuat 2001; Morgun 2002; Pavlinov 2017, 2018, 2022a; Reydon 2022]; they "are applicable in different contexts of biological investigation where they perform different roles and refer to different ontologies" [Reydon 2005b: 61]. The latter author, whose standpoint is close to the abovementioned eliminative pluralism, brings incommensurability of species conceptions to a kind of logical "semiotic" conclusion [Reydon 2004, 2005a, 2022; Reydon and Kunz 2019]. For him, "the species problem is better understood as a case of homonymy rather than pluralism. [...] That is, the term 'species' is a homonymic term that stands proxy for a number of independent scientific concepts that throughout the developmental history of biological science have come to be called by the same name" [Reydon 2005b: 60, 61]; Frank Zachos seems to agree with this interpretation [Zachos 2016]. This incommensurability hinders consideration of how and why the particular conceptions are applied in practice by different specialists [Luckow 1995; Kunz 2001, 2012; Pušić et al. 2017].

According to Jody Hey, the species problem in its current intricate state is largely accounted for by that its discussants are mentally "stuck" at that stage of its development that was achieved at the beginning of the 20th century [Hey 2001a,b]. For a further search for its possible solutions, a new understanding is needed, based on some new information about the structure and function of wildlife, and this "new information" is suggested to search in the researchers' minds, more precisely, in their comprehension of that wildlife. Perhaps this is true in part, but it should be taken into account that the species notion is applied also outside biology (see Section 6.2). In this regard, a recently raised question of what kind of "species" can be in an "artificial life" [Bedau et al. 2000; Kampis et al. 2007] seems to be quite significant. All this expands considerably the scope of relevance of the species problem and shows that biologists' intuition of what "their" species is or may be, hardly decreases in itself a passion of the discussion of the problem.

A conceptualist approach to a constructive solution of the species problem should involve such reshaping of the conceptual space of the cognitive situation that would allow preserving all the diversity of the species conceptions and, at the same time, highlighting their certain conceptual unity. Two general approaches are possible in this regard, partly complementing each other [Pavlinov 2018, 2022a]. One of them is to arrange an array of species conceptions into a kind of *facet classification* (or a relational database) equipped with a general meta-language to describe the overall structure of organismal diversity. Another is to elaborate a kind of *conceptual pyramid*, which brings together all particular conceptions into a hierarchical or second-order pluralism (see Section 6.1).

6 The Contemporary
The Newest Issues

At the end of the 20th and the beginning of the 21st centuries, discussions on the diverse manifestations of the species problem do not subside but instead intensify due to acquiring partly new content. This reflects the fact that the species notion still retains one of the key positions in biology by being included in its basic thesaurus and applied by many biological disciplines [Cracraft 1989a; Ruse 1995; LaPorte 2007; Pavlinov 2013a, 2021b; Minelli 2022]. Philosophers get actively involved in these discussions to treat the species problem in their own way (e.g., [Hull 1976, 1977, 1978, 1997; Sober 1984; Rosenberg 1987; Ereshefsky 1988, 1998, 1999, 2010b, 2017; Mahner and Bunge 1997; Dupré 1999, 2022; Slater, 2013]).

Understanding the multidimensional nature of the structure of species diversity gives rise to an aspiration to discuss, in a new guise, the old basic questions, such as: (a) whether natural species exists as a specific biological phenomenon, (b) if it exists, what are its essential properties and what is its ontological status, and (c) how to reflect the diversity of its manifestations in the conceptual space of biology. This chapter considers these and related topics.

6.1 SPECIES DEFINITION: A CONCEPTUAL PYRAMID

The core of any scientific concept is the *definition* of a conceptualized object it refers to: if there is no definition, there is no concept; it appears to be an intuitive informal understanding, which is usually taken out of the scope of rationally organized science. In our case, the purpose of the definition is to turn a certain intuitive image of the "species in general", inherited from folk systematics and ordinary knowledge, into a rationally organized cognitive model to solve the following principal tasks: (a) outlining the species notion in the conceptual space of respective cognitive situation, (b) providing the possibility of mutual interpretation of different species conceptions, and (c) substantiation of criteria for discriminating particular species in practical research [Mayr 1957a; Rosenberg 1987; Mallet 1995; Hull 1997; Wilson 1999a; Stamos 2003; Wilkins 2007, 2009b; McCabe 2008; Pavlinov 2018, 2021a, 2022a; Mishler, 2021].

The following consideration of this issue (a) is based on a realistic interpretation of the definition, according to which it is the object itself, and not a notion designating it, that is actually defined, and (b) the object is given an intensional definition by referring to its biologically sound properties [Gorsky 1974; Boylan 1983; Falcon 1997]. As noted in Chapter 1, such a definition obtains a certain meaningful interpretation depending on a given onto-epistemic context, and it can be of two basic types, *inductive* (elementaristic) or *deductive* (holistic).

In the first case, the "ascending" (down–top) definition is based on inductive methodology: it focuses on the organisms and certain relations between them.

DOI: 10.1201/9781003346296-7

For instance, if these involve reproductive interactions, the question "What and why are species?" may be answered like this: due to sexes and sexual reproduction [Ghiselin 1997; Shcherbakov 2010]. All species definitions based on phenetic and occasionally pragmatic ideas also belong here. Inductive definition yields an understanding of species as "*epiphenomena*, developed or evolved [...] from lower-level processes" ([Cracraft 1989a: 48], italics added).

In the second case, the "descending" (top–down) definition corresponds to the deductive genus-species logical scheme, in which a notion is defined as a "species" detailing a notion of inclusive "genus". This approach is most consistent with the provisions of conceptualism, which makes respective definition framework relative [Richards 2010; Bartlett 2015; Pavlinov 2021a, 2022a]. Accordingly, the species definition is a particular judgment derived from a more general one about a certain natural object, which includes species as its element. For example, the conceptions of ecospecies and phylospecies meet this condition: species is defined as a structural element of the ecosystem or as a fragment of phyletic lineage.

Keeping in mind the category-unit dualism of species (see Section 1.1.1), when discussing definitions of species, its two hypostases, as a unit and as a category, are usually treated separately to be conceptualized differently [Mayr 1982, 1996; Mishler and Brandon 1987; de Queiroz 1999; Mayden 2002; Bock 2004; Richards 2008, 2010; Wilkins 2009a, 2018; Zachos 2016]. However, this standpoint does not take into account that, in the general conceptual space shaping the species problem, the notions of species unit and species rank (category) are interrelated; therefore, when one of them is considered, the other is presumed [Bartlett 2015; Barker 2022; Pavlinov 2022a]. When defining species, the indication of species integrator serves as a connecting link between these hypostases: this allows regarding particular species conceptions on a single causal basis and allocating the objects being defined to the same level of generality, which is called species rank [Boyd 1999]. Accordingly, the task of defining biological "species in general" involves both species unit and species rank equally.

The latter thesis becomes of fundamental importance when regarding species as a natural object of quasi-individual nature (see Section 6.2). Unlike a class or a natural kind, a species-individual cannot be defined in a logical sense since only multiple objects or general notions denoting them are subject to definition [Gorsky 1974, 1983; Reydon 2006; Slater 2013]. Obviously, this clearly refers to the particular species delineated in the practical studies. However, in general, when "species in general" is considered, including its quasi-individual account, no particular species units are meant, but rather the species category to which such units belong. This means that "species in general" can be endowed with a status of natural kind, so it can be defined to the extent that its definition refers to a certain set of the natural entities allocated to the species category. In this case, "species definition" and "species concept" receive a similar meaning by referring to the same object, viz., a certain natural phenomenon of a certain level of generality [Ereshefsky 2017].

* * *

So far, a seemingly comprehensive framework-relative definition of "species in general" is being developed in the context of hierarchically organized interrelated species conceptions of various levels of generality [Kitcher 1984a; Mayden 1997, 1999, 2002; Naomi 2011; Baetu 2012; Pavlinov 2013a, 2017, 2018; 2021a, 2022a; Zachos 2016], which yields the abovementioned *hierarchical* or *second-order* pluralism [Richards 2010; Sloan 2013; Nathan and Cracraft 2020]. Such a view of the latter is forerun by the "pyramid of concepts" of medieval scholasticism [Makovelsky 2004], so it can be generalized by the notion of *conceptual pyramid* [Pavlinov 2018, 2021a]; it is offered to formulate it as a kind of "*orbital model*" to reflect its multidimensionality [Kokić 2009]. The principal meaning of this conceptual construction is that it performs a framework function by hierarchical ordering the entire array of particular conceptions on a uniform causal basis: the conceptions recognized at some levels of generality (except the highest) are inferred as particular interpretations of the conceptions of higher levels of generality.

A "pyramidal" character of this conceptual construct is determined by certain properties of the reduction cascade that generates it (see Section 1.1.1). One of these properties is an increase in the total number of conceptions as we move from top to bottom along this cascade. At a higher level of the conceptual pyramid, a cognitive model is introduced providing an understanding of what natural "species in general" is or may be. Its lower levels include more particular theoretical species conceptions, for example, evolutionary or ecological ones. They are specified at far lower levels by adding certain details in their definitions—for example, different interpretations (phylogenetic, genealogical, etc.) of the evolutionary species conception. The lowermost levels include operational conceptions to elaborate certain criteria for delineating the concrete species units. Such "pyramidal" construction of the hierarchy of species conceptions inevitably yields an acknowledgement of their plurality and shows that a monistic reduction of the species problem [Lam 2020] is untenable.

Such "pyramid of conceptions" can be considered a specific representation of the ontic component of cognitive situation that performs the following functions. First, by referring to a "top" concept, from which all judgments about species are inferred, it allows discussing an issue of why and how biological species arises as a specific natural phenomenon. Second, it provides both logically correct and biologically sound intensional definitions of species units as different hypostases of biological "species in general". Third, provided that this whole construction is scaffolded deductively in a certain natural (from some viewpoint) fashion, the inferences of particular species conceptions in such framework ensure their naturalness (from the same viewpoint). Finally, in the context of a single pyramid, different species definitions turn out to be conceptually interrelated and thus mutually interpretable.

Framework-relative definitions of species conceptions make the latter meaningful according to a certain biologically sound concept or theory of the highest level of generality. Successively, without reference to the latter, any particular conception seems to be introduced *ad hoc*, i.e., without serious biologically sound justification. This leads to an important corollary concerning operational species conceptions: these are the theoretical ones that make them biologically meaningful. This agrees with the general conceptualistic idea of the theory-ladenness of any empirical observations in natural sciences [Hempel 1965; Quine 1969; Carrier 1994].

It becomes more evident if we treat an exploration of species diversity as an elaboration of particular scientific hypotheses about the latter's structure [Pavlinov 2013a]. From a conceptualistic perspective, no such hypothesis can be meaningfully elaborated if it has no conceptual metaphysical framework in its background. In our case, for any species classification, treated as a hypothesis, to be scientifically sound and substantively meaningful, it should be developed within the framework of a certain theoretical species conception referring to a certain natural biological phenomenon.

According to the classical theory of definitions, the sought "top" concept/theory is to be treated as such "logical genus", by which division it is possible to acquire "logical species" containing a general definition of natural species that biology deals with [Pavlinov 2013a, 2021a, 2022a]. For avoiding logical "genus–species" tautology, it is necessary to define in the same general context ("logical genus") some other biological units (other "logical species") of the same level of generality as natural species proper, but which are certainly not species. Such an approach is aimed at understanding what the species phenomenon is and how it differs from any "nonspecies" phenomena (for instance, biomorph).

Thus, when constructing a conceptual pyramid, one of the principal questions involves determining properly its top: which concept/theory should be allocated to the latter to make it a "unique beginner" in the considered hierarchy (in the sense of [Berlin 1992]). Obviously, the general concept (meta-concept) of species cannot be considered in this guise: in such hierarchy, it remains without a logically consistent definition [Wierzbicka 1996] and cannot therefore be rationally distinguished from any "nonspecies" phenomena of the same level of generality. Consequently, this "top" concept should be far more general, referring to a certain natural object that includes both species and "nonspecies" as its structural elements [Sluys 1991; Hey 2001a; Pavlinov 2013a, 2021a, 2022a]. It follows from this that species should be defined not by itself but as one of the elements of the overall structure of biodiversity, considered along with its other elements (such as biomorph, guild, etc.).

Jody Hey suggests the theory of life as such "unique beginner" of the entire pyramid [Hey 2001a], and Ronald Sluys distinguishes evolutionary and ecological hierarchies considered in such vein [Sluys 1991]. In a rather constructive form, both these proposals can be concretized by the concept of *evolving structured biota*. It allows making the main emphasis on those natural causes that operate at the level of biota and structure it, as it develops and functions, thus generating and individualizing its various structural supra-organismal units [Pavlinov 2013a, 2022a]. The joint co-action of these causes yields *dynamic stability* of both the entire biota and its various structural-functional units, including species, as one of their most fundamental properties to be comprehended [Brooks and Wiley 1986].

So, to get a metaphysically consistent definition of natural "species in general", the entire conceptual pyramid should be construed as a descending cascade of the causes structuring biota, thus generating certain supra-organismal units in its structure, including species, and providing their dynamic stability. This means that the "top" concept/theory presumes defining species on a causal basis: it explains why and how the living becomes structured and, by this, why and how species arises as a particular unit of its structure. This means that any substantive species conception, to be biologically sound, is to be *cause-focused* and *explanatory* [Mayr 1982; Ereshefsky 1992b; Pavlinov 2013a; Barker 2019b, 2022]. So, if a multicausal nature of "species in general" is acknowledged, its definition should not be reductive by indicating only one particular cause (historical or else), but exhaustive and incorporating all causal factors that ensure its very existence [Zavadsky 1968; Sluys 1991; Wilkins 2009a; Baetu 2012; Pavlinov 2013a, 2022a; Kendig 2014; Nathan 2017]. According to this general conceptual construction, particular cause-focused species conceptions accentuate on particular causes (ecological, historical, etc.) that are regarded the most important for individuation of species [Ereshefsky 1992b; Wilkins 2003a, 2009a; Barker 2010, 2019b, 2022; Pavlinov 2013a, 2022a]. Successively, consideration of these causes responsible for the origin and dynamic stability of species units in different groups of organisms is to become one of the main issues in the contemporary species problematics.

* * *

The attempts to transform a general understanding of species into a more or less rigid species definition, satisfactory enough from a metaphysical standpoint, face a particular "epistemic pitfall". It arises due to the fact that every theoretical definition is *conceptually inexact* [Körner 1966], which is strengthened by the *principle of inverse relation* between rigor and meaningfulness of definition [Voyshvillo 1989]. It states that the more strictly a concept is formulated, the less likely there is something in nature to which it may correspond [Hempel 1965]. In the case of species, a negative effect of this principle is evident when different interpretations of species as a historical commonality are regarded: all attempts to make its definition more rigid lead to introduction of narrower conceptions (see Section 5.4), which are more reductionist and therefore less "natural". With this, it should be noted that the species concept is not unique in this respect: this holds true for the concepts of homology, gene, organism, ontogenesis [Ghiselin 1981; Hall 1994; Beurton et al. 2000; Brigandt 2003b, 2012; Folse and Roughgarden 2010; MacLeod 2012; Pavlinov 2012; Minelli 2014; Pradeu et al. 2016].

For this "epistemic pitfall" to be circumvented, it is to be realized that any definition of natural species, claiming to be meaningful biologically, is doomed to be nonrigid semantically [O'Hara 1993; Pavlinov 2018, 2022a]. Because of this, the epistemic component of our cognitive situation should include some elements of the abovementioned fuzzy logic, which formalizes the possibility of operating with nonrigid context-dependent definitions [Zadeh 1992; Kosko 1993]. It underlies a probabilistic model of language [Nalimov 1979], according to which the species notion, as a linguistic variable, is characterized by a certain probabilistic distribution of its possible particular meanings, which are fixed by particular context-relative conceptions. Employing such logic presumes a probabilistic context-dependent interpretation of any judgments about natural species—about their properties, characters, composition, rank, etc. Accordingly, when elaborating definition for a particular species conception, it makes sense to fix rigidly its "core" and remain its "periphery" fuzzily outlined.

6.2 SPECIES ONTOLOGY: NEW QUESTIONS

At the present stage of the conceptual history of the species problem, actively considered are the same two basic ontological interpretations of species, realistic and nominalistic, that were being discussed in medieval scholasticism. However, a conceptualism perspective sheds new light on the contemporary consideration of species ontology: it is theory-dependent, so different understandings of reality make this rigid dichotomy fallacious [Quine 1969; Hull 1977, 1989; Putnam 1991; Dupré 1993; Mahner and Bunge 1997; Psillos 1999; Swoyer 2006]. Species pluralism also brings a certain novelty to this discussion: some philosophers consider it an argument against species reality [Stanford 1995; Stamos 2003], though others do not see contradictions between pluralism and realism [Kitcher 1984a,b; Brigandt 2003a]. Besides, the contemporary consideration of this fundamental issue involves the new ways of posing species ontology, which were not previously taken into account [Ghiselin 1981; Kitcher 1987; Mayr 1987; Stamos 2003; Slater 2013; Bartlett 2015; Pavlinov 2017, 2018, 2021a, 2022a]. In general, these new ways, which all presume realistic account of natural species, may be classified into three principal groups: stationary (synchronic), evolutionary (diachronic), and mixed (synchronic-diachronic).

One of the popular stationary interpretations ascribes species the status of *natural kind*. In this capacity, the notion of species, as a universal unit of the structure of diversity, is considered in various natural sciences and humanities: for example, recognized are "species" of atoms, molecules, minerals, forms of consciousness, languages, etc. [Bunge 1977; Kitts and Kitts 1979; Ghiselin 1981; Arteca and Mezey 1989; Simões 1992; Grunwald 1997; Mahner and Bunge 1997; Allen and Bekoff 1999; Stamos 2002; Elder 2008], and diversification of chemicals is sometimes referred to as "speciation" [Ure and Davidson 2001; Clemens and Dharmendra 2015]. In one of the ontology versions, a fundamental unit of this overall structure is called *ontological species* [Mahner and Bunge 1997].

Each natural kind is outlined by a certain essential property (several conjugate properties), which is common with the necessity to some set of things and making their totality law-like; in this sense, such kind is *natural* [Quine 1969, 1994; Ruse 1987; Devitt 2008, 2011; Koslicki 2008; Ereshefsky 2009b; Reydon 2009; Hawley and Bird 2011; Slater 2013, 2015; Magnus 2015; Kendig 2016; Richards 2016; Spencer 2016; Austin 2018; Bird and Tobin 2018]. This way of defining natural kind is sometimes referred to as a *natural kind essentialism*: it implies that an essential property in question constitutes a *kindhood* (*kindness*) of the respective totality [Wilson 1999b; Rieppel 2007; Reydon 2009; Hawley and Bird 2011; Slater 2015]. With this, the general notion of essence can be filled with different contents, according to which several pairs of oppositional interpretations are distinguished: relational *vs.* intrinsic, shared-nature *vs.* typological, teleological *vs.* material, etc. [Talpsepp 2013]. Thus, a natural kind is not a something "given": according to contemporary conceptualism, an outlining of a certain natural kind, or *kinding* [Kendig 2016; Kendig and Bartley 2019], is context-relative: it depends on a particular aspect of consideration, which is set by a particular cognizing subject [Quine 1969; Mahner 1993; Shatalkin 2012; Slater 2013; Kendig 2016; Bird and Tobin 2017; Austin 2018; Onishi and Serpico 2022].

There are a lot of instances when the general notion of natural kind is effectively applied to the description of various aspects of the diversity of organisms. The examples are their grouping according to their functions in ecosystems (hetero- or autotrophs, biomorphs or guilds), anatomical body plans (uni- or multicellulates, vertebrate or invertebrate animals), social organization (solitary or colonial animals), etc. However, an application of this notion to biospecies faces certain contradictions, and their removal requests nontrivial specifications. On the one hand, biospecies is normally defined polythetically, whereas natural kind monothetically [Okasha 2002; Lewens 2012; Ludwig 2018], so polythetic definition makes biospecies thus treated a *vague biological kind* [Simons 2013; Slater 2013] (see also on homeostatic property cluster below). On the other hand, biospecies is restrained spatiotemporally, whereas natural kind is typically treated as a universal, so the former is defined as a *restricted kind* [Guyot 1986]. From the biological perspective, ecological limits should undoubtedly be added to these boundaries, although certain philosophers, prone to simplification, consider this as an "extraneous complication" (e.g., [Splitter 1988]).

Another version of stationary ontology considers species a *system* in its general system-theoretic sense [Kordonsky 1983; Shatalkin 1983a]. Two general categories of systems are proposed to distinguish the systems of objects of the same kind and the object systems, which are ordered by different system-forming properties [Urmantsev 1988]. In the first case, either common essential characteristics or certain laws of systemic isomorphism are meant: this makes such a system by its meaning close to natural kind. In the second case, the systemic integrity of a group is ensured by the internal relations among its elements, which make it close to an individual-like object. Biospecies can be treated in either of these two guises depending on its implied biological properties [Lysenko 1952; Mayr 1963, 1969, 1988; Volkova and Filyukov 1966; Sucker 1978; Reig 1982; Borghini 2000; Stamos 2003; Wilkins 2003a, 2007; Crane 2004; LaPorte 2006; Sykora 2007; Malikov and Golenishchev 2009; Rekovets and Kuzmenko 2021], and their combining yields understanding of species as a *dynamic system* [Blair 1956; Collier 1985]. At any rate, taking into account a rather weak integrity of species as a systemic object, as compared to an organism, it is assigned the status of *quasisummative system* [Shaposhnikov 1975] or *individualized class* [Hull 1976].

A more flexible interpretation of the stationary ontology of species is provided by the conception of *homeostatic property cluster* [Boyd 1999; de Queiroz 1999; Wilson 1999b; Ereshefsky 2001, 2009b; Ereshefsky and Matthen 2005; Rieppel 2005, 2006, 20110a; Wilson et al. 2007; Brigandt 2009; Barker and Wilson 2010; Slater 2013, 2015; Ereshefsky and Reydon 2015; Richards 2016]. In this case, species is understood as a composite (complex) whole, which is characterized by the following features [Wilson et al. 2007]: systemic relations between the properties of its members ensuring its homeostasis, its intrinsic heterogeneity, and the fuzziness of its boundaries. In the original version, the philosopher of biology Richard Newell Boyd (1942–2021) believes that the homeostatic property cluster is a special case of natural kind, so in fact it is a *homeostatic property cluster kind*, and that "individual species have [...] essences, so that a form of 'essentialism' is true for species", although not in the classical sense [Boyd 1999: 142]. A "nonclassicity" of Boydean understanding of essence is presumed by his interpreting the commonality of properties of cluster species polythetically, because of which its homeostasis turns out to be imperfect: "some things may display some but not all of the properties [...]; some but not all of the relevant underlying homeostatic mechanisms may be present" [*op.cit.*: 143].

Two interpretations of the ontic status of species should be mentioned particularly, which position it differently in the surrounding world [Simpson 1961; Mayr 1963, 1969, 1988]. According to one of them, biospecies can be defined *absolutely* through its internal connections. According to another, the concept of species is *relative*: it means that a species can be considered as such only with regard to another species, as, say, a brother in relation to his brother. The second understanding of species goes back to the logical genus-species scheme (see Section 3.5).

When regarding the structural organization of Nature in the context of global evolutionism, a historical interpretation of species ontology comes to the fore. In this case, it is treated not as a stationary object, but as a developing unit being treated as a *process* [Paramonov 1951; Zavadsky 1961, 1968; Rieppel 2009; Dupré 2020, 2021, 2022], or a *process-system* [Rieppel 2009, 2010a,b]. As such process is local with regard to time and space, one of the most important outcomes of such consideration is the interpretation of species as an object of *individual* nature [Ghiselin 1974, 1981, 1987, 1997; Hull 1976, 1978; Bernie 1984; Shatalkin 1984; Sober 1984; Falk 1988; Brooks and Wiley, 1986; Ereshefsky 1988; Baum 1998; Coleman and Wiley 2001; Mayden 2002; Reydon 2003; Brogaard 2004; Crane 2004; Richards 2010; Haber 2016]. Individual interpretation of species implies that its structure is described in terms of mereology as a *constitutive hierarchy* [Haber 2016], with the conspecific organisms and populations being not elements but parts of species [Hull 1977; Ghiselin 1997; Rieppel 2012].[1]

This "individuality thesis" presumes a certain conceptual framework based on the understanding of what individual is in a general ontic sense [Heise 1981; Sober 2000]. However, there is no such obvious understanding as applied to species [Lidén 1990; Huneman 2014]; one of the reasons is that species individuality can be of different causality and of various degrees [Guyot 1986; Ereshefsky 1991; Pozdnyakov 1994; Baum 1998; Rieppel 2007, 2009; Shatalkin 2012]. To highlight specifics of the "individuality thesis" with regard to biospecies, in comparison to a standard definition of individual (Latin *individuum*, indivisible), the former is treated as a *dispersed individual* [de Sousa 1989]; it is also defined as a *holon* [Wood 1994; Ghiselin 1997; Shatalkin 2012] or a *mereological sum* [Brogaard 2004] or a *quasi-individual* [Pavlinov 2018, 2022a] or just a *unit* [Van Valen 1988].

The "individuality thesis" is historically connected with Buffon's views [Lovejoy 1959; Ghiselin 1969, 1974; Sloan 1995; Gayon 1996; Stamos 2003], with which active biologization of the species problem began in the second half of the 18th century (see Section 5.1.2). Natural groups of different levels of generality, including species, were regarded in such status by organismal natural philosophy of the 19th century (Oken etc.): it is most vividly manifested in ascribing to species of a certain life cycle [Nägeli 1865;

[1] Something similar is presumed by partonomic interpretation of classification universals suggested by Neoplatonists and their early commentators (see Sections 2.2 and 3.1).

Haeckel 1866; Dillon 1966]. The philosopher Karl Raimund Popper (1902–1994) can be counted among the contemporary forerunners of this thesis: in his *Logic of Scientific Discovery*, he argues that, depending on a cognitive intention, species can be thought of as individual (concreteness) or class (abstraction) [Popper 1935; Sloan 2002].

Individual status of "vertical" species is ensured by its existence as a phyletic lineage [Hull 1976, 1977; de Queiroz 1999]. It is emphasized that it is this status of species that provides for its participation in the evolutionary process [Ghiselin 1974, 1981, 1987, 1997; Griffiths 1974a,b, 1999; Hull 1976, 1977, 1989; Bernie 1984; Rosenberg 1985; Williams 1985, 1989; Mishler and Brandon 1987; Ereshefsky 1988, 1991; Cracraft 1989a; Gayon 1996; Horvath 1997b; Coleman and Wiley 2001; Wilkins 2007; Rieppel 2010a; Neto 2016; Flematti 2018]. E. Haeckel called species thus understood "genealogical individual" [Haeckel 1866], now it is considered a *historical individual* [Ghiselin 1974, 1981; Wiley 1981; Kluge 1990; Pavlinov 2005, 2018] or a *functional unit* [Baum 2009; Barker and Velasco 2013; Barker 2022]. Reproductive conception presumes individuality of "horizontal" species due to its being a closed genetic system [Mayr 1982, 1988; Paterson 1985; Masters et al. 1987; Kunz and Werning 2009; Friedmann 2013; Masters and Génin 2018].

The principal difference between individual and natural-kind ontologies, as applied to the species issues, can be summarized as follows [Ghiselin 1974, 1981, 1987, 1997, 2002; Hull 1976, 1977, 2001; Ruse 1987; Falk 1988; Ereshefsky 1991; Coleman and Wiley 2001; Pavlinov 2018]. According to the first interpretation, supported by modern phylogenetic theory, biospecies is local for its unique historical fate: each species origins only once and dies out irrevocably. According to the second interpretation, rooted in scholastic natural philosophy, species is a universal and has no clear space-time boundaries. This second interpretation is most obviously exemplified by transmutations of chemical elements occurring unlimited number of times in any part of the universe. In the Antiquity and Middle Ages, a similar multiple generation of lower organisms was supposed, whose species belonging was provided for by the respective eternal essences embodied in them [Lunkevich 1960]. This difference underlies an important distinguishing between species and biomorph: the latter has no spatio temporal boundaries, so it is a natural kind taking a certain place in the functional structure of ecosystems [Du Rietz 1931; Brown 1977; Aleev 1986; Kirpotin 2005].

From a certain scientific-philosophical perspective, a fundamental disadvantage of the individual account of species is that the latter "falls out" of that classical context, according to which every conception is scientifically sound to the extent that an object fixed by it admits certain nomological generalizations [Ruse 1981, 1987; Caplan 1981a,b; Rosenberg 1987; Leroux 1993; Lange 1995; Mahner and Bunge 1997; Wilson 1999b; Reydon 2005a, 2006, 2009]. R. Wilson emphasizes that these are such generalizations, having the status of natural laws, that allow presuming reality (objectivity) of correctly outlined natural kinds [Wilson 1999b]. On this basis, paradoxically enough, Michael Ruse sees in the "individuality thesis" a manifestation of nominalism [Ruse 1981].

Thus, one of the hotspots of the modern species problematics, in contrast to its previous stages, seems to be a dilemma of *species-individual vs. species-kind*

[Mayr 1987; Stuessy 1989; LaPorte 2004; Sciacca 2020]. They are usually regarded as incompatible by describing ontologically different objects, but this does not prevent their joint consideration based on the onto-epistemic *principle of complementarity*, according to which they are endowed with a *composite* ontology [Armand 2008].[2] It is assumed that the dynamic stability of species-as-kind/cluster in certain spatiotemporal boundaries is provided by a specific co-action of species integrators, which has a temporal extension [Rieppel 2006, 2012; Onishi and Serpico 2022]. This agrees with the Boyd's statement that "we have no reason to deny that there can be genuine natural kinds that are historically delimited" [Boyd 1999: 155].

According to such combined consideration, species can be defined as a *historical kind*, viz., a natural kind/system endowed with a temporal extension and a certain *historical essence* [Ghiselin 1981; Cracraft 1989a; Ereshefsky 1991; Griffiths 1999; Millikan 1999; Okasha 2002; LaPorte 2004; Rieppel 2005, 2006; Wilson 2007; Wilson et al. 2007; Richards 2016; Pavlinov 2018; Maxwell et al. 2020; Wagner and Tomlinson 2020; Godman 2021], though this interpretation finds its critics [Pedroso 2012; Devitt 2018b]. Close to this is an understanding of species as a class united by its historical commonality [Gayon 1996; Quarfood 1999], which refers to Simpson's specific "evolutionary role" [Simpson 1961]. An interpretation of species as *kumatoid* [Zuev 2002, 2015] is also worthwhile mentioning here. Such combinatorics makes biospecies a very specific natural phenomenon [Mayr 1987]: "species are not either individuals, or natural kinds. Instead, species are complex wholes (particulars, individuals) that instantiate a specific natural kind" [Rieppel 2007: 373; Rieppel 2010b], or *biological natural kinds* [Wilson et al. 2007]. In an extreme guise, biospecies may deserve recognition as a natural entity endowed with its own ontic status, namely, the status of *species* [von Vaupel 1987].

In this regard, the abovementioned consideration of species as an element of developing biota, which is treated as a nonequilibrium system, looks attractive, as it makes species itself a system of such kind [Pavlinov 1992, 2013a]. An immersion of such interpretation in the context of evolutionary epistemology makes it possible to correlate biospecies with the elements of other nonequilibrium systems. From this perspective, ideas and conceptions can be regarded as "species-like" entities that are born, live, and die out like biological species [Wilson 1990; Colin and Bekoff 1997]. Highly likely is a close analogy between biospecies and ethnogroups, as well as the languages, thus understood [Gil-White 2011; Tallerman and Gibson 2012; Wagner and Tomlinson 2020].

Thus, from a conceptualist standpoint, a certain "ecumenical stance" [Casetta and Vecchi 2019] seems quite sound, which focuses on the complex nature of species integrators and their interactions, collectively ensuring the dynamic stability of species as a natural phenomenon. Therefore, as R. Boyd notes, consideration is to be given to "the relative power of these sorts of homeostatic mechanisms in sustaining the sort of homeostatic integrity characteristic of biological species"

---

[2] However, the monist philosopher Robert Wilson, not acknowledging such complementarity, believes that this brings the situation to "the absurdity of saying that one and the same thing is a natural kind and an individual" [Wilson 1996: 310]. Though, his view of wave–particle duality in quantum physics seems to be the same (https://www.nehrlich.com/blog/2003/11/17/quantum-psychology-by-robert-anton-wilson/).

[Boyd 1999: 165]. Under some assumptions, such complexly viewed species can be described by a certain *feedback model*, according to which different integrators interact to conduct a common function of ensuring species homeostasis [Barker 2019a, 2022].

Focusing on the species integrators that ensure the ontological status of natural species is attractive because of the following. It suggests that species may appear in different ontic hypostases in different groups of organisms depending on their biological organization; this is designated as *status pluralism* [Slater 2013]. A possibility of such interpretation seems to be presumed by the already mentioned division of "species in general" in euspecies and quasispecies (see Section 5.4): by their properties, the former tends to quasi-individual ontology, while the latter to natural kind ontology. This general ontic multiaspectedness is complemented by a specific biological one: perhaps, ecospecies corresponds more to a homeostatic property cluster, whereas phylospecies corresponds to a historical quasi-individual.

* * *

Because of the great attention to species as a natural entity, consideration of its ontology as a classification unit partly recedes into the background of the cognitive situation. Meanwhile, some of the respective issues constitute a significant part of the contemporary species problematics. In the context of the theme discussed in this section, the most important is a question of ontological status of classification species with regard to the ontology of natural species.

This issue brings us back to the species' natural-classificatory dualism; the following general considerations are possible here. On the one hand, taking into account the currently dominant trend of biologization of the species problematics, presuming a close correspondence between natural and classification units, it seems possible to attribute to them the same ontology [Löther 1972]. On the other hand, both elements of this dualism are usually considered as entities of different kinds: for instance, natural species is treated as a quasi-individual, while classification species as a class [Woodger 1952; Bunge 1977; Chebanov 1980; Shatalkin 1983a, 1984, 1988; Mahner and Bunge 1997; Dupré 1999, 2001]. This dual view becomes reinforced by the interpretation of classification as a specific *taxonomic hypothesis* [Panchen 1992; Pavlinov 1995, 2018; Starobogatov 1996; Baum 1998; Rasnitsyn 2007; Fitzhugh 2005, 2009; Gill 2014]: it is obvious that the natural species itself and the hypothetical inferences about it are fundamentally different in their status. However, in fact, the natural-classificatory dualism, as such, is not involved here because such hypothesis is formulated not about particular species, as sometimes supposed [Starobogatov 1996; Fitzhugh 2005, 2009; Pante et al. 2015], but about the structure of species diversity in a certain group of organisms [Pavlinov 1995, 2011, 2018]. Therefore, the natural diversity of organisms and classification as its cognitive model (hypothesis) are really quite different: the former refers to ontology and the latter to epistemology. However, this kind of dualism does not exclude a close correlation between the particular natural and classification species.

6.3 THE "NEW ESSENTIALISM": AN EVOLVING SPECIESHOOD?

The general notion of essence, developed in the second half of the 19th and early 20th centuries to become characteristic of contemporary essentialism, is the result of mixing Aristotelian *ousia* and Platonic *eidos* [Popper 1935], which is undoubtedly erroneous. At the same time, it is subjected to fundamental vilification by positivism, rooted in scholastic nominalism, and early postpositivism. In biology, an essentialist interpretation of organisms and their aggregates (including species) is rejected on the basis of the evolutionary idea: if species evolve and the properties of organisms change during evolution, there is no place for unchanging essences in the living. A special role in forming such a negative attitude to essentialism among biologists is played by a sharp juxtaposition of two thinking styles, typological and population, by E. Mayr: the first is antievolutionary and the second is an important part of evolutionary thinking [Mayr 1942, 1959a, 1982, 1988]. The fascination with this Mayrian idea yields a declaration of "death of essentialism" [Hull 1965; Ereshefsky 2001, 2017; Gould 2002], which become a "received opinion" of the vast majority of biologists.

However, a careful analysis of Mayrian juxtaposition shows it to be untenable on several points. First, his interpretation of essentialism relies on a flawed opinion of A. Cain, who drew it from an incorrect interpretation of Aristotle's ideas in one of the articles of the early 20th century [Winsor 2001, 2003, 2006a; Wilkins 2003a, 2009a; Müller-Wille 2011; Talpsepp 2013]. Second, Mayr mistakenly identifies essentialism with typology and understands the latter too simplistically: in fact, typological views of the early 19th century (Goethe, before all) imply an ordered *dynamic* diversity of organisms [Lyubarsky 1996; Riegner 2013; Vasilyeva and Stephenson 2013; Pavlinov 2018]. Finally, an emphasis by population thinking on the uniqueness of individuals and ignoring a certain population commonality actually means a "hidden nominalism": population appears as a simple sum of individuals, which clearly contradicts its understanding as a unit of evolution [Sober 1980; Ariew 2008; Nanay 2010].

The postpositivist revolution in biology in the mid-20th century directs the "historical pendulum" of the development of its onto-epistemology in the direction opposite to that initiated by positivistic ideas. One of the results of this movement becomes a *new essentialism*: it substantiates the ontology of biological supra-organismal systems, including the causes of their stability, by their endowment with certain emergent properties constituting their essences [Ellis 2001, 2002; Devitt 2008; Austin 2018]. This ontology is well compatible with the ideas developed by synergetics, which consider in this way any complex nonequilibrium systems, both material (including biota) and mental (including science as an information system) [Prigogine and Stengers 1984; Barantsev 2003].

The new essentialism closely concerns the species problem by offering an understanding of biospecies as a natural unit endowed with a certain essence [Sober 1980; Boyd 1999; Ellis 2001, 2002; Rieppel 2007, 2010b; Wilson et al. 2007; Devitt 2008, 2018a,b; Lewens 2012; LaPorte 2017; Austin 2018; Maxwell et al. 2020; Pavlinov 2022a]. The latter may be considered a manifestation of a particular intrinsic property of species that might be called, by a clear analogy with the abovementioned kindhood, *specieshood* of whatever particular content, which distinguishes species from any "nonspecies" with their own essences or "-hoods" [Pavlinov 1992, 2009, 2013a, 2018, 2021a, 2022a; Brooks and McLennan 1999; Griffiths 1999; Wilkins 2007; Hawley and Bird 2011; Kendig 2014; LaPorte 2017; Barker 2019b].

From this essentialist perspective, searching for specieshood as an intrinsic, essential property of natural species becomes one of the principal issues of the current and the nearest future species problematics. To start moving in the right direction, it is necessary to reformat properly the entire cognitive situation to make the notion of specieshood a focal point in its conceptual space. This involves consideration of the questions of a fairly general (philosophical) nature concerning the ontology of specieshood, including the following interrelations: of *kind essentialism* and *category essentialism*, keeping in mind species unit-category dualism and its interpretation as a natural kind, cluster, or quasi-individual [Wilson 2007; Wilson et al. 2007; Devitt 2008, 2011, 2018a; Barker 2010; Richards 2010; Hawley and Bird 2011]; of the just-indicated forms of "stationary" essentialism and *historical essentialism* [de Queiroz 1998; Griffiths 1999; LaPorte 2004; Pedroso 2012; Devitt 2018b]; of *mereological* (property) and *kind* (category) essentialism, keeping in mind species partonomic-taxonomic dualism [Richards 2010; Devitt 2018a]; of contingent and necessary properties of species [Ghiselin 1997; de Queiroz 1998, 1999; Devitt 2008, 2010]; etc.

For these and possibly other issues to fit the general trend of biologization of the species problem and fill them with a more tangible content, it seems crucial to link the notions of specieshood and species integrator. From this perspective, the first noticeable step toward the biologically substantive understanding of specieshood seemed to be defining species as a syngameon at the turn of the 19th to the 20th centuries (see Section 5.2.2). Viewed from a more general metaphysical perspective, it means that a hypothetical specieshood should be regarded as an integrated part of the overall natural history of organisms [Pavlinov 2013a, 2018, 2021a, 2022a]. It incorporates, in a certain (now unobvious) way, biological mechanisms responsible for the dynamic stability of species, viz., its self-reproduction and self-isolation, its place in the niche structure of ecosystems, its persistence as genealogical lineages, etc. Such understanding yields an assumption that a hypothetical specieshood may depend, to more or less degree, on various aspects of the species biology, which inevitably makes it group specific.

Taking this dependence for granted, it might be assumed that the presumed specieshood manifestations change together with other biological properties of organisms in the course of their evolution as part of the historical development of the functional-structural organization of biota in general and its various elements in particular, including species [Pavlinov 2013a, 2018, 2021a, 2022a]. So we have here something like an *evolving specieshood*, being a specification of the abovementioned "historical essence", which content changes in the course of biological evolution. This speculative assumption represents an "essentialist" development of an earlier idea that species itself, as a biological phenomenon, is a product of evolution and has its own history [Polyansky 1958, 1959; Zavadsky 1961, 1968; Dobzhansky 1970; Brooks and Wiley 1986; Ghiselin 1987, 1997].

Judging by the relevant data on the nature of species integrators in the groups of organisms of different levels of biological advancement, the differences between them may be considered as a result of the former's evolutionary enhancement. From this viewpoint, it is possible to talk about "pre-species" form of biological organization at the very first stages of its evolution (viruses), at which no particular species integrators occur. The next go fuzzily integrated quasispecies or pseudospecies of prokaryotes and lower eukaryotes, whose species integrators are still quite loose.

Finally, the more perfect intrinsic mechanisms shaping the specieshood become developed (such as the amphimictic reproductive system) in more advanced organisms that result in more cohesive and discrete euspecies. This evolution of specieshood is accompanied by a "halo" of constantly emerging "virtual" species that arise on different biological bases. Due to the improvement of species integrators and the increase of cohesion and discreteness of species, the latter's ontological status changed from natural kind to quasi-individual. Such consideration is relevant to the "*species-as-status*" conception [Reydon 2005a, 2022; Reydon and Kunz 2019]: in fact, all "around-species" units, reaching a certain level of individuality and evolutionary independence, may be ascribed such status, but their particular specieshoods would be different.[3]

The most important question in this regard is, whether quasi-, pseudo-, and euspecies can be considered different manifestations of a single species structure of biota? It is, in general, similar to the question addressed at the organismal level. This is because, according to the evolutionary model just outlined, the integrators working at both organismal and species levels are developing in parallel: a progressive evolution of biological organism, as a discrete dynamically stable system, is accompanied by an enhancement of species integrators, which entails raising of the dynamic stability of species being a similar system. A positive answer to this question may yield acknowledgement of a single "around-species" level of the structure of biota: this would mean an acknowledgement of "species in general" as a natural phenomenon, which is differently manifested by specific "modes of existence" in the groups of organisms with specific species integrators [Zavadsky 1961, 1968; Grant 1981; de Queiroz 1999; Brigandt 2003a; Wilkins 2003a, 2009a; Pavlinov 2013a, 2021a, 2022a].

To the contrary, the failure to reveal any universal natural (biological) mechanisms responsible for a certain universal specieshood would mean that there is no species as a biologically universal natural phenomenon inhered in biota. This would be tantamount to admitting that, as eliminative pluralism supposes, there are essentially different units of the functional-structural organization of biota intrinsic to the different groups of organisms with different natural histories. Accordingly, they should be formalized not as different "modes" of the same "species in general", but rather as different "species-like" entities with essentially different particular "-hoods" for which some *biodiversity unit* may serve as an "umbrella" notion. With this, it is to be emphasized that this negative conclusion will be biologically motivated to the same extent as the previous positive one

6.4 IF NOT SPECIES, THEN WHAT?

One of the most radical solutions to the species problem presumes an exclusion of the species notion/concept from the contemporary biological thesaurus altogether, which would finalize a conceptual history of this problem. This suggestion is justified from various standpoints [Burma 1949, 1954; Davidson 1954; Michener 1962;

---

[3] David Hull admits that species category may evolve, but believes (for an unclear reason) that "this evolution is conceptual, not biological" [Hull 2006: 52].

Sokal and Sneath 1963; Riddle and Hafner 1999; Hendry et al. 2000; Kober 2008], and it is summarized by an agitating for "biology without species" [Kober 2008] or "post-species biology" [Mishler 2021, 2022]. Something similar is presumed by the rejection of the species rank suggested by the adherents of the rankless taxonomic hierarchy [Michener 1962; Ereshefsky 2001; Mishler 2010, 2021, 2022; Zachos 2016; Mishler and Wilkins 2018]. Accordingly, it is suggested to recognize some other natural units for describing "biological reality", which might be amenable to a more rigid operational interpretation, and to preserve for the species notion only its classificatory function [Mahner 1993; Mahner and Bunge 1997; Dupré 2001; Rapini 2004; Baum 2009], i.e., to get back to its scholastic nominalistic account.

However, when bringing such too far-reaching suggestion, one should take into consideration that the general notion of species is sturdily embedded in the thesaurus of many fundamental and applied biological disciplines. So, implementation of this idea may request a substantial reorganization of a large part of the conceptual space of contemporary biology, for such rejection would entail serious corrections of other concepts and notions associated, one way or another, with that of species.

One of the fundamental cognitive functions conducted by the latter is presumed by a certain universally valid epistemic principle. According to it, for any differences between organisms to be explored, one must have some more or less solid unified basis of their comparison that makes them components of a single commonality. Indeed, conceptual "pluralism, after all, is a meaningful position only with respect to things that all in some respect are of the same kind" [Reydon 2005b: 58]. For many research tasks in biology, such currently acknowledged basis refers to the *conspecificity*, i.e., to the belonging of organisms to the same species as a really existing natural unit.

> For instance, in comparative analyses of different ecosystems, there is an evident need for a certain unified basis of comparison (the abovementioned "denominator") of the biomorphs (guilds etc.) recognized within each of the local ecosystems. Currently, it is the species that fulfills such function of serving as a "universal currency": for the evolutionary ecologists, biomorphs exist in the particular ecosystems not by themselves but as the local manifestations of the widespread species [Schwartz 1980].

From this perspective, it is clear that, for biology to get rid of species, it is necessary to introduce another no less general concept that might provide another solid unified basis of broad biologically sound comparisons [LaPorte 2007; Pavlinov 2013a, 2021a, 2022a]. It is evident that such replacement would require a serious justification by the reference to a certain fundamental, sufficiently ample biological theory embracing various branches of biology and not limited, say, to a cladistic section of systematics with its general lineage conception. With this, it is to be kept in mind that any new general concept, developed in the framework of a "postspecies biology", will face the same cognitive problems considered above, viz., ontic and epistemic reduction, multiplicity and incommensurability of particular interpretations, their fuzziness, etc. So the cognitive situation, now shaped by the species problem, will simply transform into another one, no less problematic. Accordingly, such a shift will certainly require the understanding and thorough analysis of a "new old" problem from the conceptualistic standpoint exposed in this book.

References

Abachiev, S. K. 2004. [*Evolutionary theory of knowledge. An experience of systemic construction.*] Moscow: Editorial URSS. (in Russian).

Abbott, R. J., M. G. Ritchie, and P. M. Hollingsworth. 2008. Introduction. Speciation in plants and animals: Pattern and process. *Philosophical Transactions of the Royal Society B* 363:2965–9.

Abramson, N. I. 2007. [Phylogeography: results, problems, prospects.] *Bulletin of the All-Russian Society of Geneticists and Breeders* 11:307–31. (in Russian).

Abruzzi, W. S. 2015. *A temporary convenience. A critical review of the species concept.* https://www.researchgate.net/publication/334577221_A_Temporary_Convenience_A_Critical_Review_of_the_Species_Concept.

Achtman, M., and M. Wagner. 2008. Microbial diversity and the genetic nature of microbial species. *Nature Reviews Microbiology* 6:431–40.

Adanson, M. 1763. *Familles des plantes*, Pt. 1. Paris: Vincent.

Agapow, P.-M., O. R. Bininda-Emonds, K. A. Crandall, et al. 2004. The impact of species concept on biodiversity studies. *The Quarterly Review of Biology* 79:161–79.

Agassiz, L. 1859. *An essay on classification.* London: Longman.

Agassiz, L. 1869. *De l'espèce et de la classification en zoologie.* Paris: G. Baillière.

Agassiz, L. 1886. *Methods of study in natural history*, 7th ed. Boston, MA: Houghton, Mifflin and Co.

Akhutin, A. V. 1988. [*The concept of Nature in Antiquity and Modern Times.*] Moscow: Nauka. (in Russian).

Albertson, D. 2014. *Mathematical theologies. Nicholas of Cusa and the legacy of Thierry of Chartres.* Oxford: Oxford University Press.

Albertus Magnus. 1867. *Alberti Magni ex Ordine Praedicatorum. De Vegetabilibus Libri VII, Historiae Naturalis Pars XVIII.* Berlin: Georg Reimer.

Albertus Magnus. 1999. *On animals. A Medieval Summa Zoologica*, Vols. I, II. Baltimore, MD: Johns Hopkins University Press.

Aldhebiani, A. Y. 2018. Species concept and speciation. *Saudi Journal of Biological Sciences* 25:437–40.

Aldrovandi, U. 1602. *De animalibus insectis libri septem, cum singulorum iconibus ad viuum expressis [...].* Bononiae: Ioan. Bapt. Bellagambam.

Aleev, Yu. G. 1986. [*Ecomorphology.*] Kiev: Naukova Dumka. (in Russian).

Alello, T. 2003. Pierre Magnol: His life and works. *Magnolia* 38:1–10.

Alexander, S. A. 2013. Infinite graphs in systematic biology, with an application to the species problem. *Acta biotheoretica* 61:181–201.

Alimov, A. F., and S. D. Stepanyanz (eds.). 2009. [*Species and speciation: An analysis of new views and trends.*]. Saint Petersburg: Zoological Institute. (in Russian).

Allen, M. J. B. 1995. *Plato's Third Eye: Studies in Marsilio Ficino's metaphysics and its sources.* Aldershot: Variorum.

Allen, C., and M. Bekoff. 1999. *Species of mind. The philosophy and biology of cognitive ethology.* Cambridge: MIT Press.

Allmon, W. D. 2016. Studying species in the fossil record: A review and recommendations for a more unified approach. In *Species and speciation in the fossil record*, eds by W. D. Allmon and M. M. Yacobucci, 59–120. Chicago, IL: University of Chicago Press.

Amadon, D. 1966. The superspecies concept. *Systematic Zoology* 153:245–9.

Amitani, Y. 2010. *The persistence question of the species problem.* Doctoral dissertation. Vancouver: University of British Columbia. https://central.bac-lac.gc.ca/.item?id=TC-BVAU-30536&op=pdf&app=Library&oclc_number=1032970076.

Amitani, Y. 2013. The communication puzzle of the species problem. *Annals of the Japan Association for Philosophy of Science* 21:1–20.

Amitani, Y. 2015. Prototypical reasoning about species and the species problem. *Biological Theory* 10:289–300.

Amitani, Y. 2017. The general concept of species. *The Journal of Philosophical Ideas, Special Issue*: 89–120.

Amitani, Y. 2022 Is the species problem *that* important. In *Species problems and beyond. Contemporary issues in philosophy and practice*, eds J. S. Wilkins, F. E. Zachos and I. Ya. Pavlinov, 39–63. Boca Raton, FL: CRC Press.

Amundson, R. 1998. Typology reconsidered: two doctrines on the history of evolutionary biology. *Biology and Philosophy* 13:153–77.

Andersen, H. 2002. The development of scientific taxonomies. In *Model-based reasoning*, eds L Magnani and N. J. Nersessian, 95–111. Boston, MA: Springer.

Andersson, L. 1990. The driving force: Species concepts and ecology. *Taxon* 39:375–82.

Andino, R., and E. Domingo. 2015. Viral quasispecies. *Virology* 479:46–51.

Antipenko, L. G. 1986. [*The problem of incompleteness of the theory and its epistemological meaning.*] Moscow: Nauka. (in Russian).

Appolonov, A. V. 2001. [John Duns Scotus and his philosophical theology.] In [*Blessed John Duns Scotus. A treatise on the Origin]*, ed. A. V. Appolonov, v–xxxvi. Moscow: Franciscan Publishing House. (in Russian).

Appolonov, A. V. 2002. [Roger Bacon.] In [*Anthology of medieval thought*] Vol. 2, ed. S. S. Neretina, 81–8. St. Petersburg: RHGI. (in Russian).

Arber, A. 1938. *Herbals: Their origin and evolution. A chapter in the history of botany, 1470–1670.* Cambridge: Cambridge University Press.

Archibald, J. D. 2014. *Aristotle's ladder, Darwin's tree. The evolution of visual metaphors for biological order.* New York, NY: Columbia University Press.

Ariew, A. 2008. Population thinking. In *Oxford handbook of philosophy of biology*, ed. M. Ruse, 64–86. Oxford: Oxford University Press.

Aristotle. 1883. *History of animals.* London: George Bell & Sons.

Aristotle. 1943. *Generation of animals.* London: William Heinemann.

Aristotle. 1962. *The categories. On interpretation. Prior analytics.* London: William Heinemann.

Aristotle. 2001. *On the parts of animals.* Oxford: Oxford University Press.

Aristotle. 2010. *De anima.* Newburyport: Focus Publishing & R. Pullins.

Arlig, A. 2011. Is there a Medieval mereology? In *Methods and methodologies. Aristotelian logic, East and West, 500–1500*, eds M. Cameron and J. Marenbon, 161–90. Leiden: E. J. Brill.

Armand, A. D. 2008. [*Two in one: The law of complementarity.*] Moscow: LKI. (in Russian).

Arnoldi, K. V. 1939. [On the question of continuous geographical variability in its general and taxonomic meaning.] *Zoological Journal* 13:685–710. (in Russian).

Arteca, G. A., and P. G. Mezey 1989. Two approaches to the concept of chemical species: Relations between potential energy and molecular shape. *International Journal of Quantum Chemistry* 36 (Supplement S23):305–20.

Artedi, P. 1738. *Ichthyologia sive opera omnia de piscibus [...] omnia in hoc genere perfection, quam antea ulla [...].* Lugduni Batavorum: Wishoff.

Ashworth, E. J. 2013. Analogy and metaphor from Thomas Aquinas to Duns Scotus and Walter Burley. In *Later medieval metaphysics. Ontology, language, and logic*, eds C. Bolyard and R. Keele, 223–48. New York, NY: Fordham University Press.

Asmus, V. F. 1976. [*Antique philosophy.*] Moscow: Vysshaya Shkola. (in Russian).

Assis, L. C. S. 2011. Species, reality and evidence: A reply to Reydon. *Cladistics* 27:1–3.
Asztalos, M. 1993. Boethius as a transmitter of Greek logic to the Latin West: The Categories. *Harvard Studies in Classical Philology* 95:367–407.
Athanasopoulos, C. 2010. Ousia in Aristotle's categories. *Logique & Analyse* 53:211–43.
Atran, S. 1987. Origin of the species and genus concepts: An anthropological perspective. *Journal of the History of Biology* 20:195–279.
Atran, S. 1990. *The cognitive foundations of natural history: Towards an anthropology of science*. New York, NY: Cambridge University Press.
Atran, S. 1998. Folk biology and the anthropology of science: Cognitive universals and cultural particulars. *Behavioral and Brain Sciences* 21:547–609.
Augustine. 2020. *The literal interpretation of Genesis: Unfinished book*. https://www.roger-pearse.com/weblog/wp-content/uploads/2020/08/Augustine-De_Genesi_Imperfectum_Image-v4.pdf.
Austin, C. J. 2018. *Essence in the age of evolution. A new theory of natural kinds*. New York: Routledge.
Avise, J. C. 2000. *Phylogeography. The history and formation of species*. Cambridge, MA: Harvard University Press.
Avise, J. C. 2009. Phylogeography: Retrospect and prospect. *Journal of Biogeography* 36:3–15.
Avise, J.C., J. Arnold, M. Ball, et al. 1987. Itraspecific phylogeography: Mitochondrial DNA bridge between population genetics and systematics. *Annual Review of Ecology and Systematics* 18:489–522.
Awbrey, F. T. 1981. Defining "kinds" do creationists apply a double standard? *Creation/Evolution Journal* 2:1–6.
Ayala, F. J., and R. Arp (eds). 2010. *Contemporary debates in philosophy of biology*. Malden, MA: John Wiley & Son.
Ayers, M. R. 1981. Locke versus Aristotle on natural kinds. *The Journal of Philosophy* 78:247–72.
Ayers, M. R. 1993. *Locke. Epistemology and ontology*. London: Routledge.
Bachmann, H. 1896. Karl Nikolaus Lang, Dr. phil. et medic., 1670–1741. *Mitteilungen des Historischen Vereins Zentralschweiz* 51:167–280.
Bachmann, H. 1906. Der Speziesbegiff. *Verhandlungen der Schweizerischen Naturforschenden Gesellschaft* 88:161–208.
Bachmann, K. 1998. Species as units of diversity: An outdated concept. *Theory in Biosciences* 117:213–30.
Bäck, A. 2008. Avicenna the commentator. In *Medieval commentaries on Aristotle's categories*, ed. L. A. Newto, 31–72. Leiden: E.J. Brill.
Baetu, T. M. 2012. Defining species: A multi-level approach. *Acta Biotheoretica* 60: 239–55.
Bailey, L. H. 1896. The philosophy of species-making. *Botanical Gazette* 22 (6): 454–62.
Baker, R.J. and R. D. Bradley. 2006. Speciation in mammals and the genetic species concept. *Journal of Mammalogy* 87:643–62.
Balakrishnan, R. 2005. Species concepts, species boundaries and species identification: A view from the tropics. *Systematic Biology* 54:689–93.
Balme, D. M. 1962. ΓΕΝΟΣ and ΕΙΔΟΣ in Aristotle's biology. *Classical Quarterly* 12:81–98.
Balme, D. M. 1980. Aristotle's biology was not essentialist. *Archiv für Geschichte der Philosophie* 62:1–12.
Baranov, P. A. (ed.). 1958. [*The species problem in botany*]. Moscow–Leningrad: USSR Academy of Sciences.
Barantsev, R. G. 1983. [System triads and classification.] In *[Theory and methodology of biological classifications]*, eds Yu. A.Shreyder and B. S. Sornikov, 81–89. Moscow: Nauka. (in Russian).
Barantsev, R. G. 2003. [*Synergetics in modern natural science.*] Moscow: URSS. (in Russian).

Barberousse, A., and S. Samadi. 2010. Species from Darwin onward. *Integrative Zoology* 5:187–97.

Barker, M. J. 2010. Specious intrinsicalism. *Philosophy of Science* 77:73–91.

Barker, M. J. 2019a. Species and other evolving lineages as feedback systems. *Philosophy Theory and Practice in Biology* 11:13.

Barker, M. J. 2019b. Eliminative pluralism and integrative alternatives: The case of species. *The British Journal for the Philosophy of Science* 70:657–81.

Barker, M. J. 2022. We are nearly ready to begin the species problem. In *Species problems and beyond. Contemporary issues in philosophy and practice*, eds J. S. Wilkins, F. E. Zachos, and I. Ya. Pavlinov, 3–38. Boca Raton, FL: CRC Press.

Barker, M. J. and J. D. Velasco. 2013. Deep conventionalism about evolutionary groups. *Philosophy of Science* 80:971–82.

Barker, M. J., and R. A. Wilson 2010. Cohesion, gene flow, and the nature of species. *The Journal of Philosophy* 107:61–79.

Barraclough, T. G. 2019. *The evolutionary biology of species.* Oxford: Oxford University Press.

Barsanti, G. 2011. Linnaeus: The order of nature, the nature of man, and evolution. *Journal of the Siena Academy of Sciences* 3:61–4.

Bartlett, S. J. 2015. *The species problem and its logic: Inescapable ambiguity and framework-relativity.* http://philsci-archive.pitt.edu/11655/1/Bartlett_The_Species_Problem_and_Its_Logic.pdf.

Bateson, W. 1894. *Materials for the study of variation treated with especial regard to discontinuity of the origin of species.* London: Macmillan & Co.

Bateson, W. 1899 (2009). Hybridisation and cross-breeding as a method of scientific investigation. In *William Bateson, F.R.S. Naturalist*, ed. B. Bateson, 161–170. Cambridge: Cambridge University Press.

Bateson, W. 1913. *Problems of genetics.* New Haven, CT: Yale University Press.

Bauhin, C. 1623. *ΠΙΝΑΞ Theatri botanici Caspari Bauhini [...].* Basileae Helvet.: Ludovici Regii.

Baum D.A. 2009. Species as ranked taxa. *Systematic Biology* 58:74–86.

Baum, D. A. 1998. Individuality and the existence of species through time. *Systematic Biology* 47:641–53.

Baum, D. A., and M. J. Donoghue 1995. Choosing among alternative phylogenetic species concepts. *Systematic Botany* 20:560–73.

Baum, D. A., and K. L. Shaw 1995. Genealogical perspectives on the species problem. In *Experimental and molecular approaches to plant biosystematics*, eds P. C. Hoch and A.G. Stephenson, 289–303. St. Louis, MO: Missouri Botanical Garden.

Baxter, R. 1998. *Bestiaries and their users in the Middle Ages.* Stroud: Sutton Publishing.

Beardsmore, R. W. 1992. The theory of family resemblance. *Philosophical Investigations* 15:131–46.

Beatty, J. 1983. What's in a word? Coming to terms in the Darwinian revolution. In *Nature animated*, Vol. 2., ed. M. Ruse, 79–100. Dordrecht: Reidel.

Beatty, J. 1985. Speaking of species: Darwin's strategy. In *The Darwinian heritage*, ed. D. Kohn, 265–81. Princeton, NJ: Princeton University Press.

Beckner, M. 1959. *The biological way of thought.* New York, NY: Columbia University Press.

Bedau, M. A., J. S. McCaskill, and N. H. Packard 2000. Open problems in artificial life. *Artificial Life* 6:363–76.

Benjamin, A. C. 1955. *Operationism.* Springfield, IL: Charles C. Thomas Publ.

Bentham, G. 1827. *Outline of a new system of logic: With a critical examination of Dr. Whately.* London: Hunt and Clarke.

Bentham, G. 1875. On the recent progress and present state of systematic botany. *Report of the British Association for the Advancement of Science*, pp. 27–54.

Berlin, B. 1992. *Ethnobiological classification: Principles of categorization of plants and animals in traditional societies*. Princeton, NJ: Princeton University Press.

Berndt, C. 2000. *Biblical classification of life: A framework and reference for authentic biblical biology*. Wildomar: Elihu Books.

Bernie, R. 1984. The species as an individual: Facing essentialism. *Systematic Zoology* 33:460–9.

Berti, E. 2016. Aristotle's concept of Nature: Traditional interpretation and results of recent studies. *Pontifical Academy of Sciences* 23. www.pas.va/content/dam/accademia/pdf/acta23/acta23-berti.pdf.

Besnard, A. F. 1864. Altes und Neues zur Lehre über die organische Art (Spezies). *Abhandlungen des zoologisch-mineralogischen Vereines in Regensburg* 9:1–72.

Bessey, C. E. 1908. The taxonomic aspect of the species question. *The American Naturalist* 42: 218–24.

Beurton, P. 2002. Ernst Mayr through time on the biological species concept a conceptual analysis. *Theory in Biosciences* 121:81–98.

Beurton, P., R. Falk, and H.-J. Rheinberger. 2000. *The concept of the gene in development and evolution*. Cambridge: Cambridge University Press.

Bianchi V. 1916. [Species and its subordinate taxonomic forms.] *Russian Zoological Journal* 1:287–97. (in Russian).

Bird, A., and E. Tobin. 2018. Natural kinds. In *Stanford encyclopedia of philosophy*, ed. E. N. Zalta. https://plato.stanford.edu/archives/spr2018/entries/natural-kinds/.

Blackman, R. 1995. What's in a name. Species concepts and realities. *Bulletin of Entomological Research* 85:1–4.

Blackwelder, R. E. 1967. *Taxonomy. A text and reference book*. New York, NY: John Wiley & Sons.

Blackwelder, R. E., and A. Boyden. 1951. The nature of systematics. *Systematic Zoology* 1:26–33.

Blair, W. F. 1956. The species as a dynamic system. *The Southwestern Naturalist* 1: 1–5.

Blaxter, M., J. Mann, T. Chapman, et al. 2005. Defining operational taxonomic units using DNA barcode data. *Philosophical Transactions of the Royal Society B* 360:1935–43.

Blumenbach, J. F. 1782. *Handbuch der Naturgeschichte*. Göttingen: Johann Christian Dieterich.

Boas, F. 1898. A precise criterion of species. *Science* 7:860–1.

Bobay, L.-M., and H. Ochman 2018. Biological species in the viral world. *Proceedings of the National Academy of Sciences* 115:6040–5.

Bobay, L.-M., and H. Ochman. 2017. Biological species are universal across life's domains. *Genome Biology and Evolution* 9:491–501.

Bobrov, E. G. 1957. [*Linnaeus, his life and works*.] Moscow–Leningrad: Publishing House of the USSR Academy of Sciences. (in Russian).

Bobrov, E. G. 1970. [*Carl Linnaeus. 1707–1778*.] Leningrad: Nauka. (in Russian).

Bock, W. J. 2004. Species: The concept, category and taxon. *Journal of Zoological Systematics and Evolutionary Research* 42:178–90.

Boecklen, W. J. 2017. Topology of syngameons. *Ecology and Evolution* 7: 10486–91.

Boehner, P. (ed.). 1964. *William of Ockham: Philosophical writings, a selection*. Indianapolis, IN: Bobbs-Merrill.

Boero, F. 2015. From Darwin's *Origin of Species* toward a theory of natural history. *F1000Prime Reports* 7:49. https://www.ncbi.nlm.nih.gov/pmc/articles/PMC4447030/.

Boëthius. 1891. *Anicii Manlii Severini Boethii in Categorias Aristotelis Commentaria*. http://www.logicmuseum.com/wiki/Authors/Boethius/categories/incategorias. (Russian edition of 2001, Moscow: Nauka).

Boëthius. 1906. *Anicii Manlii Severini Boethii in Isagogen Porphyrii commentorum editionis secundae*. Vienna: F. Tempsky/Leipzig: G. Freitag. http://www.logicmuseum.com/wiki/Authors/Boethius/isagoge/CPorII. (Russian edition of 1990, Moscow: Nauka.).

Bond, J. E., and A. K. Stockman. 2008. An integrative method for delimiting cohesion species: Finding the population-species interface in a group of californian trapdoor spiders with extreme genetic divergence and geographic structuring. *Systematic Biology* 57:628–46.

Bonnet, C. 1769. *Contemplation de la Nature*, 2d ed. Vol. 1. Amsterdam: Marc-Michel Rey.

Bonneuil, C. 2002. The manufacture of species: Kew gardens, the empire, and the standardisation of taxonomic practices in late 19th century botany. In *Instruments, travel and science. Itineraries of precision from the 17th to the 20th century*, eds M.-N. Bourguet, C. Licoppe, O. Sibum, 189–215. London & New York, NY: Routledge.

Borghini, A. 2000. The relational species concept. *Academia Letters*. https://www.academia.edu/18825648/The_Relational_Species_Concept.

Borgmeier, T. 1957. Basic questions of systematics. *Systematic Zoology* 6:53–69.

Botha, M. E. 1989. Theory development in perspective: The role of conceptual frameworks and models in theory development. *Journal of Advanced Nursing* 14:49–55.

Bowler, P. J. 1973. Bonnet and Buffon: Theories of generation and the problem of species. *Journal of the History of Biology* 6:259–81.

Bowler, P. J. 1975. The changing meaning of "evolution". *Journal of the History of Ideas* 36:95–114.

Bowler, P. J. 2003. *Evolution: The history of an idea*, 3d ed. Berkeley, CA: University of California Press.

Bowler, P. J. 2005. Variation from Darwin to the Modern Synthesis. In *Variation: A central concept in biology*, eds B. Hallgrimsson and B. K. Hall, 9–27. San Diego, CA: Academic Press.

Boyd, R. 1999. Homeostasis, species, and higher taxa. In *Species: New interdisciplinary essays*, ed. R. A. Wilson, 141–85. Cambridge, MA: MIT Press.

Boylan, M. 1983. *Method and practice in Aristotle's biology*. Lanham, MD: University Press of America.

Bradie, M., and W. Harms 2016. Evolutionary epistemology. In *Stanford encyclopedia of philosophy*, ed. E. N. Zalta. https://plato.stanford.edu/archives/spr2016/entries/epistemology-evolutionary/.

Brasier, C. M. 1997. Fungal species in practice: Identifying species units in fungi. In *Species: The units of biodiversity*, eds M. F. Claridge, H. A. Dawah, and M. R. Wilson, 135–70. London: Chapman & Hall.

Bréhier, É. 1958. *Philosophy of Plotinus*. Chicago, IL: University of Chicago Press.

Breidbach, O. and M. Ghiselin. 2006. Baroque classification: A missing chapter in the history of systematics. *Annals of the History and Philosophy of Biology* 11:1–30.

Bremekamp, C. E. B. 1953. A re-examination of Cesalpino's classification. *Acta Botanica Neerlandica* 1:580–93.

Brenton, L. C. L. 1980. *The Septuagint with apocrypha: Greek and English*. Grand Rapids, MI: Zondervan Publ.

Brigandt, I. 2003a. Species pluralism does not imply species eliminativism. *Philosophy of Science* 70:1305–16.

Brigandt, I. 2003b. Homology in comparative, molecular, and evolutionary developmental biology: The radiation of a concept. *Journal of Experimental Zoology. Pt. B: Molecular and Developmental Evolution* 299B:9–17.

Brigandt, I. 2009. Natural kinds in evolution and systematics: Metaphysical and epistemological considerations. *Acta Biotheoretica* 57:77–97.

Brigandt, I. 2012. The dynamics of scientific concepts: the relevance of epistemic aims and values. In *Scientific concepts and investigative practice*, eds U. Feest and F. Steinle, 75–104. Baltimore, MD: John Hopkins University Press.

Brigandt, I., and P. E. Griffiths 2007. The importance of homology for biology and philosophy. *Biology and Philosophy* 22:633–41.

Brogaard, B. 2004. Species as individuals. *Biology and Philosophy* 19:223–42.
Bronowski, A. 2007. The stoic view on universlas. *Documenti e studi sulla tradizione filosofica medievale* 18:71–87.
Brooks, D. R., and D. A. McLennan. 1999. Species: Turning a conundrum into a research program. *Journal of Nematology* 31:117–33.
Brooks, D. R., and E. O. Wiley. 1986. *Evolution as entropy*. Chicago, IL: University Chicago Press.
Brothers, D. J. 1985. Species concepts, speciation, and higher taxa. In: *Species and speciation*, ed. E. S. Vrba, 35–42. Pretoria: Transvaal Museum.
Brower, A. V. Z. 2021. A slippery reality: The epistemological shifting sands of tokogeny, phylogeny, lineages and species delimitation. *Systematics and Biodiversity* 19:782–96.
Brower, J. E. 2016. Aquinas on the problem of universals. *Philosophy and Phenomenological Research* 92:715–35.
Brower, J. E., and K. Guilfoy (ed.). 2004. *The Cambridge companion to Abelard*. Cambridge (UK): Cambridge University Press.
Brown, C. H. 1977. Folk botanical life-forms: Their universality and growth. *American Anthropologist* 79:317–42.
Brown, D. E. 1991. *Human universals*. Philadelphia, PA: Temple University Press.
Brunfels, O. 1530–1536. *Herbarum vivae eicones ad naturae imitationem summa cum diligentia et artificio effigiatse [...]*, T. I–III. Argentorati: Ioannem Scottü.
Bryan, M. 2005. *John Ray (1627–1705), Pioneer in the natural sciences. A celebration and appreciation of his life and work*. Braintree: John Ray Trust.
Buffon, G.-L. 1749. *Histoire naturelle, générale et particulière, avec la description du Cabinet du Roy*, T. 1. Paris: L'Imprimere royale.
Buffon, G.-L. 1755a. *Histoire naturelle, générale et particulière, avec la description du Cabinet du Roy*, T. 5. Paris: L'Imprimere royale.
Buffon, G.-L. 1755b. Espèce (Hist. Nat.). In *Encyclopedie, ou Dictionaaire raisonné, des arts et métiers*, T. 5, ed. D. Diderot, 956–7. Paris: Briasson, David, Le Breton & Durand.
Buffon, G.-L. 1756. *Histoire naturelle, générale et particulière, avec la description du Cabinet du Roy*, T. 6. Paris: L'Imprimere royale.
Buffon, G.-L. 1766. *Histoire naturelle, générale et particulière, avec la description du Cabinet du Roy*, T. 14. Paris: L'Imprimere royale.
Buffon, G.-L. 1769. *Histoire naturelle, générale et particulière*. Nouvelle édition, T. 10. Paris: L'Imprimere royale.
Buffon, G.-L. 1770. *Histoire naturelle des Oiseaux*, T. 1. Paris: L'Imprimere royale.
Buffon, G.-L. 1843. De la Nature. Second vue. In *Oeuvres choisies de Buffon [...]*, T. 1, 52–64. Paris: Pourrat Fréres.
Buffon, G.-L. 1954. *Oeuvres philosophiques*. Paris: Presses Universitaires de France.
Buhle, J.-G. 1816. *Histoire de la philosophie moderne, depuis la Renaissance des lettres jusqu'a Kant, Tome deuxième, 1re partie*. Paris: F. J. Fournier.
Bunge, M. 1977. *Ontology, 1: The furniture of the world*. Dordrecht: Reidel.
Burbrink, F. T., B. I. Crother, C. M. Murray, et al. 2022. Empirical and philosophical problems with the subspecies rank. *Ecology and Evolution* 12:1–17.
Burkhardt, R. W. 1977. *The spirit of system: Lamarck and evolutionary biology*. Cambridge, MA: Harvard University Press.
Burkhardt, R. W. 1987. Lamarck and species. In *Histoire du concept d'espece dans les sciences de la vie*, eds. S. Atran, J.-L. Fischer, and J. Roger, 161–80. Paris: Fondation Singer-Polignac.
Burma, B. H. 1949. The species concept: A semantic review. *Evolution* 3:369–70.
Burma, B. H. 1954. Reality, existence, and classification: A discussion of the species problem. *Madroño* 12:193–209.
Bykhovsky, B. E. (ed.). 1972. [*The development of a concept of structural levels in biology*.] Moscow: Nauka. (in Russian).

Bzovy, J. 2016. *Species pluralism: Conceptual, ontological, and practical dimensions*. PhD Thesis. London: University of Western Ontario. https://ir.lib.uwo.ca/cgi/viewcontent.cgi?article=5976&context=etd.

Caesalpino A. 1571. *Andreae Caesalpini Aretini, Quaestionum Peripateticarum Lib.V. [...]*. Venetiis: Apud Iuntas.

Cesalpino A. 1583. *De plantis libri XVI Andreae Cesalpini Aretini [...]*. Florentiae: Georgium Marescottum.

Cain, A. J. 1954. *Animal species and their evolution*. Princeton, NJ: Princeton University Press.

Cain, A. J. 1958. Logic and memory in Linnaeus' system of taxonomy. *Proceedings of the Linnean Society of London* 169:144–63.

Cain, A. J. 1994a. Rank and sequence in Caspar Bauhin's Pinax. *Botanical Journal of the Linnean Society* 114:311–56.

Cain, A. J. 1994b. Numerus, figura, proportio, situs; Linnaeus's definitory attributes. *Archives of Natural History* 21:17–36.

Cain, A. J. 1996. John Ray on "accidents". *Archives of Natural History* 23:343–68.

Cain, A. J. 1997. John Locke on species. *Archives of Natural History* 24:337–60.

Cain, A. J., and G. A. Harrison. 1958. An analysis of the taxonomist's judgment of affinity. *Proceedings of the Zoological Society of London* 131:85–98.

Camardi, G. 2001. Richard Owen, morphology and evolution. *Journal of the History of Biology* 34:481–515.

Cameron, M. 2007. Abelard (and Heloise?) on intention. *American Catholic Philosophical Quarterly* 81:323–38.

Camp, W. H. 1951. Biosystematy. *Brittonia* 7:113–27.

Camp, W. H., and C. L. Gilly. 1943. The structure and origin of species, with a discussion of intraspecific variability and related nomenclatural problems. *Brittonia* 4:323–85.

Cantino, P. D., and K. de Queiroz. 2020. *International code of phylogenetic nomenclature (PhyloCode), Version 6*. Boca Raton, FL: CRC Press.

Caplan, A. L. 1980. Have species become déclassé? In *Proceedings of the Biennial Meeting of the Philosophy of Science Association*, Vol. 1, 71–82.Chicago:University of Chicago Press

Caplan, A. L. 1981a. Pick your poison: Historicism, essentialism, and emergentism in the definition of species. *The Behavioral and Brain Sciences* 4:285–286.

Caplan, A. L. 1981b. Back to class: a note on the ontology of species. *Philosophy of Science* 48:130–40.

Carnap, R. 1969. *The logical structure of the world: And pseudoproblems in philosophy*. Chicago, IL: Open Court Publ.

Carr, D. C. 1923. *Linnaeus and Jussieu, or The rise and progress of systematic botany*. London: John W. Parker.

Carr, J. 1997. *Aristotle's use of "genus" in logic, philosophy, and science*. Doctoral dissertation. Edinburgh: University of Edinburgh.

Carrier, M. 1994. *The completeness of scientific theories*. Norwell, MA: Kluwer Academic Publ.

Caruel, T. 1883. Pensées sur la taxinomie botanique. *Botanische Jahrbücher für Systematik, Pflanzengeschichte und Pflanzengeographie* 4:549–616.

Casacci, L. P., F. Barbero and E. Balletto. 2014. The "evolutionary significant unit" concept and its applicability in biological conservation. *Italian Journal of Zoology* 81:182–93.

Casetta, E. 2010. Categories, taxa, and chimeras. In *New essays in logic and philosophy of science*, eds M. D'Agostino, G. Giorello, F. Laudisa, et al., 264–78. London: College Publications.

Casetta, E. 2012. The temporal boundaries of biological species. In *Open problems in philosophy of sciences*, eds P. Graziani, L. Guzzardi, and M. Sangoi, 35–45. London: College Publications.

Casetta, E., and D. Vecchi. 2019. Species are, at the same time, kinds and individuals: A causal argument based on an empirical approach to species identity. *Synthese* 198:3007–25.

Cassirer, E. 1950. *Problem of knowledge. Philosophy, science, and history since Hegel.* New Haven, CT: Yale University Press.

Chambers, G. 2012. The species problem: Seeking new solutions for philosophers and biologists. *Biology and Philosophy* 27:755–65.

Chambers, R. 1844. *Vestiges of the natural history of creation.* London: John Churchill.

Chandler, C. R., and M. H. Gromko. 1989. On the relationship between species concepts and speciation processes. *Systematic Zoology* 38:116–25.

Charles, D. 1997. Aristotle and the unity and essence of biological kinds. In *Aristotelische Biologie. Intentionen, Methoden, Ergebnisse*, eds W. Kullmann and S. Follinger, 27–42. Stuttgart: Pranz Steiner Verlag.

Charlesworth, M. J. 1982. *St. Anselm's Proslogion.* Oxford: Clarendon Press.

Chase, J. M., and M. A. Leibold. 2003. Ecological niches: *Linking classical and contemporary approaches.* Chicago, IL: University of Chicago Press. 216 p.

Chebanov, S. V. 1980. [Internal and external systems in classification theory.] In [*Systems research,*] 140–46. Moscow: Nauka. (in Russian).

Chebanov, S. V. 2007. [To what extent did Linnaeus dealt with classification?] [*Linnaean collection],* ed. I. Ya. Pavlinov, 437–54. Moscow: Moscow State University Publ. (in Russian).

Chernov, Yu. I. 1991. [Biological diversity, its essence, and related problems.] *Uspehi Sovremennoj Biologii* 111:499–507. (in Russian).

Chiaradonna, R., and G. Galluzzo (eds). 2013. *Universals in ancient philosophy.* Pisa: Scuola Normale Superiore Pisa.

Chiaradonna, R. 2008. What is Prophyry's Isagoge? *Documenti E Studi Sulla Tradizione Filosofica Medievale* 19:1–30.

Chiaradonna, R. 2017. Definitions, species, wholes, parts. Clem. Strom. VIII. 6.19.2-20.2. In *Liber Logicus (Stromata VIII) by Clement of Alexandria'*, ed. M. Havrda, 1–13. Prague. https://www.academia.edu/31560456/_Liber_logicus_by_Clement_of_Alexandria_Proof_inquiry_scepticism_causation_in_an_early_Christian_text_An_international_workshop_Prague_April_20-22_2017.

Chung, C. 2003. On the origin of the typological/population distinction in Ernst Mayr's changing views of species, 1942–1959. *Studies in History and Philosophy of Biological and Biomedical Sciences* 34: 277–96.

Claridge, M. F. 2010. Species are real biological entities. In *Contemporary debates in philosophy of biology*, eds F. J. Ayala and R. Arp, 91–109. Malden, MA: John Wiley & Son.

Claridge, M. F., H. A. Dawah, and M. R. Wilson (eds). 1997. *Species: The units of biodiversity.* London: Chapman & Hall.

Clausen, R. T. 1941. On the use of the terms "subspecies" and "variety". *Rhodora* 43:157–67.

Clausen, J., D. D. Keck, and W. M. Hiesey. 1939. The concept of species based on experiment. *American Journal of Botany* 26:103–6.

Clemens, W., and K. G. Dharmendra (ed.). 2015. *Radionuclides in the environment. Influence of chemical speciation and plant uptake on radionuclide migration.* Heidelberg: Springer.

Clements, F. E. 1908. An ecologic view of the species conception. *The American Naturalist* 191:253–64.

Clucas, S. 2010. Scientia and inductio scientifica in the *Logica Hamburgensis* of Joachim Jungius. In *Scientia in early modern philosophy*, eds T. Sorell, G. A. J. Rogers, and J. Kraye, 53–70. Dordrecht: Springer.

Clusius, C. 1601. *Atrebatis rariorum Plantarum Historia [...].* Antverpiae: Ioannem Moretum.

Cockerell, T. D. A. 1897. Physiological species. *Entomological News* 8: 234–6.

Cohan, F. M. 2002. What are bacterial species? *Annual Review of Microbiology*, 56:457–87.

Coleman, W. 1962. Lyell and the "reality" of species: 1830–1833. *Isis* 53:325–38.

Coleman, W. 1964. *Georges Cuvier, zoologist: A study in the history of evolution theory.* Cambridge, MA: Harvard University Press.

Coleman, K. A., and E. O. Wiley. 2001. On species individualism: A new defense of the species-as-individuals hypothesis. *Philosophy of Science* 68:498–517.

Colin, A., and M. Bekoff. 1997. *Species of mind: The philosophy and biology of cognitive ethology.* Cambridge, MA: MIT Press.

Colless, D. H. 1967. An examination of certain concepts in phenetic taxonomy. *Systematic Zoology* 16:6–27.

Colless, D. H. 2006. Taxa, individuals, clusters and a few other things. *Biology and hilosophy* 21:353–67.

Collier, J. 1985. *The reality of species.* Thesis Doctorate in Philosophy. London: University of London. https://repository.royalholloway.ac.uk/file/73428683-ea4a-4855-a8be-e422b3a953e3/1/10098511.pdf.

Collier, J. 1998. *A unified approach to species.* https://www.academia.edu/2986055/A_Unified_Approach_to_Species.

Conix, S. 2019a. Radical pluralism, classificatory norms and the legitimacy of species classifications. *Studies in History and Philosophy of Biological and Biomedical Sciences.* 73: 27–34.

Conix, S. 2019b. Measuring evolutionary independence: A pragmatic approach to species classification. *Biology and Philosophy* 34:53.

Connell, S. M. 2001. Toward an integrated approach to Aristotle as a biological philosopher. *The Review of Metaphysics* 55:297–322.

Cook, O. F. 1906. Factors of species-formation. *Science* 23:506–7.

Cope, E. D. 1887. *The origin of the fittest. Essays of evolution.* New York, NY: Appleton & Co.

Copleston, F. 1993a. *A history of philosophy, Vol. 2. Medieval philosophy.* Doubleday, NY: An Image Book.

Copleston, F. 1993b. *A history of philosophy, Vol. 3. Late Medieval and Renaissance philosophy.* Doubleday, NY: An Image Book.

Corazzon, R. 2020. Plotinus' criticism of Aristotle's categories. *Theory and History of Ontology.* https://www.ontology.co/pdf/categories-plotinus.pdf.

Corrigan, K. 2005. *Reading Plotinus. A Practical introduction to neoplatonism.* West Lafayette, IN: Purdue University Press.

Cowles, H. C. 1908. An ecological aspect of the conception of species. *The American Naturalist* 42:265–71.

Coyne, J. A., and H. A. Orr. 1998. The evolutionary genetics of speciation. *Philosophical Transactions: Biological Sciences* 353:287–305.

Coyne, J. A., and H. A. Orr. 2004. *Speciation.* Sunderland, MA: Sinauer Assoc.

Coyne, J. A., and H. A. Orr. 2010. Speciation: A catalogue and critique of species concepts. In *Philisophy of biology. An anthology*, eds A. Rosenberg and R. Arp, 272–91. Malden, MA: Blackewell Publ.

Coyne, J. A., H. A. Orr, and D. J. Futuyma 1988. Do we need a new species concept? *Systematic Zoology* 37:190–200.

Cracraft, J. 1983. Species concepts and speciation analysis. *Current Ornithology* 1:159–87.

Cracraft, J. 1987. Species concepts and the ontology of evolution. *Biology and philosophy* 2:329–46.

Cracraft, J. 1989a. Species as entities of biological theory. In *What the philosophy of biology is: Essays dedicated to David Hull*, ed. M. Ruse, 31–52. Dordrecht: Kluwer Academic. Publ.

Cracraft, J. 1989b. Speciation and its ontology: The empirical consequences of alternative species concepts for understanding patterns and processes of differentiation. In *Speciation and its consequences*, eds D. Otte and J. Endler, 28–59. Sunderland, MA: Sinauer Assoc.

Cracraft, J. 1993. What, if anything, is a species? In *Species, species concepts, and primate evolution*, eds W. H. Kimbel and L. B. Martin, 3–20. Berlin: Springer Science+Business Media.

Cracraft, J. 1997. Species concepts in systematics and conservation biology: An ornithological viewpoint. In *Species: The units of biodiversity*, eds M. F. Claridge, H. A. Dawah, and M. R. Wilson, 325–39. London: Chapman & Hall.

Cracraft, J. 2000. Species concepts in theoretical and applied biology: A systematic debate with consequences. In *Species concepts and phylogenetic theory: A debate*, eds Q. D. Wheeler and R. Meier, 3–14. New York, NY: Columbia University Press.

Crane, J. K. 2003. Locke's theory of classification. *British Journal for the History of Philosophy* 11:249–59.

Crane, J. K. 2004. On the metaphysics of species. *Philosophy of Science* 71:156–73.

Crane, J. K., and R. Sandler. 2011. Species concepts and natural goodness. In *Carving nature at its joints: Natural kinds in metaphysics and science*, eds J. K. Campbell, M. O'Rourke, and M. H. Slater, 289–311. Cambridge, MA: MIT Press.

Crisp, M. D., and G. T. Chandler. 1996. Paraphyletic species. *Telopea* 6:813–44.

Cronquist, A. 1978. Once again, what is a species? *Beltsville Agricultural Research Center* 2:3–20.

Crowe, T. M. 1999. Species as multifaceted entities. In *Proceedings of the 22th International Ornithological Congress, Durban*, eds Adams N. J. and R. H. Slotow, 1490–5. Johannesburg: BirdLife South Africa.

Cuénot, L. 1936. *L'espèce. (Encyclopédie scientifique. Bibliothèque de biologie générale)*. Paris: G. Doin & Cie Éditeurs.

Cuvier, G. 1805. *Leçons d'anatomie comparée*. Paris: Baudouin.

Cuvier, G. 1834. *Cuvier's Animal Kingdom [...]*. London: Bradbury and Evans, Printers, Whitefriars.

D'Onofrio, S. R. 2008. *Aquinas as representationalist: The ontology of the Species intelligibilis*. Doctoral dissertation. Buffalo, NY: State University of New York at Buffalo. https://www.academia.edu/11058110/Aquinas_as_Representationalist_The_ontology_of_the_Species_Intelligibilis.

Danser, B. 1929. Über die bergriffe Komparium, Kommiskuum und Konvivium und üeber die Entstehungsweise der Konvivien. *Genetica* 11:399–450.

Dardağan, A. 2018. Neoplatonic "Tree of Life" (Arbor Porphyriana: A diagram of logic and mystical theology). *STANAK, Society for Research of Bosnian Medieval History*. https://osf.io/preprints/socarxiv/g2qxe/.

Darlington, C. D. 1940. Taxonomic species and genetic systems. In *The new systematics*, ed. J. Huxley, 137–60. London: Oxford University Press.

Darlington, P. J. 1971. Modern taxonomy, reality, and usefulness. *Systematic Zoology* 20:341–65.

Darwin, C. 1859. *On the origin of species by means of natural selection, or the preservation of favoured races in the struggle for life*. London: John Murray.

Daston, L. J., and P. Galison. 2007. *Objectivity*. New York, NY: Zone Books.

Daudin, H. 1927. De Linné à Jussieu. Méthode de la classification et l'idée de la série en bota nique et en zoologie (1740–1790). *Revue Philosophique de la France et de l'Etranger* 103:469–73.

Davidson, J. F. 1954. A dephlogisticated species concept. *Madroño*, 12:246–51.

Davies, B., and G. R. Evans (eds). 1998. *Anselm of Canterbury. The major works*. Oxford: Oxford University Press.

Dayrat, B. 2005. Towards integrative taxonomy. *Biological Journal of the Linnean Society* 85:407–15.

de Gandolle, A. 1855. *Géographie botanique raisonnée, ou Exposition des faits principaux et des lois concernant la distribution geographique des plantes de l'époque actuelle*. Paris: V. Masson / Gèneve: J. Kessmann.

de Candolle, A. 1867. *Lois de la nomenclature botanique adoptées par le Congrès international de botanique tenu à Paris en août 1867 [...]*. Paris: J. B. Baillere et fils.

de Candolle, A. P. 1819. *Théorie élémentaire de la botanique, ou, Exposition des principes de la classification naturelle et de l'art de décrire et d'étudier les végétaux*, 2nd ed. Paris: Deterville.

de Candolle, A., and C. A. Cogniaux 1876. Quelques points de nomenclature botanique. *Bulletin de la Société Royale de Botanique de Belgique* 15:477–85.

de Jussieu, A. L. 1789. *Genera plantarum secundum ordines naturales disposita [...]*. Parisiis: Viduam Herissant.

de Jussieu, A. L. 1824. *Principes de la method naturelle des végétaux*. Paris: F. G. Levrault.

de Meeûs, T., P. Durand, and F. Renaud. 2003. Species concepts: what for? *Trends in Parasitology* 19:425–7.

De Necker, N. J. 1790. *Phytozoologie philosophique [...]*. Neuwied sur la Rhin: Société typographique.

de Pinna, M. C. C. 2000. Species concepts and phylogenetics. *Reviews in Fish Biology and Fisheries* 9:353–73.

de Queiroz, K. 1998. The general lineage concept of species, species criteria, and the process of speciation. In *Endless forms: Species and speciation*, eds D. J. Howard and S. H. Berlocher, 57–75. Oxford: Oxford University Press.

de Queiroz, K. 1999. The general lineage concept of species and the defining properties of species category. In *Species: New interdisciplinary essays*, ed. R. A. Wilson, 49–89. Cambridge, MA: MIT Press.

de Queiroz, K. 2005a. A unified concept of species and its consequences for the future of taxonomy. *Proceedings of the California Academy of Sciences*, ser. 4, 56 (Suppl. 1):196–215.

de Queiroz, K. 2005b. Different species problems and their resolution. *BioEssays* 27:1263–9.

de Queiroz, K. 2005c. Ernst Mayr and the modern concept of species. *Proceedings of the National Academy of Sciences* 102 (Suppl. 1):6600–7.

de Queiroz, K. 2007. Species concepts and species delimitation. *Systematic Biology* 56:879–86.

de Queiroz, K., and M. J. Donoghue. 1988. Phylogenetic systematics and the species problem. *Cladistics* 4:317–38.

de Queiroz, K., and M. J. Donoghue. 1990. Phylogenetic systematics and species revisited. *Cladistics* 6:83–90.

de Queiroz, K., and J. Gauthier. 1992. Phylogenetic taxonomy. *Annual Review of Ecology, Evolution and Systematics* 23:449–80.

de Sousa, R. 1989. Kinds of kinds: Individuality and biological species. *International Studies in the Philosophy of Science* 3:119–35.

De Vries, H. 1904. *Species and varieties: Their origin by mutation*. Chicago, IL: Open Court Publ.

Dean, J. 1979. Controversy over classification: A case study from the history of botany. In *Natural order. Historical studies of scientific culture*, eds B. Barnes and S. Shapin, 211–30. London: Sage Publ.

Dean, J. P. 1980. *A naturalistic model of classification and its relevance to some controversies in botanical systematics, 1900–1950*. Doctoral dissertation. University of Edinburgh. https://era.ed.ac.uk/handle/1842/17590.

DeCandolle, A. P., and K. Sprengel. 1821. *Elements of the philosophy of plants, containing the principles of scientific botany [...]*. Edinburgh: Blackwood.

Demarest, B. 2017. Kant's epigenesis: specificity and developmental constraints. *History and Philosophy of Life Sciences* 39: 3.

Descartes, R. 1982. *Principles of philosophy*. Dodrecht: Springer.

Del Soldato, E. 2020. Natural philosophy in the Renaissance. In *Stanford encyclopedia of philosophy*, ed. E. N. Zalta. https://plato.stanford.edu/archives/fall2020/entries/natphil-ren/.

Deverreux, D., and P. Pellegrin (eds). 1990. *Biologie, logique et metaphysique chez Aristote.* Paris: CNRS.

Devitt, M. 2008. Resurrecting biological essentialism. *Philosophy of Science*, 75:344–82.

Devitt, M. 2011. Natural kinds and biological realisms. In *Carving nature at its joints: Natural kinds in metaphysics and science*, eds J. K. Campbell, M. O'Rourke, and M. H. Slater, 155–73. Cambridge, MA: MIT Press.

Devitt, M. 2018a. Individual essentialism in biology. *Biology and Philosophy* 33: 39.

Devitt, M. 2018b. Historical biological essentialism. *Studies in History and Philosophy of Biological and Biomedical Sciences* 71:1–7.

DeYoung, R. K., C. McCluskey, and C. Van Dyke. 2009. *Aquinas's ethics. Metaphysical foundations, moral theory, and theological context.* Notre Dame, IN: University of Notre Dame Press.

Di Liscia, D.A., E. Kessler, and C. Methuen (eds). 1997. *Method and order in Renaissance philosophy of nature. The Aristotle commentary tradition.* Brookfield, VT: Ashgate Publishing.

Dillon, J. 1996. *The Middle Platonists. 80 B.C. to A.D. 220.* 2nd ed. Ithaca, NY: Cornell Universitv Press.

Dillon, L. S. 1966. The life cycle of the species: an extension of current concepts. *Systematic Zoology* 15:112–26.

Dobzhansky, T. 1935. A critique of the species concept in biology. *Philosophy of Science* 2:344–55.

Dobzhansky, T. 1937. *Genetics and the origin of species.* New York, NY: Columbia University Press.

Dobzhansky, T. 1970. *Genetics of evolutionary process.* New York, NY: Columbia University Press.

Dobzhansky, T. 1973. Nothing in biology makes sense except in the light of evolution. *The American Biology Teacher* 35:125–9.

Dodoens, R. 1553. *Trium priorum de stirpium historia commentariorum imagines ad vivum expressae [...].* Antverpiae: Joanni Leoi.

Domingo, E. 2002. Quasispecies theory in virology. *Journal of Virology* 76:463–5.

Donegan, T. M. 2018. What is a species? A new universal method to measure differentiation and assess the taxonomic rank of allopatric populations, using continuous variables. *ZooKeys* 757:1–67.

Donoghue, M. J. 1985. A critique of the biological species concept and recommendations for a phylogenetic alternative. *The Bryologist* 88:172–85.

Doolittle, W. F. 2018. Speciation without species: A final word. *Philosophy and Theory in Biology* 11:14.

Doolittle, W. F., and R. T. Papke 2006. Genomics and the bacterial species problem. *Genome Biology* 7:116.

Dorofeeva, D. Yu. (ed.). 2001. *The book of Angels: An anthology.* Saint Petersburg: Amphora.

Doyen, J. T., and C. N. Slobodchikoff. 1974. An operational approach to species classification. *Systematic Biology* 23:239–47.

Dress, A., V. Moulton, M. Steel, et al. 2010. Species, clusters and the "Tree of life": A graph-theoretic perspective. *Journal of Theoretical Biology* 265:535–42.

Drobner, H. R. 2000. Gregory of Nyssa as philosopher: De anima et resurrectione and De hominis opificio. *Dionysius* 18:69–102.

Drooger, C. W. 1954. Remarks on the species concept in paleontology. *The Micropaleontologist* 8:23–6.

Drouin, J.-M. 2001. Principles and uses of taxonomy in the works of Augustin-Pyramus de Candolle. *Studies in History and Philosophy of Biological and Biomedical Sciences* 32: 255–75.

Du Rietz, G. E. 1930. The fundamental units of biological taxonomy. *Svensk Botanisk Tidskrift*, 24:333–428.

Du Rietz, G. E. 1931. Life forms of terrestrial flowering plants. *Acta Phytogeographica Suecica* 3:1–95.

Dubois, A. 2011. Species and "strange species" in zoology: Do we need a "unified concept of species"? *Comptes rendus de l'Académie des sciences, Series IIA, Earth and Planetary Science* 10:77–94.

Dupré, J. 1981. Natural kinds and biological taxa. *The Philosophical Review* 90:66–90.

Dupré, J. 1993. *The disorder of things: Metaphysical foundations of the disunity of science.* Cambridge, MA: Harvard University Press.

Dupré, J. 1999. On the impossibility of a monistic account of species. In *Species: New interdisciplinary essays*, ed. R. A. Wilson, 3–21. Cambridge, MA: MIT Press.

Dupré, J. 2001. In defence of classification. *Studies in History and Philosophy of Biological and Biomedical Sciences* 32:203–19.

Dupré, J. 2020. Life as process. *Epistemology and Philosophy of Science* 57:96–113.

Dupré, J. 2021. *The metaphysics of biology.* New York, NY: Cambridge University Press.

Dupré, J. 2022. (Some) species are processes. In *Species problems and beyond. Contemporary issues in philosophy and practice*, eds J. S. Wilkins, F. E. Zachos, and I. Ya. Pavlinov, 279–92. Boca Raton, FL: CRC Press.

Dzik, J. 2005. The concept of chronospecies in ammonites. *Atti II Conv. Int. F.E.A. Pergola* 87: 273–89.

Echelle, A. A. 1990. In defense of the phylogenetic species concept and the ontological status of hybridogenetic taxa. *Herpetologica* 46:109–13.

Efremov, V. A. 2009. [A theory of concept and the conceptual space.] *Herzen University Journal of Humanities & Sciences* 104:96–106. (in Russian).

Ehrlich, P. R. 1961. Has the biological species concept outlived its usefulness? *Systematic Biology* 10:167–76.

Eigen, M. 1983. Viral quasispecies. *Scientific American* 269:42–9.

Elder, C. L. 2008. Biological species are natural kinds. *The Southern Journal of Philosophy* 46:339–62.

Eldredge, N. 1993. What, if anything, is a species? In *Species, species concepts and primate evolution*, eds W. H. Kimbel and L. B. Martin, 3–20. Boston, MA: Springer.

Eldredge, N. 1995. Species, selection, and Paterson's concept of the specific-mate recognition system. In *Speciation and the recognition concept*, eds D. Lambert and H. G. Spencer, 464–77. Baltimore, MD: Johns Hopkins University Press.

Eldredge, N., and J. Cracraft. 1980. *Phylogenetic patterns and the evolutionary process.* New York, NY: Columbia University Press.

Eldredge, N., and S. N. Salthe. 1984. Hierarchy and evolution. In *Oxford surveys in evolutionary biology*, eds R. Dawkins and M. Ridley, 184–208. Oxford: Oxford University Press.

Ellis, B. 2001. *Scientific essentialism.* Cambridge: Cambridge University Press.

Ellis, B. 2002. *The philosophy of nature. A Guide to the new essentialism.* Chesham: Acumen Publ.

Ellis, B. 2011. The problem with the species problem. *History and Philosopy of Life Science* 33:343–64.

Endersby, J. 2008. *Imperial nature, Joseph Hooker and the practice of Victorian science.* Chicago, IL: University of Chicago Press.

Engels, F. 1964. [*Dialectics of Nature.*] Moscow: Political Literature Publ. (in Russian).

Ereshefsky, M. (ed.). 1992a. *The units of evolution. Essays on the nature of species.* Cambridge: MIT Press.

Ereshefsky, M. 1988. *The ontological status of species: A study of individuality and its role in evolutionary theory.* Ph.D. thesis. Madison: University of Wisconsin. https://philpapers.org/rec/ERETOS.

Ereshefsky, M. 1991. Species, higher taxa, and the units of evolution. *Philosophy of Science* 58:84–101.

Ereshefsky, M. 1992b. Eliminative pluralism. *Philosophy of Science* 59:671–90.

Ereshefsky, M. 1998. Species pluralism and anti-realism. *Philosophy of Science* 65:103–20.

Ereshefsky, M. 1999. Species and the Linnaean hierarchy. In *Species: New interdisciplinary essays*, ed. R. A. Wilson, 285–305. Cambridge: MIT Press.

Ereshefsky, M. 2001. *The poverty of the Linneaean hierarchy: A philosophical study of biological taxonomy*. New York, NY: Cambridge University Press.

Ereshefsky, M. 2009a. Darwin's solution to the species problem. *Synthese* 175:405–25.

Ereshefsky, M. 2009b. Natural kinds in biology. *Routledge Encyclopedia of Philosophy*. https://www.rep.routledge.com/articles/thematic/natural-kinds-in-biology/v-1#.

Ereshefsky, M. 2010a. What's wrong with the new biological essentialism. *Philosophy of Science* 77:674–85.

Ereshefsky, M. 2010b. Microbiology and the species problem. *Biology and Philosophy* 25:553–68.

Ereshefsky, M. 2011. Mystery of mysteries: Darwin and the species problem. *Cladistics* 27:67–79.

Ereshefsky, M. 2017. Species. In *Stanford encyclopedia of philosophy*, ed. E. N. Zalta. https://plato.stanford.edu/entries/species/.

Ereshefsky, M., and M. Matthen. 2005. Taxonomy, polymorphism, and history: An introduction to population structure theory. *Philosophy of Science* 72:1–21.

Ereshefsky, M., and T. A. C. Reydon. 2015. Scientific kinds. *Philosophical Studies* 172:969–86.

Eriksson, G. 1983. Linnaeus the botanist. In *Linnaeus, the man and his work*, ed. T. Frängsmyr, 63–109. Berkeley: University of California Press.

Erismann, C. 2007. The logic of being: Eriugena's dialectical ontology. In *The many roots of Medieval logic: The Aristotelian and the non-Aristotelian traditions*, ed. J. Marenbon, 73–88. Leiden: E. J. Brill.

Ernst, T. 2011. *Quidam homo est asinus: The originality and influence of Peter Abelard upon Medieval thought*. Doctoral dissertation. Louisville: University of Louisville. https://ir.library.louisville.edu/cgi/viewcontent.cgi?article=1408&context=etd.

Esper, E. I. C. 1781. *De varietatibus specierum in Naturae productis, Disqvisitio philosophica. Sectio I.* Erlangen: typis Elrodtianis.

Evangeliou, C. 1982. The ontological basis of Plotinus' criticism of Aristotle's theory of categories. In *The structure of being*, ed. R. B. Harris, 73–82. Albany, NY: State University of New York Press.

Evangeliou, C. 1996. *Aristotle's categories and Porphyry*. Leiden: E. J. Brill.

Evenhuis, N. L. 2008. The "Mihi itch" a brief history. *Zootaxa* 1890:59–68.

Evola, J. 1995. *The Hermetic tradition. Symbols and teachings of the royal art*. Rochester, VT: Inner Traditions.

Faegri, K. 1937. Some fundamental problems of taxonomy and phylogenetics. *The Botanical Review* 3:400–23.

Faith, D. P. 2003. Biodiversity. In *Stanford encyclopedia of philosophy*, ed. E. N. Zalta. http://plato.stanford.edu/archives/sum2003/entries/biodiversity/.

Falcon, A. 1997. Aristotle's theory of division. *Bulletin of the Institute of Classical Studies*, Suppl. 68: 127–46.

Falcon, A., and D. Lefebvre (eds). 2018. *Aristotle's Generation of animals. A critical guide*. Cambridge: Cambridge University Press.

Falk, R. 1988. Species as individuals. *Biology and Philosophy* 3:455–62.

Farber, P. L. 1972. Buffon and the concept of species. *Journal of the History of Biology* 5:259–84.

Farndell, F. 2010. *All things natural. Ficino on Plato's Timaeus*. London: Shepheard-Walwyn Publ.

Fausek, V. A. (ed.). 1904. [*Theory of development*.] Saint Petersburg: Brockhaus-Efron Publ. (in Russian).

Faust, D. 1987. *The limits of scientific reasoning*, 2nd ed. Minneapolis, MN: University of Minnesota.

Filipchenko, Yu. A. 1915. [*Variability and evolution.*] Petrograd: A. S. Panafidina Publ. (in Russian).

Fine, G. 1993. *On ideas: Aristotle's criticism of Plato's theory of forms.* Oxford: Oxford University Press.

Fischer, G. 1811. [*The system of fossils serving as the basis of the order in which they are placed in the Museum of the Imperial Moscow University [...].*] Moscow: University typogr. (in Russian).

Fitzhugh, K. 2005. The inferential basis of species hypotheses: The solution to defining the term 'species'. *Marine Ecology* 26:155–65.

Fitzhugh, K. 2009. Species as explanatory hypotheses: Refinements and implications. *Acta Biotheoretica* 57:201–48.

Fleck, L. 1979. *Genesis and development of a scientific fact.* Chicago, IL: University of Chicago Press.

Flematti, J. 2018. The ontology of evolutionary theory. Are species individuals rather than classes? *Crítica: Revista Hispanoamericana de Filosofía* 15:63–73.

Fleming, J. 1822. *The philosophy of zoology; or a general view of the structure, functions, and classification of animals*, Vol. 2. Edinburgh: Archibald Constable.

Folse, H. J., and J. Roughgarden. 2010. What is an individual organism? A multilevel selection perspective. *The Quarterly Review of Biology* 85:447–72.

Foucault, M. 1970. *The order of things. An archaeology of the human sciences.* New York, NY: Pantheon Books.

Frair, W. 2000. Baraminology: Classification of created organisms. *Creation Research Society Quarterly* 37:82–91.

Frankham, R., J. D. Ballou, M. R. Dudash, et al. 2012. Implications of different species concepts for conserving biodiversity. *Biological Conservation* 153:25–31.

Franklin, J. 1986. Aristotle on species variation. *Philosophy* 61:245–52.

French, R. 1994a. *Ancient natural history: Histories of nature.* London: Routledge.

French, R. 1994b. *William Harvey's natural philosophy.* Cambridge: Cambridge University Press.

Freudenstein, J. V., M. B. Broe, R. A. Folk, et al. 2017. Biodiversity and the species concept—lineages are not enough. *Systematic Biology* 66:644–56.

Friedmann, V. S. 2007. [Systems of recognition of "friend–foe", and the renaissance of the biological species concept.] In *XXI Lyubishchev readings*, 201–15. Ulyanovsk: Ulyanovsk State Pedagogical University Publ. (in Russian).

Friedmann, V. S. 2013. Conspecific recognition systems and the rehabilitation of the biological species concept in ornithology. In *The species problem: Ongoing issues*, ed. I. Ya. Pavlinov, 193–224. Rijeka: InTech Open Access Publ.

Frigg, R., and J. Nguyen. 2016. Scientific representation. In *Stanford encyclopedia of philosophy*, ed. E. N. Zalta. https://plato.stanford.edu/entries/scientific- representation/.

Frost, D. R. and D. M. Hillis. 1990. Species in concept and practice: Herpetological applications. *Herpetologica* 46:87–104.

Frost, D. R., and A. G. Kluge. 1994. A consideration of epistemology of systematic biology, with special reference to species. *Cladistics* 10:259–94.

Funk, D. J., and K. E. Omland. 2003. Species-level paraphyly and polyphyly: Frequency, causes, and consequences, with insights from animal mitochondrial DNA. *Annual Review of Ecology, Evolution and Systematics* 34:397–423.

Fürst von Lieven, A., and M. Humar. 2008. A cladistic analysis of Aristotle's animal groups in the Historia animalium. *History and Philosophy of Science* 30:227–62.

Futuyma, D. J. 1987. On the role of species in anagenesis. *American Naturalist* 130:465–73.

Gabbay, D. M., and J. Woods (eds). 2008. *Handbook of the history of logic, Vol. 2. Mediaeval and Renaissance logic.* Oxford: Elsevier.

Gärdenfors, P. 2004. Conceptual spaces as a framework for knowledge representation. *Mind and Matter* 2:9–27.

Garveo, G. 1651. *Exercitationes de Generatione Animalium [...].* Londini: Du-Gardianis.

Gaydenko, P. P., and V. V. Petrov (eds). 2002. [*Philosophy of Nature in Antiquity and in the Middle Ages,*] Part 3. Moscow: Progress-Tradition. (in Russian).

Gaydenko, P. P. 2000. [*The history of New European philosophy in its connection with science.*] Moscow: PER SE. (in Russian).

Gaydenko, P. P. 2003. [*Scientic rationality and philosophical mind.*] Moscow: Progress-Traditsia. (in Russian).

Gaydenko, P. P. 2008. [*The One.*] In [*Ancient philosophy: An Encyclopedic dictionary.*] 375–79. Moscow: Progress-Tradition. (in Russian).

Gaydenko, V. P. 2001. [On the basic concepts of the doctrine of Thomas Aquinas.] In [*Thomas Aquinas. Ontology and theory of knowledge: Fragments of essays,*] 3–32. Moscow: Insitute of Philosophy.

Gaydenko, V. P., and G. A. Smirnov. 1989. [*Western European science in the Middle Ages: General principles and the doctrine of movement.*] Moscow: Nauka. (in Russian).

Gayon, J. 1996. The individuality of the species: A Darwinian theory?—from Buffon to Ghiselin, and back to Darwin. *Biology and Philosophy* 11:215–44.

Gesner, C. 1560. *Nomenclator aquatilium animantium. Icones animalium aquatilium in mari [...].* Zürich: Christoph. Froschoverum.

Ghiselin, M. T. 1966. On psychologism in the logic of taxonomic controversies. *Systematic Zoology* 15:207–15.

Ghiselin, M. T. 1974. A radical solution to the species problem. *Systematic Zoology* 23:536–44.

Ghiselin, M. T. 1981. Categories, life, and thinking. *The Behavioral and Brain Sciences* 4:269–313.

Ghiselin, M. T. 1987. Species concepts, individuality, and objectivity. *Biology and Philosophy* 2:127–43.

Ghiselin, M. T. 1997. *Metaphysics and the origin of species.* New York, NY: State University of New York Press.

Ghiselin, M. T. 2002. Species concepts: the basis for controversy and reconciliation. *Fish and Fisheries* 3:151–60.

Ghiselin, M. T. 1969. The principles and concepts of systematic biology. *Systematic Biology: Proceedings of the International Conference*, 1692, 45–66. Washington: National AcadEmy of Sciences.

Gill, F. B. 2014. Species taxonomy of birds: Which null hypothesis? *Auk* 131:150–61.

Gilmour, J. S. L. 1940. Taxonomy and philosophy. In *The new systematics*, ed. J. Huxley, 461–74. Oxford: Oxford University Press.

Gilmour, J. S. L. 1961. Taxonomy. In *Contemporary botanical thought*, eds A. M. MacLeod and L. S. Cobley, 27–45. Chicago, IL: Quadrangle Book.

Gilson, E. 1955. *History of Christian philosophy in the Middle Ages.* New York, NY: Random House.

Gilson, E. 1960. *The Christian philosophy of Saint Augustine.* New York, NY: Random House. 598 p.

Gilson, E. 1984. *From Aristotle to Darwin and back again. A Journey in final causality, species, and evolution.* Notre Dame, IN: University of Notre Dame Press.

Gilson, E. 1991. *The spirit of Mediaeval philosophy.* Notre Dame, IN: University of Notre Dame Press.

Gilson, E. 2002. *Thomism. The philosophy of Thomas Aquinas*, 6th ed. Toronto: Pontifical Institute of Mediaeval Studies.

Gilson, E. 2010. [*The philosophy of the Middle Ages (from origine of patristique to the end of the 14th century).]* Moscow: Respublika. (in Russian).

Gilson, E. 2018. *John Duns Scotus: Introduction to his fundamental positions.* London: T&T Clark. 632 p.

Gil-White, F. J. 2011. Are ethnic groups biological "species" to the human brain? Essentialism in our cognition of some social categories. *Current Anthropology* 42:515–54.

Giray, E. F. 1976. An integrated biological approach to the species problem. *The British Journal for the Philosophy of Science* 27:317–28.

Giseke, P. D. 1792. *Praelectiones in ordines naturales plantarum [...]*. Hamburgi: Benj. Gottl. Hoffmanni.

Gish, D. T. 1993. *Creation scientists answer their critics*. El Cajon, CA: Institute for Creation Research.

Gishlick, A. 2006. Baraminology. *Reports of the National Center for Science Education* 26. https://ncse.ngo/baraminology.

Glanzberg, M. 2013. Truth. In *Stanford encyclopedia of philosophy*, ed. E. N. Zalta. https://plato.stanford.edu/entries/truth/.

Glass, B. 1959. The germination of the idea of biological species. In *Forerunners of Darwin: 1745–1859*, eds B. Glass, O. Temkin, and W. L. Straus, 30–48. Baltimore, MD: Johns Hopkins Press.

Glover, G. J. 2007. *Beyond the firmament: Understanding science and the theology of creation*. Chesapeake, VA: Watertree.

Gmelin, F. 1792. *Animal Kingdom, or Zoological System, of the celebrated sir Charles Linnaeus. Class I. Mammalia [...]*. Edinburgh: A. Strahan and T. Cadell.

Godman, M. 2021. Scientific realism with historical essences: The case of species. *Synthese* 198:3041–57.

Goldschmidt, R. 1940. *The material basis of evolution*. New Haven, CT: Yale University Press.

Golovin, P. N. 1958. [The concept of a species in mycology.] In [*The species problem in botany*, Vol. 1,] ed. P. A. Baranov, 34–67. Moscow–Leningrad: USSR Academy of Sciences. (in Russian).

Gontier, N. 2011. Depicting the tree of life: The philosophical and historical roots of evolutionary tree diagrams. *Evolution, Education and Outreach* 4:515–38.

Gontier, N., and M. Bradie. 2017. Acquiring knowledge on species-specific biorealities: The applied evolutionary epistemological approach. In *Routledge handbook of evolution and philosophy*, ed. R. Joyce,136–152. New York, NY: Routledge.

González-Forero, M. 2008. Removing ambiguity from the biological species concept. *Journal of Theoretical Biology* 256:76–80.

Goodin, S. 1999. Locke and Leibniz and the debate over species. In *New essays on the rationalists*, eds R.J. Germaro and C. Huenemann, 163–78. Oxford: Oxford University Press.

Gorman, J. 1992. *Understanding history: An introduction to analytical philosophy of history*. Ottawa: University of Ottawa Press.

Gorsky, D. P. 1974. [*Definition (logical and methodological issues)*.] Moscow: Mysl. (in Russian).

Gorsky, D. P. 1983. [*Logic*.] Moscow: Uchpedgiz. (in Russian).

Gotthelf, A., and J. Lennox (eds). 1987. *Philosophical issues in Aristotle's biology*. Cambridge: Cambridge University Press.

Gotthelf, A. 1991. Division and explanation in Aristotle's Parts of Animals. *The Society for Ancient Greek Philosophy Newsletter* 241. https://orb.binghamton.edu/cgi/viewcontent.cgi?article=1240&context=sagp.

Gotthelf, A. 1999. Darwin on Aristotle. *Journal of the History of Biology* 32:3–30.

Gotthelf, A. 2012. *Teleology, first principles, and scientific method in Aristotle's biology*. Oxford: Oxford University Press.

Gottlieb, G. 1992. *Individual development and evolution: The genesis of novel behavior*. New York, NY: Oxford University Press.

Gould, S. J. 2002. *The structure of evolutionary theory*. Cambridge, MA: Harvard University Press. 1436 p.

Gould, S. J., and N. Eldredge. 1977. Punctuated equilibria: The tempo and mode of evolution reconsidered. *Paleobiology*, 3(2): 115–51.

Goulding, T. C., and B. Dayrat. 2016. Integrative taxonomy: Ten years of practice and looking into the future. In *Aspects of biodiversity*, eds I. Ya. Pavlinov, M. V. Kalyakin, and A. V. Sysoev, 116–33. Moscow: KMK Sci. Press.

Grant, R. 2011. Naturalizing the metaphysics of species: A perspective on the species problem. *South African Journal of Philosophy* 30:63–9.

Grant, V. 1957. The plant species in theory and practice. In *The Species problem. A Symposium presented at the Atlanta Meeting of the American Association for the Advancement of Science*, ed. E. Mayr, 39–80. Washington, DC: American Association for the Advancement of Science.

Grant, V. 1981. *Plant speciation*, 2nd ed. New York, NY: Columbia University Press.

Grantham, T. A. 1993. Beyond "individuality" and "pluralism": A review of Ereshefsky's *Units of evolution: Essays on the nature of species. Biology and Philosophy* 8:457–68.

Gray, A. 1876. *Darwiniana. Essays and reviews pertaining to Darwinism.* New York, NY: D. Appleton & Co.

Greene, E. L. 1909. *Landmarks of botanical history. A study of certain epochs in the development of the science of botany.* Pt 1. *Prior to 1562.* Washington, DC: Smithsonian Miscellaneous Collections.

Greene, E. L. 1910. Certain aspects of the species question. *The American Midland Naturalist* 1:245–63.

Greene, J. C. 1992. From Aristotle to Darwin: Reflections on Ernst Mayr's interpretation in *The Growth of Biological Thought. Journal of the History of Biology* 25:257–84.

Gregory, M. E. 2006. *Diderot and the metamorphosis of species.* New York, NY & London: Routledge.

Grene, M. 1974a. *The understanding of nature: Essays in the philosophy of biology.* Dordrecht: D. Reidel.

Grene, M. 1974b. Is genus to species as matter to form? Aristotle and taxonomy. *Synthese* 28:51–69.

Grene, M., and D. Depew. 2004. *The philosophy of biology. An episodic history.* Cambridge: Cambridge University Press.

Griffiths, G. C. D. 1974a. On the foundations of biological systematics. *Acta Biotheoretica* 23:85–131.

Griffiths, G. C. D. 1974b. Some fundamental problems in biological classification. *Systematic Zoology* 22:338–43.

Griffiths, P. E. 1999. Squaring the circle: Natural kinds with historical essences. In *Species: New interdisciplinary essays*, ed. R. A. Wilson, 209–228. Cambridge, MA: MIT Press.

Gross, A. G. 1988. Philosophy versus science: The species debate and the practice of taxonomy. *Proceedings of the Biennial Meeting of the Philosophy of Science Association* 1:223–30.

Grunwald, E. 1997. *Thermodynamics of molecular species.* New York, NY: John Wiley.

Guedes, M. 1967. La méthode taxonomique d'Adanson. *Revue d'Histoire des Sciences et de leurs applications* 20:361–86.

Guyot, K. 1986. Specious individuals. *Philosophica* 37:101–26.

Haack, S. 1979. Epistemology with a knowing subject. *The Review of Metaphysics* 33:309–35.

Haber, M. H. 2016. The individuality thesis (3 ways). *Biology and Philosophy* 31:913–30.

Hackett, J. 1992. *Roger Bacon (circa 1214/1220–1292).* In *Medieval philosophers*, ed. J. Hackett, 90–103. Detroit, MI: A Bruccoli Clark Layman Book.

Hackett, J. 1994. Albert the Great (b. 1200; d. 1280). In *Individuation in Scholasticism. The Later Middle Ages and the Counter-Reformation 1150–1650*, ed. J. J. E. Gracia, 97–116. Albany, NY: State University of New York Press.

Hackett, J. 2020. Roger Bacon. In *Stanford encyclopedia of philosophy*, ed. E. N. Zalta. https://plato.stanford.edu/archives/sum2020/entries/roger-bacon/.

Hadot, I. 2015. *Athenian and Alexandrian Neoplatonism and the harmonization of Aristotle and Plato.* Leiden: E. J. Brill.

Haeckel, E. 1866. *Generelle Morphologie der Organismen: Allgemeine Grundzüge der organischen Formen-Wissenschaft,* Vol. 2. Berlin: Georg Reimer.

Haeckel, E. 1874. Anthropogenie: *oder, Entwickelungsgeschichte des Menschen [...].* Leipzig: W. Engelmann.

Haffer, J. H. 2006. The history of the biological species concept. *Acta Zoologica Sinica,* 52 (Supplement):415–20.

Hagen, J. B. 1982. *Experimental taxonomy, 1930-1950: The impact of cytology, ecology, and genetics on ideas of biological classification.* Doctoral dissertation. Corvallis: Oregon State University. https://ir.library.oregonstate.edu/concern/graduate_thesis_or_dissertations/sb397c26n?locale=de.

Hall, B. K. (ed.) 1994. *Homology, the hierarchical basis of comparative biology.* San Diego, CA: Academic Press.

Hall, B. K., and W. M. Olson (eds). 2006. *Keywords and concepts in evolutionary developmental biology.* Cambridge, MA: Harvard University Press.

Hall, H. M., and F. E. Clements. 1923. *The phylogenetic method in taxonomy: The North American species of Artemisia, Chrysothamnus, and Atriplex.* Washington, DC: Publications of Carnegie Instinstitue.

Hampton, A. J. B., and J. P. Kenney. 2021. *Christian Platonism: A history.* Cambridge: Cambridge University Press.

Hanage, W. P. 2013. Fuzzy species revisited. *BMC Biol.* 11: 41. https://bmcbiol.biomedcentral.com/articles/10.1186/1741-7007-11-41.

Hanage, W. P., C. Fraser, and B. G. Spratt. 2005. Fuzzy species among recombinogenic bacteria. *BMC Biology* 3:6. http://www.biomedcen.tral.com/1741-7007/3/6.

Harrison, P. 2009. Linnaeus as a second Adam? Taxonomy and the religious vocation. *Zygon* 44:879–93.

Harrison, R. G. 1998. Linking evolutionary pattern and process: The relevance of species concepts for the study of speciation. In *Endless forms: Species and speciation.* eds D. J. Howard and S. H. Berlocher, 19–31. Oxford: Oxford University Press.

Hatch, M. H. 1941. The logical basis of the species concept. *The American Naturalist* 75:193–212.

Hathaway, R. F. 1969. *Hierarchy and the definition of order in the Letters of Pseudo-Dionysius.* Hague: Marlinus Nijhott.

Hausdorf, B. 2011. Progress toward a general species concept. *Evolution* 65:923–31.

Häuser, C. L. 1987. The debate about the biological species concept a review. *Journal of Zoological Systematics and Evolutionary Research* 25:241–57.

Hawkins, H. L. 1935. The species problem. *Nature,* 136:574–5.

Hawley, K., and A. Bird. 2011. What are natural kinds? *Philosophical Perspectives* 25:205–21.

He, F., and P. Legendre. 2002. Species diversity patterns derived from species-area models. *Ecology* 83:1185–98.

Hebert, P. D. N., A. Cywinska, S. L. Ball, et al. 2003. Biological identifications through DNA barcodes. *Proceedings of the Royal Society B* 270:313–21.

Heider, D. 2014. *Universals in Second Scholasticism.* Amsterdam: John Benjamins.

Heise, H. 1981. Universals, particulars, and paradigms. *The Behavioral and Brain Sciences* 4:289–90.

Helbig, A. J., A. G. Knox, D. T. Parkin, et al. 2002. Guidelines for assigning species rank. *Ibis* 144:518–25.

Hempel, C. G. 1965. *Aspects of scientific explanation and other essays in the philosophy of science.* New York, NY: The Free Press.

Hendry, A. P., M. S. Vamosi, S. J. Latham, et al. 2000. Questioning species realities. *Conservation Genetics* 1:67–76.

Hennig, W. 1966. *Phylogenetic systematics.* Urbana, IL: University Illinois Press.

Henry, D. 1991. *Medieval mereology*. Amsterdam: B. R. Grüner.

Henry, D. 2011. Aristotle's pluralistic realism. *The Monist* 94:197–220.

Heptner, V. G. 1947. [The problem of species in modern zoology.] In Mayr E. *Systematics and the origin of species, from the viewpoint of zoologist*, 5–22. Moscow: Inostrannaya Literatura. (in Russian).

Herber, E. C. (ed.). 1963. *Correspondence between Spencer Fullerton Baird and Louis Agassiz – two pioneer American naturalists*. Washington, DC: Smithsonian Institution. 237 p.

Heslop-Harrison, J. 1960. *New concepts in flowering-plant taxonomy*. Cambridge: Harvard University Press.

Hetzenauer, P. M. (ed.). 1914. *Biblia Sacra Vulgatae Editionis [...]*. Romae: Fr. Pustet & Co.

Hey, J. 2001a. *Genes, categories, and species. The evolutionary and cognitive cause of the species problem*. New York, NY: Oxford University Press.

Hey, J. 2001b. The mind of the species problem. *Trends in Ecology and Evolution* 16:326–9.

Hey, J. 2006. On the failure of modern species concepts. *Trends in Ecology and Evolution* 21:447–50.

Hey, J., R. Waples, M. Arnold, et al. 2003. Understanding and confronting species uncertainty in biology and conservation. *Trends in Ecology and Evolution* 18:597–603.

Heywood, V. H. 1963. The "species aggregate" in theory and practice. *Regnum Vegetabile* 27:26–37.

Heywood, V. H. 1989. Nature and natural classification. *Plant Systematics and Evolution* 167:87–92.

Hoch, P. C. and A. G. Stephenson (eds). 1995. *Experimental and molecular approaches to plant biosystematics*. St. Louis, MO: Missouri Botanical Garden.

Hofsten, N., von. 1958. Linnaeus's conception of Nature. Uppsala: *Kungliga Vetenskaps-societetens Årsbok* (1957):65–105.

Hogben, L. 1940. Problems of the origins of species. In *The new systematics*, ed. J. Huxley, 270–86. London: Oxford University Press.

Holmes, J. A. 2000. *Species pluralism*. PhD. East Lansing: Michigan State University. https://philpapers.org/rec/HOLSP-3.

Holopainen, T. 1996. *Dialectic and theology in the eleventh century*. Leiden: E. J. Brill.

Holsinger, K. E. 1987. Pluralism and species concepts, or when must we agree with one another? *Philosophy of Science* 54:480–5.

Holton, G. 1973. *Thematic origins of scientific thought: Kepler to Einstein*. Cambridge, MA: Harvard University Press.

Holton, G. 1996. The role of themata in science. *Fomldations of Physics* 26:453–65.

Hopkins, J. 1978. *A concise introduction to the philosophy of Nicholas of Cusa*. Minneapolis, MN: University of Minnesota Press.

Hopkins, W. 1860. Physical theories of the phenomena of life, Pt. 1. *Fraser's Magazine for Town and Country* 61:739–52.

Hoque, T. 2008. Buffon: from natural history to the history of nature? *Biological Theory* 2:413–9.

Horvath, C. D. 1997a. Phylogenetic species concept: Pluralism, monism, and history. *Biology and Philosophy* 12:225–32.

Horvath, C. D. 1997b. Some questions about identifying individuals: Failed intuitions about organisms and species. *Philosophy of Science* 64(4): 654–68.

Huang, Y. 2010. A heuristic view about the evolution and species. *ArXiv*:1004.3228v1. https://doi.org/10.48550/arXiv.1004.3228.

Hubbs, C. L. 1934. Racial and individual variation in animals, especially fishes. *The American Naturalist* 68:115–28.

Hudson, R. R. and J. A. Coyne. 2002. Mathematical consequences of the genealogical species concept. *Evolution* 56:1557–65.

Hull D. L. 2002. Recent philosophy of biology: A review. *Acta biotheoretica* 50:117–28.
Hull D. L. 2006. Essentialism in taxonomy: Four decades later. *Annals of the History and Philosophy of Biology* 11:47–58.
Hull, D. L. 1965. The effect of essentialism on taxonomy: Two thousand years of stasis. *British Journal for the Philosophy of Science* 15:314–26.
Hull, D. L. 1969. What philosophy of biology is not. *Synthese* 20:157–84.
Hull, D. L. 1976. Are species really individuals? *Systematic Zoology* 25:174–91.
Hull, D. L. 1977. The ontological status of species as evolutionary unit. In *Foundational problems in the special sciences*, eds R. Butts and J. Hintikka, 91–102. Dordrecht: D. Reidel.
Hull, D. L. 1978. A matter of individuality. *Philosophy of Science* 45:335–60.
Hull, D. L. 1985. Linné as an Aristotelian. In *Contemporary perspectives on Linnaeus*, ed. J. Weinstock, 37–54. Lanham: University Press of America.
Hull, D. L. 1988. *Science as a process*. Chicago, IL: University Chicago Press.
Hull, D. L. 1989. *Metaphysics of evolution*. Albany, NY: State University of New York Press.
Hull, D. L. 1997. The ideal species concept—and why we can't got it. In *Species. The units of biodiversity*, eds M. F. Claridge, A. H. Dawah, and M. R. Wilson, 357–80. London: Chapman & Hall.
Hull, D. L. 1999. On the plurality of species: Questioning the party line. In *Species: New interdisciplinary essays*, ed. R. A. Wilson, 23–48. Cambridge, MA: MIT Press.
Hull, D. L. 2001. *Science and selection: Essays on biological evolution and the philosophy of science*. New York, NY: Cambridge University Press.
Humboldt, A. 1806. *Ideen zur einer physiognomik der Gewachse*. Tubingen: Cotta.
Huneman, P. 2014. Individuality as a theoretical scheme. I. Formal and material concepts of individuality. *Biological Theory* 9:361–73.
Huxley, J. S. (ed.). 1940. *The new systematics*. London: Oxford University Press.
Huxley, J. S. 1939. Clines: An auxiliary method in taxonomy. *Bijdragen tot de Dierkunde* 27:491–520.
Huxley, J. S. 1942. *Evolution: The modern synthesis*. London: G. Allen & Unwin.
Ibrahim, A. 1987. La notion de moule intérieur dans les théories de la génération au XVIIIe siècle. *Archives de Philosophie* 50:555–80.
Illiger, J. K. W. 1800. *Versuch einer Systematischen vollstäandigen Terminglogie für das Thierreich und Pflanzenreich*. Helmstadt: C. G. Fleckeiseh.
Inge, W. R. 1900. The permanent influence of Neoplatonism upon Christianity. *The American Journal of Theology* 4(2): 328–44.
Irmscher, C. 2013. *Louis Agassiz: Creator of American science*. New York, NY: Houghton Mifflin Harcourt.
Isaac, N. J. B., and A. Purvis. 2004. The 'species problem' and testing macroevolutionary hypotheses. *Diversity and Distributions* 10:275–81.
Iwatsuki, K., P. H. Raven, and W. J. Bock (eds.). 1986. *Modern aspects of species*. Tokyo: Univ of Tokyo Press.
Jamniczky, H. A. 2005. Biological pluralism and homology. *Philosophy of Science* 72:687–98.
Janick, J. 2003. Herbals: The connection between horticulture and medicine. *Hort Thechnology* 13:229–38.
Jardine, N. 1967. The concept of homology in biology. *British Journal for the Philosophy of Science* 18:125–39.
Johnson, D. S. 1908. Aspects of the species question. *The American Naturalist* 42: 217–81.
Jones, A. J. 1972a. A general analysis of the biblical "kind" (min). *Creation Research Society Quarterly* 9:53–7.
Jones, A. J. 1972b. Boundaries of the Min: An analysis of the Mosaic lists of clean and unclean animals. *Creation Research Society Quarterly* 9:114–23.
Jones, A. J. 1973. How many animals on the Ark? *Creation Research Society Quarterly* 10:102–8.

Jones, J.-E. 2020. Locke on real essence. In *Stanford encyclopedia of philosophy*, ed. E. N. Zalta. https://plato.stanford.edu/archives/spr2020/entries/real-essence/.
Jordan, A. 1847. *Observation sur plusieurs plantes nouvelles ou critiques de la France. Premier fragment.* Paris: J.-B. Baillière.
Jordan, A. 1873. Remarques sur le fait de l'existence en société à l'état sauvage des espèces affines. *Annales de la Société Linnéenne de Lyon, 2e série* 20:195–213.
Jorge, G., and Newton L. 2016. Medieval theories of the categories. In *Stanford encyclopedia of philosophy*, ed. E. N. Zalta. https://plato.stanford.edu/archives/win2016/entries/medieval-categories/.
Jungius, J. 1747. *Opuscula botanico-physica.* Coburgi: Georgii Ottonis.
Kamelin, R. V. 2004. [*Lectures on plant systematics. Chapters of theoretical systematics of plants.*] Barnaul: Azbuka. (in Russian).
Kämpfer, P., and R. Rosselló-Mora. 2004. The species concept for prokaryotic microorganisms—An obstacle for describing diversity? *Poiesis Prax* 3:62–72.
Kampis, G., L. Gulyás, and S. Soós. 2007. The species problem in artificial life. *Proceedings of the 2007 IEEE Symposium on Artificial Life (CI-ALife 2007)*, 139–146. Piscataway, NJ: IEEE.
Kanaev, I. I. 1966. [*Georges-Louis Leclerc de Buffon (1707–1788).*] Moscow: Nauka. (in Russian).
Kangro, H. 1969. Joachim Jungius und Gottfried Wilhelm Leibniz: Ein Beitrag zum geistigen Verhältnis Beider gelehrten. *Studia Leibnitiana* 1:175–207.
Kant, I. 2007. Of the different races of human beings. In *Anthropology, history, and education*, eds G. Zoller, and R. B. Louden, 84–97. Cambridge: Cambridge University Press.
Kant, I. 2004. *Kritik der Urteilskraft.* Leipzig: Reclam-Verlag GmbH.
Karamanolis, G. E., and A. D. R. Sheppard. 2007. *Studies on Porphyry.* London: University of London.
Kedar, Y. 2017. Roger Bacon (c. 1220–1292) and his system of laws of Nature: Classification, hierarchy and significance. *Perspectives on Science* 25:719–45.
Kedar, Y. 2021. *Virtus* and *species* in the philosophy of Nature of Roger Bacon (ca. 1220–1293). *Revista Española de Filosofía Medieval* 28:43–66.
Kendig, C. (ed.). 2016. *Natural kinds and classification in scientific practice.* Abingdon: Routledge. *267 p.*
Kendig, C. 2010. An ontogenetic-ecological conception of species: A new approach to an old idea. http://philsci-archive.pitt.edu/5267/1/KendigEPSA09.pdf.
Kendig, C. 2014. Towards a multidimensional metaconception of species. *Ratio* 27 (2):155–72.
Kendig, C. 2022. Metaphysical presuppositions about species stability: Problematic and unavoidable. In *Species problems and beyond. Contemporary issues in philosophy and practice*, eds J. S. Wilkins, F. E. Zachos, and I. Ya. Pavlinov, 293–314. Boca Raton, FL: CRC Press.
Kendig, C., and B. Bartley. 2019. Synthetic kinds: Kind-making in synthetic biology. In *Perspectives on classification in synthetic sciences: Unnatural kinds*, ed. J. R. S. Bursten, 78–96. London: Taylor & Francis.
Kerner, A. 1866. *Gute und schlechte Arten.* Innsbruck: Wagner.
Khlebosolov, E. I. 2003. [Functional concept of species in biology.] In [*Ecology and evolution of animals.*] 3–22. Ryazan: Ryazan State Pedagogical University Publ. (in Russian).
Khokhryakov, M. K. 1955. [On the species in fungi.] *Botanical Journal* 40:33–45. (in Russian).
Kholodkovsky, N. A. 1910. [On biological species.] *Izvestia of Imperial. Academy of Sciences*4:751–70. (in Russian).
Kholodnaya, M. A. 2004. [*Cognitive styles. About the nature of the individual mind.*] 2nd ed. Saint Petersburg: Peter Publishing. 384 p.
Khramov, A. V. 2021. [The first book about evolution in Russian: Translation and reception of "*Vestiges of the natural history of creation*" in Russia.] *Historical and Biological research* 13:28–56. (in Russian).

King, P. 1992. Duns Scotus on the common nature and the individual differentia. *Philosophical Topics* 20(2): 51–76.

King, P. 2004. Metaphysics. In *The Cambridge companion to Abelard*, eds J. E. Brower and K. Guilfoy, 65–125.Cambridge: Cambridge University Press.

Kirpotin, S. N. 2005. [Life forms of organisms as organization patterns and spatial environmental factors.] *Zhurnal Obshchei Biologii* 66:239–50. (in Russian).

Kitchell, K. F., and I. M. Resnick. 1999. Introduction: The life and works of Albert The Great. In *Albertus Magnus On Animals. A Medieval Summa Zoologica*, Vol. 1, eds K. F. Kitchell and I. M. Resnick, 1–42. Baltimore, MD: Johns Hopkins University Press.

Kitcher, P. 1984a. Species. *Phylosophy of Science* 51:308–33.

Kitcher, P. 1984b. Against the monism of the moment: A reply to Elliott Sober. *Philosophy of Science*, 51:616–30.

Kitcher, P. 1987. Ghostly whispers: Mayr, Ghiselin, and the "philosophers" on the ontological status of species. *Biology and Philosophy* 2:184–92.

Kitts, D. B. 1987. Plato on kinds of animals. *Biology and Philosophy* 2:315–28.

Kitts, D. B., and D. J. Kitts. 1979. Biological species as natural kinds. *Philosophy of Science* 46:613–22.

Klebahn, H. 1904. *Die wirtswechselnden Rostpilze: Versuch einer Gesamtdarstellung ihrer biologischen Verhält.* Berlin: Gebrüder Borntraeger.

Kleinschmidt, O. 1900. Arten oder Formenkreise? *Journal für Ornithologie* 48:134–39.

Kleinschmidt, O. 1926. Der weitere Ausbau der Formenkreislehre. *Journal für Ornithologie* 74:405–8.

Kleisner, K. 2007. The formation of the theory of homology in biological sciences. *Acta Biotheoretica* 55:317–40.

Klima, G. 2017. The Medieval problem of universals. In *Stanford encyclopedia of philosophy*, ed. E. N. Zalta. https://plato.stanford.edu/entries/universals-medieval/.

Klocker, H. 1992. *William of Ockham and the Divine freedom*, 2nd ed. Milwaukee, WI: Marquette University Press.

Kluge, A. G. 1990. Species as historical individuals. *Biology and Philosophy* 5:417–31.

Kluge, E.-H. W. 1976. Roscelin and the Medieval problem of universals. *Journal of the History of Philosophy* 14:405–14.

Knight, C. 2012. Unit-ideas unleashed: A reinterpretation and reassessment of Lovejovian methodology in the history of ideas. *Journal of the Philosophy of History* 6:195–217.

Knyazeva, E. N. 2006. [Epistemological constructivism.] *Philosophia Nauki & Tekhniki* 12:133–53. (in Russian).

Kober, G. 2008. *Biology without species: A solution to the species problem.* Doctoral dissertation. Boston: Boston University. https://philpapers.org/rec/KOBBWS.

Koerner, L. 1996. Carl Linnaeus in his time and place. In *Cultures of natural history*, eds N. Jardine, J. A. Secord, and E. C. Spary, 145–63. Cambridge: Cambridge University Press.

Kokić, T. 2009. Orbital model of species. Symposium: Panel "Darwin Today", 17: 157–72. Sarajevo: Academy of Sciences and Arts of Bosnia and Herzegovina. https://www.academia.edu/561309/Orbital_Model_of_Species.

Komarov, V. L. 1927. [*Flora of the Kamchatka Peninsula*, Vol. 1.] Leningrad: USSR Academy of Sciences. in Russian).

Komarov, V. L. 1940. [*The doctrine of species in plants.*] Moscow: Akademia Nauk Publ. (in Russian).

Komarov, V. L. 1949. [*Selected essays. Flora of Manchuria, Pt. 1.*] Moscow–Leningrad: USSR Academy of Sciences. (in Russian).

Kordonsky, S. G. 1983. [Taxa and analytical objects as systems.] In [*System method and modern science.*], 141–50. Novosibirsk: Novosibirsk State University. (in Russian).

Körner, S. 1959. *Conceptual thinking: A logical inquiry.* New York, NY: Dover.

Körner, S. 1966. *Experience and theory: An essay in the philosophy of science*. London: Routledge.
Kornet, D. J. 1993. Permanent splits as speciation events: A formal reconstruction of the internodal species concept. *Journal of Theoretical Biology* 164:407–35.
Kornet, D. J., and J. W. McAllister. 2005. The composite species concept: A rigorous basis for cladistic perspective. In *Current themes in theoretical biology. A Dutch perspective*, eds T. A. C. Reydon and L. Hemerik, 95–127. Amsterdam: Springer.
Kornet, D. J., J. A. J. Metz, and H. A. J. M. Schellinx. 1995. Internodons as equivalence classes in genealogical networks: Building-blocks for a rigorous species concept. *Journal of Mathematical Biology* 34:–22.
Korzhinsky, S. I. 1893. [Flora of the East of European Russia in its systematic and geographical rations, Pt. 1.] *Izvestia Tomsk State Univerersity* 5: 71–299. (in Russian).
Kosko, B. 1993. *Fuzzy thinking: The new science of fuzzy logic*. New York, NY: Hyperion.
Koslicki, K. 2008. Natural kinds and natural kind terms. *Philosophy Compass*, 3/4: 789–802.
Koyré, A. 1973. *Éludes d'histoire de la pensée scientifique*. Paris: Gallimard.
Kozo-Polyansky, B. M. 1937. [A. Jussieu and C. Linnaeus (On the centenary of the death of A. L. Jussieu).] *Bulleten Moskovskogo obshestva ispytatelej prirody, Nov. ser.* 46:249–58. (in Russian).
Krell, F.-T. 2004. Parataxonomy *vs.* taxonomy in biodiversity studies pitfalls and applicability of 'morphospecies' sorting. *Biodiversity and Conservation* 13:795–812.
Kremyansky, V. I. 1969. [*Structural levels of living matter*]. Moscow: Nauka. (in Russian).
Kristeller, P. O. 1964. *The philosophy of Marsilio Ficino*, 2nd ed. Gloucester: Peter Smith.
Kuhn, T. S. 1962. *The structure of scientific revolutions*. Chicago, IL: University of Chicago Press.
Kuklok, A. 2013. *Conceptualism and objectivity in Locke's account of natural kinds*. Doctoral dissertation. Cambridge, MA: Harvard University. https://dash.harvard.edu/handle/1/11181128.
Kull, K. 2009. Umwelt and modelling. In *The Routledge companion to semiotics*, ed. P. Cobley, 43–56. London: Routledge.
Kullmann, W., and S. Föllinger (eds). 1997. *Aristotelische Biologie*. Stuttgart: Franz Steiner.
Kunz, W. 2001. Taking more care in using different species concepts. *Parasitology Research* 87:413–6.
Kunz, W. 2002. When is a parasite species a species? *Trends in Parasitology* 18:121–4.
Kunz, W. 2012. *Do species exist? Principles of taxonomic classification*. Weinheim: Wiley-Blackwell.
Kunz, W., and M. Werning. 2009. The biological species as a gene-flow community. Species essentialism does not imply species universalism. *2nd Conference of the European Philosophy of Science Association*. Chicago, IL: PhilSci Archive. https://www.academia.edu/33486287/The_Biological_Species_as_a_Gene_Flow_Community_Species_Essentialism_Does_Not_Imply_Species_Universalism.
Kuprevich, V. F. 1949. [*The species problem in heterotrophic and autotrophic plants*.] Leningrad: USSR Academy of Sciences. (in Russian).
Kuprevich, V. F. 1950. [Species as a stage in the evolution of heterotrophic and autotrophic plants.] In [*Problems of botany*,] 1:149–62. Leningrad: USSR Academy of Sciences. (in Russian).
Kuraev, V. I., and F. V. Lazarev. 1988. [*Precision, truth, and growth of knowledge*.] Moscow: Nauka. (in Russian).
Kurantov A. P., Styazhkin N. I. 1978. [*William Occam*.] Moscow: Mysl. 191 p. (in Russian).
Kursanov L. I. 1945. [The notion of species in lower plants.] *Microbiology* 14:4–7. (in Russian).
Kuzin, B. S. 1962. [On lower taxonomic categories.] In [*Issues in general zoology and medical parasitology*,] 138–54. Moscow: State Publishing of Medicine. (in Russian).
Kuznetsova, N. I. 1998. [Philosophy of science and the history of science: The evolution of relationships against the background of the twentieth century.] *Philosophia Nauki & Tekhniki* 4:61–72. (in Russian).

Kuznetsova, N. I. 2009. [Presentism and antiquarism—two pictures of the past.] *Arbor Mundi* 15:164–96. (in Russian).
Laërtius, D. 1853. *The lives and opinions of prominent philosophers.* London: Henry G. Bohn.
Lakatos, I. 1978. *The methodology of scientific research programmes.* Cambridge: Cambridge University Press.
Lam, K. H. 2020. The realism of taxonomic pluralism. *Metaphysics* 3:1–16.
Lamarck, J.-B. 1984. *Zoological philosophy: An exposition with regard to the natural history of animals by Jean-Baptiste Lamarck.* New York, NY: Hafner Publ.
Lamoureux, D. O. 2008. *Evolutionary creation: A Christian approach to evolution.* Eugene (OR): Wipf and Stock.
Lang, K. N. 1722. *Methodus testacea marina nova et facilis [...] in suas debitas et distinctas classes, genera et species distribuendi [...].* Lucernae: Henrici Rennwardi Wyssing.
Lang, P. 1964. *De Speusippi Academici Scriptis.* Frankfurt: Minerva.
Lange, M. 1995. Are there natural laws concerning particular biological species? *The Journal of Philosophy* 92:430–51.
Lansing, J. 1995. Genus, species, and vantages. In *Language and the cognitive construal of the world*, eds J. R. Taylor and R. E. MacLaury, 365–75. Berlin: Mouton de Gruyter.
LaPorte, J. 2004. *Natural kinds and conceptual change.* Cambridge: Cambridge University Press.
LaPorte, J. 2006. Species as relations: Examining a new proposal. *Biology and Philosophy* 21:381–93.
LaPorte, J. 2007. In defense of species. *Studies in History and Philosophy of Biological and Biomedical Sciences* 38:255–69.
LaPorte, J. 2017. Modern essentialism for species and its animadversions. In *The Routledge handbook of evolution and philosophy*, ed. R. Joyce, 182–93. New York, NY: Routledge.
Larson, J. L. 1967. Linnaeus and the natural method. *Isis* 58(3): 304–20.
Larson, J. L. 1968. The species concept of Linnaeus. *Isis* 59 (3): 291–9.
Larson, J. L. 1971. *Reason and experience: The representation of natural order in the work of Carl von Linné.* Berkely, CA: University California Press.
Laubichler, M. D., and J. Maienschein (ed.). 2007. *From embryology to evo-devo: A history of developmental evolution.* Cambridge, MA: MIT Press.
Leavitt, R. 1909. A vegetative mutant, and the principle of homoeosis in plants. *Botanical Gazette* 47:30–68.
Lee, M. S. Y. 1995. Species concepts and the recognition of ancestors. *Historical Biology* 10:329–9.
Lee, M. S. Y. 2003. Species concepts and species reality: salvaging a Linnaean rank. *Journal of Evolutionary Biology* 16:179–88.
Lee, M., and M. Wolsan. 2002. Integration, individuality and species concepts. *Biology and Philosophy* 17:651–60.
Lehman, H. 1967. Are biological species real? *Philosophy of Science* 34:157–67.
Leibniz, G. 2018. *The Monadology.* New York, NY: GlobalGrey.
Leibniz, G. W. 1996. *New essays on human understanding.* Cambridge: Cambridge University Press.
Lektorsky, V. A. (ed.). 1995. *Historical types of rationality*, Vol. 1. Moscow: Institute of Philosophy. (in Russian).
Lennox, J. G., and R. Bolton (eds). 2010. *Being, nature, and life in Aristotle: Essays in honor of Allan Gotthelf.* Cambridge: Cambridge University Press.
Lennox, J. G. 1980. Aristotle on genera, species, and "the more and the less". *Journal of the History of Biology* 13:321–46.
Lennox, J. G. 1985. Are Aristotelian species eternal? In *Aristotle on nature and living things*, ed. A. Gotthelf, 67–94. Pittsburgh, PA: Mathesis Publ.

Lennox, J. G. 2001. *Aristotle's philosophy of biology: Studies in the origins of life science.* Cambridge: Cambridge University Press.

Lennox, J. G. 2006. Aristotle's biology and Aristotle's philosophy. In *The Cambridge companion to ancient philosophy*, eds M. L. Gill and P. Pellegrin, 292–315. Hoboken, NJ: Blackwell Publ.

Lennox, J. G. 2009. Form, essence, and explanation in Aristotle's biology. In *A companion to Aristotle*, ed. G. Anagnostopoulos, 348–67. Hoboken, NJ: Blackwell Publ.

Lennox, J. G. 2017. Aristotle's biology. In *Stanford encyclopedia of philosophy*, ed. E. N. Zalta. https://plato.stanford.edu/entries/aristotle-biology/.

Lepenies, W. 1985. Linnaeus' Nemesis Divina and the idea of retaliation in the eighteenth century. In *Contemporary perspectives on Linnaeus*, ed. J. Weinstock, 85–115. Lanham, MD: University Press of America.

Leroux, N. 1993. *What are biological species? The impact of the current debate in taxonomy on the species problem.* Thesis Masters of Arts. Montréal: McGill University. https://escholarship.mcgill.ca/concern/theses/9306t007d.

Levin, D. A. 1979. The nature of plant species. *Science* 204:381–4.

Levin, D. A. 2000. *The origin, expansion, and demise of plant species.* Oxford: Oxford University Press.

Levin, G. D. 2011. [*Truth and rationality.*] Moscow: Canon+. (in Russian).

Levushkin, S. I. 1976. [The notion of species and ecology.] [*Species and its productivity in the area*], 83–5. Vilnius: Vilnius State University.

Lewens, T. 2012. Species, essence and explanation. *Studies in History and Philosophy of Biological and Biomedical Sciences* 43:751–7.

Lidén, M. 1990. Replicators, hierarchy, and the species problem. *Cladistics* 6:183–6.

Lidén, M. 1992. Species: Where's the problem? *Taxon* 41:315–7.

Lidén, M., and B. Oxelman. 1989. Species. Pattern or process? *Taxon* 38:228–32.

Lightner, J. 2013. An initial estimate of avian Ark kinds. *Answers Research Journal* 6:409–66.

Lindberg, D. C. 1983. *Roger Bacon's philosophy of Nature [...].* Oxford: Clarendon Press.

Lindeboom, G. A. 1982. Jan Swammerdam (1637–1680) and his Biblia Naturae. *Clio Medica* 17: 113–31.

Lindley, J. 1832. *An introduction to botany.* London: Longman, Rees, Orme, Brown, Green, & Longman.

Lindley, J. 1835. *A key to structural, physiological, and systematic botany [...].* London: Longman, Rees, Orme, Brown, Green, & Longman.

Lindley, J. 1836. *The vegetable kingdom: or The structure, classification, and uses of plants, illustrated upon the natural system.* London: Bradbury & Evans.

Linnaeus, C. 1735. *Systema Naturae, sive Regna tria Naturae systematice proposita per classes, ordines, genera, & species.* Lugduni Batavorum: Haak.

Linnaeus, C. 1736. *Fundamenta botanica quae majorum operum prodromi instar theoriam scientiae botanices [...].* Amstelodami: Salomonem Schouten.

Linnaeus, C. 1737a. *Critica botanica in qua nomina plantarum generica, specifica, & variantia [...].* Lugduni Batavorum: Conradum Wishoff.

Linnaeus, C. 1737b. *Genera plantarum eorumque characteres naturales secundum numerum, figuram, situm [...].* Lugduni Batavorum: Conradum Wishoff.

Linnaeus, C. 1749. *Amoenitas academiae, seu, dissertationes variae physicae, medicae botanicae [...].* Holmiae et Lipsiae: apud Godofredum Kiesewetter.

Linnaeus, C. 1751. *Philosophia botanica in qua explicantur fundamenta botanica [...].* Stockholmiæ: apud Godofredum Kiesewetter.

Linnaeus, C. 1753. *Species plantarum, exhibentes plantas rite cognitas, ad genera relatas [...]*, T. I. Holmiae: imp. Laurentii Salvii. 560 p.

Linnaeus, C. 1762. *Fundamentum fructificationis [...].* Upsaliae.

Linnaeus, C. 1766–8. *Systema naturae per regna tria naturae [...], Editio duodecima reformata*, T. I–III. Holmiae: imp. Laurentii Salvii.
Linnaeus, C. 1792. *Praelectiones in Ordines naturales plantarum [...].* Hamburgi: Benj. Gottl. Hoffmanni.
Linnaeus, C. 2003. *Philosophia botanica.* Oxford: Oxford University Press.
Liu, J. 2016. The integrative species concept and species on the speciation way. *Biodiversity Science* 24:1004–8.
Lloyd, A. C. 1962. Genus, species and ordered series in Aristotle. *Phronesis* 7:67–90.
Lloyd, A. C. 1981. *Form and universal in Aristotle.* Liverpool: Francis Cairns.
Lloyd, A. C. 1990. *The anatomy of Neoplatonism.* Oxford: Clarendon Press.
Lloyd, G. E. R. 1961. The development of Aristotle's theory of classification of animals. *Phronesis* 6:59–81.
Locke, J. 1999. *An essay concerning human understanding.* University Park, PA: Penn State University Press.
Locy, W. A. 1915. *Biology and its makers*, 3rd ed. New York, NY: Henry Holt and Co.
Lones, T. E. 1912. *Aristotle's researches in natural science.* London: West, Newman & Co.
Look, B. C. 2009. Leibniz and Locke on natural kinds. In *The early moderns in quest for the unity of knowledge*, ed. V. Alexandrescu, 380–409. Bucharest: Zeta Books.
Look, B. C. 2020. Gottfried Wilhelm Leibniz, In *Stanford encyclopedia of philosophy*, ed. E. N. Zalta. https://plato.stanford.edu/archives/fall2020/entries/leibniz/.
Losev, A. F. 1960. [*History of Greek literature.*] Moscow: Nauka. (in Russian).
Losev, A. F., and A. A. Takho-Godi. 1993. [*Plato. Aristotle.*] Moscow: Molodaya Gvardiya. (in Russian).
Löther, R. 1972. *Die Beherrschung der Mannigfaltigkeit. Philosophische Grundlagen der Taxonomie.* Jena: Gustav Fischer.
Lotsy, J. P. 1916. *Evolution by means of hybridization.* Hague: M. Nijhoff.
Lotsy, J. P. 1931. On the species of the taxonomist in its relation to evolution. *Genetica* 13:1–16.
Lovejoy, A. O. 1936. *The Great Chain of Being: A study of the history of an idea.* Cambridge, MA: Harvard University Press.
Lovejoy, A. O. 1959 (1911). Buffon and the problem of species. In *Forerunners of Darwin: 1745–1859*, eds B. Glass, O. Temkin, and W. L. Sraus, 84–112. Baltimore, MD: John Hopkins Press.
Løvtrup, S. 1979. The evolutionary species: Fact or fiction? *Systematic Zoology* 28:386–92.
Løvtrup, S. 1987. On the species problem. *Environmental Biology of Fishes* 20:3–9.
Luckow, M. 1995. Species concepts: Assumptions, methods, and applications. *Systematic Botany* 20:589–605.
Ludwig, D. 2018. Letting go of "natural kind". Towards a multidimensional framework of non-arbitrary classification. *Philosophy of Science* 85:31–52.
Lunkevich, V. V. 1960. [*From Heraclitus to Darwin. Essays on the history of biology.*] Vol. 1. 2nd ed. Moscow: Uchpedgiz. (in Russian).
Lurie, E. 1960. *Louis Agassiz. A life in science.* Chicago, IL: University of Chicago Press.
Lysenko, T. D. 1952. [*New in the science of biological species.*] Moscow: State Publishing of Agricultural Literature. (in Russian).
Lyubarsky, G. Yu. 1996. [*Archetype, style and rank in biological systematics.*] Moscow: KMK Sci. Pres. (in Russian).
Lyubarsky, G. Yu. 2015. [*The birth of science. Analytical morphology, classification system, scientific method.*] Moscow: YaSK Publ. (in Russian).
Lyubarsky, G. Yu. 2018. [*Origins of hierarchy: The history of taxonomic rank.*] Moscow: KMK Sci. Press. (in Russian).
Lyubishchev, A. A. 1972. To the logic of systematics. In *Problems of evolution*, Vol. 2, 45–68. Novosibirsk: Nauka.

Maat, J. 2004. *Philosophical languages in the seventeenth century: Dalgarno, Wilkins, Leibniz*. Dordrecht: Kluwer.

MaCarthur, R. 1965. Patterns of species diversity. *Biological Reviews* 40:510–33.

Mace, G. M. 2004. The role of taxonomy in species conservation. *Philosophical Transactions of the Royal Society of London, Ser. B* 359:711–19.

MacLeod, M. 2012. Rethinking scientific concepts for research contexts: the case of the classical gene. In *Scientific concepts and investigative practice*, eds U. Feest and F. Steinle, 47–74. Baltimore, MD: John Hopkins University Press.

Madigan, A. S. J. 1994. Alexander on Aristotle's species and genera as principles. In *Aristotle in Late Antiquity*, ed. L. Schrenk, 76–91. Washington, DC: Catholic University of America Press.

Magee, J. 1998. *Anicii Manlii Severini Boethii De divisione liber: Critical edition, translation, prolegomena, and commentary*. Leiden: E. J. Brill.

Magnol, P. 1689. *Prodromus historiae generalis plantarum in quo Familiae plantarum [...]*. Monspelij: Gabrielis & Honorati Pech.

Magnus, D. 1996. Theory, practice, and epistemology in the development of species concepts. *Studies in History and Philosophy of Science* 21:521–45.

Magnus, P. D. 2015. John Stuart Mill on taxonomy and natural kinds. *The Journal of the International Society for the History of Philosophy of Science* 5:269–80.

Mahner, M. 1993. What is a species? *Journal for General Philosophy of Science* 24:103–26.

Mahner, M., and M. Bunge. 1997. *Foundations of biophilosophy*. Frankfurt: Springer Verlag.

Maienschein, J. 2000. Why study history for science? *Biology and Philosophy* 15:339–48.

Makovelsky, A. O. 2004. [*The history of logic*.] Moscow: Kuchkovo Pole. (in Russian).

Malikov, V. G., and F. N. Golenishchev. 2009. [A systemic concept of speciation and the species problem.] In [*Species and speciation: An analysis of new views and trends*], eds A. F. Alimov and S. D. Stepanyanz, 117–40. Saint Petersburg: Zoological Institute. (in Russian).

Malikov, V. G., and F. N. Golenishchev. 2016. [Reality and relativity of species.] In *Aspects of biodiversity*, Vol. 1, eds I. Ya. Pavlinov, M. V. Kalyakin, and A. V. Sysoev, 134–45. Moscow: KMK Sci. Press. (in Russian).

Mallet, J. 1995. A species definition for the modern synthesis. *Trends in Ecology and Evolution* 10:294–99.

Mallet, J. 2001. Species, concept of. In *Encyclopedia of biodiversity*, Vol. 5, ed. S. Levin, 427–40. London: Academic Press.

Mallet, J. 2004. Species problem solved 100 years ago. *Nature* 430:503.

Mallet, J. 2008. Hybridization, ecological races and the nature of species: Empirical evidence for the ease of speciation. *Philosophical Transactions of the Royal Society B* 363:2971–86.

Mallet, J. 2010a. Why was Darwin's view of species rejected by twentieth century biologists? *Biology and Philosophy* 25:497–527.

Mallet, J. 2010b. Group selection and the development of the biological species concept. *Philosophical Transactions of the Royal Society* 365:1853–63.

Mallet, J. 2013a. Darwin and species. In *The Cambridge encyclopedia of Darwin and evolutionary though*, ed. M. Ruse, 109–15. New York, NY: Cambridge University Press.

Mallet, J. 2013b. Subspecies, semispecies, superspecies. In *Encyclopedia of biodiversity*, 2d ed. Vol. 5, ed. S. Levin, 45–48. London: Academic Press.

Maloney, T. S. 1985. The extreme realism of Roger Bacon. *The Review of Metaphysics* 38:807–37.

Mamchur, E. A. 2004. [*Objectivity of science and relativism (To discussions in modern epistemology)*.] Moscow: Institute of Philosophy RAN. (in Russian).

Manceau, M., and A. Lambert. 2019. The species problem from the modeler's point of view. *Bulletin of Mathematical Biology* 81:878–98.

Mandelbaum, M. 1965. The history of ideas, intellectual history, and the history of philosophy. *History and Theory* 5:33–66.
Marenbon, J. 1988. *Early Medieval philosophy (480–1150): An introduction*, 2nd ed. London: Routledge.
Marenbon, J. 1997. *The philosophy of Peter Abelard.* Cambridge: Cambridge University Press.
Marenbon, J. 2003. *Boethius.* New York, NY: Oxford University Press.
Margolis, E. and S. Laurence. 2012. Concepts. In *Stanford encyclopedia of philosophy*, ed. E. N. Zalta. https://plato.stanford.edu/entries/concepts/.
Marradi, A. 1990. Classification, typology, taxonomy. *Quality & Quantity* 24:129–57.
Marsh, F. L. 1941. *Fundamental biology.* Lincoln, OR: Self-published.
Marsh, F. L. 1944. *Evolution, creation, and science.* Washington, DC: Review & Herald Publ.
Marsh, F. L. 1960. The Genesis kinds in our modern world. *The Journal of the American Scientific Affiliation* 12:4–8.
Martijn, M. 2010. *Proclus on Nature. Philosophy of Nature and its methods in Proclus' Commentary on Plato's Timaeus.* Leiden: E. J. Brill.
Martin, C. J. 2009. The logical textbooks and their influence. In *The Cambridge companion to Boethius*, ed. J. Marenbon, 56–84. Cambridge: Cambridge University Press.
Marx, W. 1977. *Introduction to Aristotle's theory of being as being.* Hague: Springer Science.
Marza, V. D., and N. Cerchez. 1967. Charles Naudin, a pionneer of contemporary biology. *Journal d'agriculture traditionnelle et de botanique appliquée* 14:369–401.
Mason, H. L. 1950. Taxonomy, systematic botany and biosystematics. *Madroño* 10:193–208.
Masters, J. C., and F. Génin. 2018. The recognition concept of species, decades on. *Rivista di Biologia* 111:93–7.
Masters, J. C., R. J. Rayner, I. J. McKay, et al. 1987. The concept of species: Recognition versus isolation. *South African Journal of Science* 83:534–7.
Masters, J. C., and H. G. Spencer 1989. Why we need a new genetic species concept. *Systematic Zoology* 38:270–9.
Maurer, A. 1982. *Medieval philosophy: An introduction*, 2nd ed. Toronto: Pontifical Institute of Mediaeval Studies.
Maurer, A. 1999. *The philosophy of William of Ockham in the light of its principles.* Toronto: Pontifical Institute of Mediaeval Studies.
Maxwell, S. J., B. C. Congdon, and T. L. Ryme 2020. Essentialistic pluralism: The theory of spatio-temporal positioning of species using integrated taxonomy. *Proceedings of the Royal Society of Queensland* 124:1–17.
Mayden, R. L. 1997. A hierarchy of species concepts: The denouement in the saga of the species problem. In *Species. The units of biodiversity*, eds M. F. Claridge, A. H. Dawah, and M. R. Wilson, 381–424. London: Chapman & Hall.
Mayden, R. L. 1999. Consilience and a hierarchy of species concepts: Advances toward closure on the species puzzle. *Journal of Nematology* 31:95–116.
Mayden, R. L. 2002. On biological species, species concepts and individuation in the natural world. *Fish and Fisheries* 3:171–96.
Mayden, R. L., and R. M. Wood. 1995. Systematics, species concepts, and the evolutionarily significant unit in biodiversity and conservation biology. *American Fisheries Society Symposium* 17:58–113.
Mayr, E. (ed.). 1957b. *The species problem. A symposium presented at the Atlanta Meeting of the American Association for the Advancement of Science.* Washington, DC: American Association for the Advancement of Science.
Mayr, E. 1942. *Systematics and the origin of species, from the viewpoint of zoologist.* New York, NY: Columbia University Press.
Mayr, E. 1949. The species concept: Semantics versus semantics. *Evolution* 3:371–2.

Mayr, E. 1957a. Species concepts and definitions. In *The species problem. A symposium presented at the Atlanta Meeting of the American Association for the Advancement of Science*, ed. E. Mayr, 1–22. Washington, DC: American Association for the Advancement of Science.

Mayr, E. 1957c. Difficulties and importance of the biological species. In *The species problem. A symposium presented at the Atlanta Meeting of the American Association for the Advancement of Science*, ed. E. Mayr, 371–388. Washington, DC: American Association for the Advancement of Science.

Mayr, E. 1959a. Typological versus population thinking. In *Evolution and anthropology: A centennial appraisal*, 409–12. Washington, DC: Anthropological Society of Washington.

Mayr, E. 1959b. Agassiz, Darwin, and evolution. *Hamard Libran Bulletin* 8:165–94.

Mayr, E. 1959c. Darwin and the evolutionary theory in biology. In *Evolution and anthropology: A centennial appraisal*, ed. B. J. Meggers, 1–10. Washington, DC: Anthropological Society of Washington.

Mayr, E. 1963. *Animal species and evolution*. Cambridge, MA: Harvard University Press.

Mayr, E. 1968. Illiger and the biological species concept. *Journal of the History of Biology* 1:163–78.

Mayr, E. 1969. *Principles of systematic zoology*. New York, NY: McGrow Hill Book.

Mayr, E. 1976. *Evolution and the diversity of life: Selected essays*. Cambridge, MA: Belknap Press.

Mayr, E. 1982. *The growth of biological thought*. Cambridge, MA: Belknap Press.

Mayr, E. 1987. The ontological status of species: Scientific progress and philosophical terminology. *Biology and Philosophy* 2:145–66.

Mayr, E. 1988. *Toward a new philosophy of biology, observations of an evolutionist*. New York, NY: Cambridge University Press.

Mayr, E. 1991. *One long argument: Charles Darwin and the genesis of modern evolutionary thought*. Cambridge, MA: Harvard University Press.

Mayr, E. 1996. What is a species, and what is not? *Philosophy of Science*, 63:262–77.

Mayr, E. 1997. *This is biology: The science of the living world*. Cambridge, MA: Harvard University Press.

Mayr, E. 1998. The role of systematics in evolutionary synthesis. In *The evolutionary synthesis: Perspectives on the unification of biology*, eds E. Mayr and W. B. Provine, 123–36. Cambridge, MA: Harvard University Press.

Mayr, E. 2000. The biological species concept. In *Species concepts and phylogenetic theory: A debate*, eds Q. D. Wheeler and R. Meier, 17–29. New York, NY: Columbia University Press.

Mayr, E., and W. J. Bock. 2002. Classifications and other ordering systems. *Journal of Zoological Systematics and Evolutionary Research* 40:169–94.

McCabe, T. 2008. Studying species definitions for mutual nonexclusiveness. *Zootaxa* 1939:1–9.

McCabe, T. 2011. Reconciling evolutionary theory and taxonomic practice. *Journal of Zoological Systematics and Evolutionary Research* 49:257–65.

McCord Adams, M. 1989. *William Ockham*, Pt. 2. Notre Dame, IN: University of Notre Dame Press.

McCray, A.T. 2006. Conceptualizing the world: Lessons from history. *Journal of Biomedical Informatics* 39:267–73.

McDade, L. A. 1995. Species concepts and problems in practice: Insight from botanical monographs. *Systematic Botany* 20:606–22.

McGinley, D. 2008. Duns Scotus's theory of Common Natures. *Filosofia Unisinos* 9:64–83.

McGrade, A. S. (ed.). 2003 *The Cambridge companion to Medieval philosophy*. Cambridge: Cambridge University Press.

McIntosh, R.P. 1986. *The background of ecology: Concept and theory*. New York, NY: Cambridge University Press.
McKelvey, B. 1982. *Organizational systematics. Taxonomy, evolution, classificaiton*. Berkeley, CA: University of California Press.
McKitrick, M. C., and R. M. Zink. 1988. Species concepts in ornithology. *The Condor* 90:1–14.
McOuat, G. 1996. Species, rules and meaning: The politics of language and the ends of definitions in 19th century natural history. *Studies in History and of Biological and Biomedical Science* 21:473–519.
McOuat, G. 2001. From cutting nature at its joints to measuring it: New kinds and new kinds of people in biology. *Studies in History and Philosophy of Biological and Biomedical Sciences* 32:613–45.
McOuat, G. 2009. The origins of 'natural kinds': Keeping 'essentialism' at bay in the age of reform. *Intellectual History Review* 19:211–30.
Meier, R., and R. Willmann. 2000. The Hennigian species concept. In *Species concepts and phylogenetic theory: A debate*, eds Q. D. Wheeler and R. Meier, 30–43. New York, NY: Columbia University Press.
Meikle, R. D. 1957. What is the subspecies? *Taxon* 6:102–5.
Mensch, J. 2013. *Kant's organicism. Epigenesis and the development of critical philosophy*. Chicago, IL: University of Chicago Press.
Merlan, P. 1946. Aristotle's unmoved movers. *Traditio* 4:1–30.
Meyen, S. V. 1977. [Taxonomy and meronomy.] In *[Methodological issues in geological sciences]*, 25–33. Kïev: Naukova Dumka. (in Russian).
Meyen, S. V., and Yu. A. Shreyder. 1976. [Methodological issues of the theory of classification.] *Voprosy Filosofii* 12:67–79. (in Russian).
Meyer-Abich, A. 1949. *Biologie der Goethe-Zeit [...]*. Stuttgart: Hippokrates-Verlag Marquardt.
Michener, C. D. 1962. Some future developments in taxonomy. *Systematic Zoology* 12:151–72.
Mikeshina, L. A. 2005. *[Philosophy of science: Modern epistemology.]* Moscow: Progress-Tradition. (in Russian).
Mikeshina, L. A. 2007. [Epistemology and cognitive science: Basic categories and principles of interaction.] *[Cognitive approach,]* ed. V. A. Lectorsky, 20–57. Moscow: Canon. (in Russian).
Mikhailov, K. E. 2003. [Typological comprehension of "biological species" and the ways to stabilize an "around-species" taxonomy in birds.] *Ornithology* 30:9–24. (in Russian).
Mill, J. S. *1843. A system of logic, ratiocinative and inductive [...]*, Vol. 1. London: John W. Parker.
Miller, G. A. 1996. Contextuality. In *Mental models in cognitive science: Essays in honour of Phil Johnson-Laird*, eds A. Garnham and J. Oakhill, 1–18. Erlbaum: Psychology Press.
Miller, H. 1849. *Foot-prints of the Creator, or, the Asterolepis of Stromness*. London: Johnstone & Hunter.
Miller, W. 2001. The structure of species, outcomes of speciation, and the 'species problem': ideas for paleobiology. *Palaeogeography, Palaeoclimatology, Palaeoecology* 176:1–10.
Miller, W. 2006. What every paleontologist should know about species: New concepts and questions. *Neues Jahrbuch für Geologie und Paläontologie* 228:557–76.
Miller, W. 2016. The species problem: Concepts, conflicts, and patterns preserved in the fossil record. In *Species and speciation in the fossil record*, eds W. D. Allmon and M. M. Yacobucci, 28–58. Chicago, IL: University of Chicago Press.
Millikan, R. G. 1999. Historical kinds and the "special sciences". *Philosophical Studies* 95:45–65.
Mina, M. V. 2007. [The species—ideal, real, and delineated by agreement.] In *[Linnaean collection]*, ed. I. Ya. Pavlinov, 305–17. Moscow: Moscow State University Publ. (in Russian).

Mina, M. V. 2010. [Evolution of the species concept from Darwin to the present days: Progress or wandering?] In *[Charles Darwin and modern biology,]* 228–36. Saint Petersburg: Nestor-History. (in Russian).

Minelli, A., and G. Fusco (eds). 2008. *Evolving pathways: Key themes in evolutionary developmental biology*. Cambridge: Cambridge University Press.

Minelli, A. 2003. *The development of animal form: Ontogeny, morphology, and evolution*. Cambridge: Cambridge University Press.

Minelli, A. 2014. Developmental disparity. In *Towards a theory of development*, eds A. Minelli and T. Pradeu, 227–45. Oxford: Oxford University Press.

Minelli, A. 2015. Taxonomy faces speciation: The origin of species or the fading out of the species? *Biodiversity Journal* 6:123–38.

Minelli, A. 2016. Tracing homologies in an ever-changing world. *Rivista di estetica* 62:40–55.

Minelli, A. 2019. The galaxy of the non-Linnaean nomenclature. *History and Philosophy of the Life Sciences* 41:31.

Minelli, A. 2020. Taxonomy needs pluralism, but a controlled and manageable one. *Megataxa* 1:9–18.

Minelli, A. 2022. The species before and after Linnaeus – tension between disciplinary nomadism and conservative nomenclature. In *Species problems and beyond. Contemporary issues in philosophy and practice*, eds J. S. Wilkins, F. E. Zachos, and I. Ya. Pavlinov, 197–226. Boca Raton, FL: CRC Press.

Minelli, A., and G. Fusco. 2013. Homology. In *The philosophy of biology: A companion for educators*, ed. K. Kampourakis, 289–322. Dordrecht: Springer Science+Business Media.

Mirbel, C.-F. 1806. *Histoire naturelle, générale et particulière, des plantes [...]*, T. 1. Paris: F. Dufart.

Mirkin, B. M., and L. G. Naumova. 2005. *[Fundamentals of general ecology.]* Moscow: University Book. (in Russian).

Mishler, B. D. 1999. Getting rid of species? In *Species: New interdisciplinary essays*, ed. R. A. Wilson, 307–15. Cambridge, MA: MIT Press.

Mishler, B. D. 2010. Species are not uniquely real biological entities. In *Contemporary debates in philosophy of biology*, eds. F. J. Ayala and R. Arp, 110–22. Hoboken, NJ: Wiley-Blackwell.

Mishler, B. D. 2021. *What, if anything, are species?* Boca Raton, FL: CRC Press.

Mishler, B. D. 2022. Ecology, evolution, and systematics in a post-species world. In *Species problems and beyond. Contemporary issues in philosophy and practice*, eds J. S. Wilkins, F. E. Zachos, and I. Ya. Pavlinov, 177–90. Boca Raton, FL: CRC Press.

Mishler, B. D., and R. N. Brandon. 1987. Individuality, pluralism, and the phylogenetic species concept. *Biology and Philosophy* 2:397–414.

Mishler, B. D., and R. N. Brandon. 1989. Sex and the individuality of species: A response to Ghiselin. *Biology and Philosophy* 4:77–9.

Mishler, B. D., and M. J. Donoghue. 1982. Species concepts: A case for pluralism. *Systematic Zoology* 31:491–503.

Mishler, B. D., and E. C. Theriot. 2000. The phylogenetic species concept: monophyly, apomorphy, anf phylogenetic species concept. In *Species concepts and phylogenetic theory: A debate*, eds Q. D. Wheeler, and R. Meier, 44–54. New York, NY: Columbia University Press.

Mishler, B. D., and J. S. Wilkins. 2018. The hunting of the SNaRC: A snarky solution to the species problem. *Philosophy, Theory, and Practice in Biology* 10:1–18.

Mivart, G. J. 1871. *On the genesis of species*. London: Macmillan and Co. 314 p.

Mlikovsky, J. and K. Zemek. 1983. On the history and presebt state of the species concept. In *General question of evolution, Proceedings of International Colloquium*, eds V. J. A. Novák and K. Zemek, 281–304. Praha: ČSAV.

Morgun, D. V. 2002. *[Epistemological foundations of the species problem in biology.]* Moscow: Moscow State University Publ. (in Russian).

Morgun, D. V. 2005. [*The species problem in the context of the formation of non-classical biology.*] Doctoral dissertation. Moscow: Moscow State University. http://www.dslib.net/filosofia-texniki/problema-vida-v-kontekste-formirovanija-neklassicheskoj-biologii.html. (in Russian).

Morison R. 1672. *Plantarum umbelliferarum distributio nova, per Tabulas cognationis et affinitatis [...].* Oxonii: Theatro Sheldoniano.

Morison, R. 1669. *Praeludia botanica.* Londini: Tho. Roycroft.

Moritz, C. 1994. Defining 'evolutionarily significant units' for conservation. *Trends in Ecology and Evolution* 9:373–5.

Morris, H. M., and J. D. Morris. 1997. *The modern creation trilogy.* Green Forest, AR: Master Books.

Motroshilova, N. V. (ed.). 2000. *[History of philosophy. West-Russia-East. The first book. Philosophy of antiquity and the Middle Ages,]* 3rd ed. Moscow: Greek-Latin Cabinet. (in Russian).

Müller-Wille, S. 2001. History redoubled: the synthesis of facts in Linnaean natural history. In *Philosophies of technology: Francis Bacon and his contemporaries*, eds C. Zittel, G. Engel, R. Nanni, et al., 515–38. Leiden: E. J. Brill.

Müller-Wille, S. 2007. Collection and collation: theory and practice of Linnaean botany. *Studies in History and Philosophy of Biological and Biomedical Sciences* 38:541–62.

Müller-Wille, S. 2011. Making sense of essentialism. *Critical Quarterly* 53:61–7.

Müller-Wille, S., and V. Orel. 2007. From Linnaean species to Mendelian factors: Elements of hybridism, 1751–1870. *Annals of Science* 64:171–215.

Müller-Wille, S., and S. Scharf. 2009. Indexing Nature: Carl Linnaeus (1707–1778) and his fact-gathering strategies. *The nature of evidence: How well do 'facts' travel?* 36:1–39. London: London School of Economics. http://eprints.lse.ac.uk/47386/1/indexing%20nature(author).pdf.

Murphy, G. L. 2002. *The big book of concepts.* Cambridge, MA: MIT Press.

Murray, I. A. 1784. *Caroli a Linne equitis Systema Vegitabilium secundum classes ordines genera species [...]*, editio decima quarta. Gottingae: Jo. Christ. Dieterich.

Myers, G. S. 1952. The nature of systematic biology and of a species description. *Systematic Zoology* 1:106–11.

Nägeli, K. 1865. *Entstehung und Begriff der naturhistorischen Art.* Munich: Königliche Akademie.

Nalimov, V. V. 1979. [*A probabilistic model of language. On relation between natural and artificial languages.*] Moscow: Nauka (in Russian).

Nanay, B. 2010. Population thinking as trope nominalism. *Synthese* 177:91–109.

Naomi, S.-I. 2011. On the integrated frameworks of species concepts: Mayden's hierarchy of species concepts and de Queiroz's unified concept of species. *Journal of Zoological Systematics and Evolutionary Research* 49:177–84.

Nathan, M. J. 2017. Pluralism is the answer! What is the question? *Philosophy, Theory, and Practice in Biology* 11:1–14.

Nathan, M. J., and J. Cracraft. 2020. The nature of species in evolution. In *The theory of evolution: Principles, concepts, and assumptions*, eds S. M. Scheiner and D. P. Mindell, 102–22. Chicago, IL: University of Chicago Press.

Naudin, Ch. 1852. Considérations philosophiques sur l'espèce et la variété. *Revue horticole*, 4-e série 1:102–9.

Naudin, Ch. 1862. Nouvelles recherches sur l'hybridité dans les végétaux. *Nouvelles Archives du Musèum d'Histoire Naturelle* 1:1–176.

Nauta, L. 2021. Lorenzo Valla. In *Stanford encyclopedia of philosophy*, ed. E. N. Zalta. https://plato.stanford.edu/archives/fall2021/entries/lorenzo-valla.

Nawar, T. 2021. Augustine on active perception, awareness, and representation. *Phronesis* 66:84–110.

Nazarenko, A. A. 2001. [Is a single concept of species possible in ornithology? (A viewpoint of a practicing taxonomist).] *Zurnal Obshchei Biologii* 62:180–6. (in Russian).

Nazarov, V. I. 1991. [*The doctrine of macroevolution: Towards a new synthesis.*] Moscow: Nauka.

Nelson, G. 1979. Cladistic analysis and synthesis: Principles and definitions, with a historical note on Adanson's Familles des Plantes (1763–1764). *Systematic Zoology* 28:1–21.

Nelson, G. 1994. Homology and systematics. In *Homology: The hierarchical basis of comparative morphology*, ed. B. K. Hall, 101–49. New York, NY: Academic Press.

Neretina, S. S. 1994. [*Word and text in medieval culture. Abelard's conceptualism.*] Moscow: Gnosis. (in Russian).

Neretina, S. S. 2006. Boethius: The Tree of creation. *Orientiry* 3:98–132. https://iphras.ru/uplfile/root/biblio/orientiry/Or_3/5.pdf. (in Russian).

Neretina, S. S. 2013. [Is Christian Platonism possible?] *Electronic Philosophical magazine Vox/Voice* 15:1–39. http://vox-journal.org. (in Russian).

Neto, C. 2016. Rethinking cohesion and species individuality. *Biological Theory* 11:1–12.

Newton, J. 2014. Locke on the 'species' of substances. *Locke Studies* 14:105–20.

Nikolskiy, G. V. 1962. [*Species and speciation.*] Moscow: Znanye. (in Russian).

Nitecki, M. H. 1957. What is a paleontological species? *Evolution* 11:378–80.

Nixon, K. C., and Q. D. Wheeler. 1990. An amplification of the phylogenetic species concept. *Cladistics* 6:211–23.

Noone, T. B. 2003. *Universals and individuation.* In *The Cambridge Companion to Duns Scotus.* ed. T. Williams, 100–28. Cambridge: Cambridge University Press.

Noor, M. 2002. Is the biological species concept showing its age? *Trends in Ecology and Evolution* 17:153–4.

Novick, A., and W. F. Doolittle. 2021. 'Species' without species. *Studies in History and Philosophy of Science* 87:72–80.

Nowak, M. A. 1992. What is a quasispecies. *Trends in Ecology and Evolution* 7:118–21.

O'Hara, R. J. 1993. Systematic generalization, historical fate, and the species problem. *Systematic Zoology* 42:231–46.

O'Hara, R. J. 1994. Evolutionary history and the species problem. *American Zoologist,* 34:12–22.

O'Meara, J. J. O. (ed.). 1987. *Eriugena. Periphyseon (the Division of Nature).* Montréal: Bellarmin.

Odum, E. C. 1953. *Fundamentals of ecology.* Philadelphia, PA: Saunders.

Ogilvie, B. W. 2006. *The science of describing: Natural history in Renaissance Europe.* Chicago, IL: University Chicago Press.

Ogurtsov, A. P. 1993. [*Philosophy of science of the Enlightenment.*] Moscow: Iinstitue of Phylosophy RAS. (in Russian).

Okasha, S. 2002. Darwinian metaphysics: Species and the question of essentialism. *Synthese* 131:191–213.

Oliver, I., and A. J. Beattie. 1996. Invertebrate morphospecies as surrogates for species: A case study. *Conservation Biology* 10:99–109.

Olmstead, R. G. 1995. Species concepts and plesiomorphic species. *Systematic Botany* 20:623–30.

Ong, W. J. 1958. *Ramus, method, and the decay of dialogue. From the art of discourse to the art of reason.* Cambridge, MA: Harvard University Press.

Onishi, Y., and D. Serpico. 2022. Homeostatic property cluster theory without homeostatic mechanisms: Two recent attempts and their costs. *Journal for General Philosophy of Science* 53:61–82.

Oren, A. 2004. Prokaryote diversity and taxonomy: Current status and future challenges. *Philosophical Transactions of the Royal Society B* 359: 623–38.

Orlov, E. V. 2011. [*The philosophical language of Aristotle.*] Novosibirsk: Siberian Branch of USSR Academy of Sciences. (in Russian).

Osborn, H. F. 1913. *From the Greeks to Darwin: An outline of the development of the evolution idea*, 5th ed. New York NY: The Macmillan Company.

Oskolsky, A. A. 2001. [On the phenomenology of biological similarity.] In [*Homology in botany: Experience and reflection,*] 100–18. Saint Petersburg: Saint Petersburg Union of Scientists. (in Russian).

Osler, M. J. 1997. Mixing metaphors: Science and religion or natural philosophy and theology in early modern Europe. *History of Science* 36:91–113.

Otte, D., and J. Endler (eds). 1989. *Speciation and its consequences.* Sunderland, MA: Sinauer Associates.

Oxenham, M. F. 2015. Lamarck on species and evolution. In *Taxonomic tapestries: The threads of evolutionary, behavioural and conservation research*, eds A. M. Behie and M. F. Oxenham, 155–70. Canberra: ANU Press.

Ozaki, K., M. Isono, T. Kawahara, et al. 2006. A Mechanistic approach to evaluation of umbrella species as conservation surrogates. *Conservation Biology* 20:1507–15.

Paasch, J. T. 2014. Scotus and Ockham on universals and individuation. In *Debates in Medieval philosophy. Essential readings and contemporary responses*, ed. J. Hause, 371–94. London: Routledge.

Padial, J.M., and I. De la Riva. 2020. A paradigm shift in our view of species drives current trends in biological classification. *Biological Reviews*, 96:731–51.

Pagel, M., C. Venditti, and A. Meade. 2006. Large punctuational contribution of speciation to evolutionary divergence at the molecular level. *Science* 314:119–21.

Panaccio, C. 2004. *Ockham on concepts.* Aldershot: Ashgate Publishing.

Panchen, A. L. 1992. *Classification, evolution, and the nature of biology.* Cambridge: Cambridge University Press.

Panchuk, M. L. 2016. *Divine ideas for metaphysical realism.* Doctoral dissertation. Columbia: University of South Carolina. 186 p. https://scholarcommons.sc.edu/etd/3365.

Panov, E. N. 1993. [Species boundary and hybridization in birds.] In [*Hybridization and the species problem in vertebrates,*] 53–96. Moscow: Moscow State University Publ. (in Russian).

Pante, E., N. Puillandre, A. Viricel. et al. 2015. Species are hypotheses: Avoid connectivity assessments based on pillars of sand. *Molecular Ecology* 24:525–44.

Paramonov, S. 1951. What is a species? The essence, the extent, and the definition of the species concept. *The Annals of the Ukrainian Academy of Arts and Science* 1:138–49.

Parmasto, E. (ed.). 1986a. [*The problem of species and genus in fungi.*] Tallinn: Institute of Zoology EAS. (in Russian).

Parmasto, E. 1986b. [The problem of the species in fungi.] In [*The problem of species and genus in fungi*] ed. E. Parmasto, 9–28. Tallinn: Institute of Zoology EAS. (in Russian).

Pasnau, R. 1997. *Theories of cognition in the later Middle Ages.* Cambridge: Cambridge University Press.

Pásztor, L., Z. Botta-Duká, T. Czárán, et al. 2016. *Theory-based ecology: A Darwinian approach.* Oxford: Oxford University Press.

Paterson, H. E. H. 1985. The recognition concept of species. In *Species and speciation*, ed. E. S. Vrba, 21–9. Pretoria: Transvaal Museum.

Pavlinov, I. Ya. (ed.). 2013b. *The species problem: Ongoing issues.* Rijeka: InTech Open Access Publ.

Pavlinov, I. Ya. 1992. [If there is the biological species, or what is the "harm" of taxonomy.] *Zurnal Obshchei Biologii* 53:757–67. (in Russian).

Pavlinov, I. Ya. 1995. [Classification as a hypothesis: Entering the problem.] *Zhurnal Obshchei Biologii* 56:411–24. (in Russian).

Pavlinov, I. Ya. 2005. [*An introduction to contemporary phylogenetics.*] Moscow: KMK Sci. Press. (in Russian).

Pavlinov, I. Ya. 2007. [Scientific pluralism and the species problem in biology. [*Carl Linnaeus in Russia*,] 127–33. Saint Petersburg: Saint Petersburg Center for the History of Ideas. (in Russian).

Pavlinov, I. Ya. 2009a. [The species problem: Another look.] In [*Species and speciation: An analysis of new views and trends*], eds A. F. Alimov and S. D. Stepanyanz, 250–71. Saint Petersburg: Zoological Institute. (in Russian).

Pavlinov, I. Ya. 2010. [Remarks on biomorphics (ecomorphological systematics).] *Zurnal Obshchei Biologii* 71:187–92. (in Russian).

Pavlinov, I. Ya. 2011. How is it possible to build a taxonomic theory. *Zoological Research* 10:45–100.

Pavlinov, I. Ya. 2012. The contemporary concepts of homology in biology: A theoretical review. *Biology Bulletin Reviews* 2:36–54.

Pavlinov, I. Ya. 2013a. The species problem: Why again? In *The species problem: Ongoing issues*, ed. I. Ya. Pavlinov, 3–37. Rijeka: InTech Open Access Publ.

Pavlinov, I. Ya. (ed.). 2013b. *The species problem: Ongoing issues.* Rijeka: InTech Open Access Publ.

Pavlinov, I. Ya. 2013c. [*Taxonomic nomenclature. 1. From Adam to Linnaeus.*] Moscow: KMK Sci. Press. (in Russian).

Pavlinov, I. Ya. 2015. [*Nomenclature in systematics. History, theory, practice.*] Moscow: KMK Sci. Press. (in Russian).

Pavlinov, I. Ya. 2017. [The species problem in biology, its roots and present.] In [*Species concept in fungi: A new look at old problem*], 5–19. Moscow: Moscow State University Press. (in Russian).

Pavlinov, I. Ya. 2018. [*Foundations of biological systematics: Theory and history.*] Moscow: KMK Scientific Press. (in Russian).

Pavlinov, I. Ya. 2021a. Conceptualization of the species problem. *Novitates Theriologicae* 12:69–85.

Pavlinov, I. Ya. 2021b. *Biological systematics: History and theory.* Boca Raton, FL: CRC Press.

Pavlinov, I. Ya. 2022a. The species problem from a conceptualist's standpoint. In *Species problems and beyond. Contemporary issues in philosophy and practice*, eds J. S. Wilkins, F. E. Zachos, and I. Ya. Pavlinov, 245–78. Boca Raton, FL: CRC Press.

Pavlinov, I. Ya. 2022b. *Taxonomic nomenclature. What's in a name – theory and history.* Boca Raton, FL: CRC Press.

Pavlinov, I. Ya., and G. Yu. Lyubarsky. 2011. [*Biological systematics: Evolution of ideas.*] Moscow: KMK Sci. Press. (in Russian).

Pavlov, N. V. 1951. [*Vladimir Leontievich Komarov.*] Moscow: USSR Academy of Sciences. (in Russian).

Payne, J. B. 1958. The concept of 'kinds' in Scripture. *Journal of the American Science American Scientific Affiliation* 10:17–20.

Pedersen, O. 1997. The first universities: *Studium Generale and the origins of university education in Europe*, Cambridge: Cambridge University Press. 310 p.

Pedroso, M. 2012. Essentialism, history, and biological taxa. *Studies in History and Philosophy of Biological and Biomedical Sciences* 43:182–90.

Pellegrin, P. 1985. Aristotle: A zoology without species. In *Aristotle on the nature of living things*, ed. A. Gotthelf, 95–116. Bristol: Bristol Classical Press.

Pellegrin, P. 1986. *Aristotle's classification of animals. Biology and the conceptual unity of Aristotelian corpus.* Berkeley, CA: University California Press.

Pellegrin, P. 1990. Taxonomie, moriologie, division. In *Biologie, Logique et Métaphysique chez Aristote*, eds D. Deverreux and P. Pellegrin, 37–48. Paris: CNRS.

Pena, P., and M. Lobelius. 1570. *Stirpium Adversaria nova: perfacilis Vestigatio, luculentaque accessio ad Priscorum [...].* Londini: Thomae Purfoetii.

Perfetti, S. 2000. *Arisostles' zoology and its Reneaissance commentators (1521–1601).* Leuven: Leuven University Press.

Persoon, C. H. 1805. *Synopsis plantarum, seu Enchiridium botanicum, complectens enumerationem systematicam specierum hucusque cognitarum*, Vol. 1. Parisiis: C. F. Cramerum.

Peterson, T.A., and A. G. Navarro-Siguenza. 1999. Alternate species concepts as bases for determining priority conservation areas. *Conservation Biology* 13:427–31.

Petry, M. J. 2001. Introduction. In *Linne C. Nemesis Divina*, ed. M. J. Petry, 3–79. Dordrecht: Springer Science + Business Media.

Pfander, P. V. 2018. [A tragedy of around-species systematics.] *Russian Ornithological Journal* 27:301–35. (in Russian).

Pietsch, T. W. 2012. Introduction. In *Cuvier's History of the natural sciences: twenty-four lessons from Antiquity to the Renaissance*, ed. T. W. Pietsch, 19–29. Paris: Muséum national d'Histoire naturelle.

Pigliucci, M. 2003. Species as family resemblance concepts: The (dis-)solution of the species problem? *BioEssays* 25:596–602.

Pigliucci, M., and J. Kaplan. 2006. *Making sense of evolution. The conceptual foundations of evolutionary biology*. Chicago, IL: University of Chicago Press.

Pini, G. 1999. Species, concept, and thing: Theories of signification in the second half of the Thirteenth century. *Medieval Philosophy and Theology* 8:21–52.

Pini, G. 2005. Scotus' realist conception of the categories: His legacy to late Medieval debates. *Vivarium* 43:63–110.

Planchon, J.-E. 1874. Le morcellement de l'espèce en botanique. *Revue des Deux Mondes (1829–1971). Troisième Période* 5:389–416.

Plate, L. 1914. Prinzipien der Systematik mit besonderer Berucksichtigung des Systems der Tiere. In *Die Kultur der Gegenwart*, T. 3, eds R. Hertwig and R. von Wettstein, 92–164. Leipzig: B. G. Teubner.

Pleijel, F. 1999. Phylogenetic taxonomy, a farewell to species, and a revision of Heteropodarke (Hesionidae, Polychaeta, Annelida). *Systematic Biology* 48:755–89.

Pleijel, F., and G. W. Rouse. 2000. Least-inclusive taxonomic unit: A new taxonomic concept for biology. *Proceedings of the Royal Society, Ser. B* 267:627–30.

Plotinus. 2018. *The Enneads*. Cambridge: Cambridge University Press.

Pokrovsky, M. P. 2014. [*An introduction to classiology*.] Ekaterinburg: Institute of Geology and Geochemistry. (in Russian).

Polansky, R. 2007. *Aristotle's De Anima: A Critical Commentary*. Cambridge: Cambridge University Press.

Politis, V. 2004. *Aristotle and the metaphysics*. London: Routledge.

Polyansky, V. I. 1956. [*On the species of lower algae*.] Moscow–Leningrad: USSR Academy of Sciences. (in Russian).

Polyansky, V. I. 1958. [The notion of species in algology.] In [*The species problem in botany*, Vol. 1,] ed. P. A. Baranov, 7–33. Moscow–Leningrad: USSR Academy of Sciences. (in Russian).

Polyansky, V. I. 1959. [On the integrity of the species and its evolution.] *Bulleten Moskovskogo obshestva ispytatelej prirody, Biol.* 64:5–13. (in Russian).

Popper, K. R. 1935. *Logik der Forschung. Zur Erkenntnistheorie der modernen Naturwissenschaft*. Wien: Springer-Verlag. 248 S.

Popper, K. R. 1974. *Objective knowledge: An evolutionary approach*. Oxford: Clarendon Press.

Popper, K. R. 1984. Evolutionary epistemology. In *Evolutionary theory: Paths into the future*, ed. J. W. Pollard, 239–55. London: John Wiley & Sons.

Porphyry. 2003. *Introduction*. Oxford: Clarendon Press.

Poulton, E. B. 1904. What is a species? *Proceedings of the Entomological Society of London (1903)*: lxxvii–cxvi.

Powers, J. 2013. Finding Ernst Mayr's Plato. *Studies in History and Philosophy of Biological and Biomedical Sciences* 44:714–23.

Pozdnyakov, A. A. 1994. [On individual nature of species.] *Zhurnal Obshchei Biologii* 44:389–97. (in Russian).
Pozdnyakov, A. A. 2007. [The ontological status of taxa from a traditional point of view.] In [*Linnaean collection*], ed. I. Ya. Pavlinov, 261–304. Moscow: Moscow State University Publ. (in Russian).
Pozdnyakov, A. A. 2015. [*Philosophical foundations of classical biology: Mechanicism in evolutionism and systematics.*] Moscow: LENAND. (in Russian).
Pozdnyakov, A. A. 2018. [*Philosophical foundations of classical biology: An introduction to organic biology.*] Moscow: URSS. (in Russian).
Pradeu, T., L. Laplane, K. Prevot, et al. 2016. Defining "development". *Current Topics in Developmental Biology* 117:171–83.
Pratt, V. 1982. Aristotle and the essence of natural history. *History and Philosophy of the Life Sciences* 4:203–23.
Pratt, V. 1984. The essence of Aristotle's zoology. *Phronesis* 29:267–78.
Prigogine, I., and I. Stengers. 1984. *Order out of chaos. Man's new dialogue with nature.* London: Heinemann.
Protasov, A. A. 2021. [Species from the point of view of biocenology.] *Species in biology: Theory and practice*, ed. I. Zagorodniuk, 86–96. Kyiv: NMNH NAS of Ukraine. (in Russian).
Psillos, S. 1999. *Scientific realism: How science tracks truth.* London: Routledge.
Puillandre, N., A. Lambert, S. Brouillet, and G. Achaz. 2012. ABGD, automatic barcode gap discovery for primary species delimitation. *Molecular Ecology* 21:1864–77.
Pušić, B., P. Gregorić, and D. Franjević. 2017. What do biologists make of the species problem? *Acta biotheoretica* 65:179–209.
Putnam, H. 1991. *Representation and reality.* Cambridge, MA: MIT Press.
Quarfood, M. 1999. The individuality of species: Some reflections on the debate. *Synthese* 120:89–94.
Quine, W. V. 1969. *Ontological relativity and other essays.* New York, NY: Columbia University Press.
Quine, W. V. 1994. Natural kinds. In *Grue! The new riddle of induction*, ed. D. Stalker, 42–56. La Salle: Open Court.
Quinn, A. 2020. Biological kinds at the turn of the 20th century: Characters, genes, and species as theoretical elements. In *Perspectives on classification in synthetic sciences: Unnatural kinds*, ed. J. R. S. Bursten, 43–60. Abingdon: Routledge.
Raikov, B. E. 1969. [*German evolutionary biologists before Darwin.*] Leningrad: Nauka. (in Russian).
Rakauskas, R. 2003.The concept of species and practical taxonomy: What is in-between? *Ekologija (Vilnius)* 1:3–7.
Ramsbottom, J. 1938. Linnaeus and the species concept. *Proceedings of the Linnean Society of London, 150th Session (1937–38)* 150:192–219.
Rapini, A. 2004. Classes or individuals? The paradox of systematics revisited. *Studies in History and Philosophy of Biological and Biomedical Sciences* 35:675–95.
Rasnitsyn, A. P. 2007. The problem of species revisited. *Paleontological Journal* 41:1151–5.
Raven, C. E. 1950. *John Ray naturalist: His life and works.* New York, NY: Cambridge University Press. 506 p.
Ray, J. 1686. *Historia plantarum species de Plantis in genere, aerùmque Partibus, Accidentibus & Differentiis […]*, T. Primus. Londoni: Mariae Clark.
Ray, J. 1693. *Synopsis Methodica Animalium Quadrupedum et Serpentini generis […].* Londini: S. Smith & B. Walford.
Ray, J. 1696. *De variis Plantarum Methodus Dissertation brevis […].* Londoni: S. Smith & B. Walford.
Ray, J. 1714. *The wisdom of God manifested in the works of the Creation […]*, 6th ed. London: William Innys.

Ray, J. 1733. *Methodus plantarum emendate et aucta [...].* Londoni: Christianum Andream Myntsing.

Ray, J. 1928a. A discourse on the specific differences in plants. In *Further correspondence of John Ray*, ed. R. W. T. Gunther, 77–83. London: The Ray Society.

Ray, J. 1928b. A discourse on the seeds of plants. In *Further correspondence of John Ray*, ed. R. W. T. Gunther, 70–7. London: The Ray Society.

Regel, R. E. 1917. [On the question of speciation.] *Trudy Bureau po Prikladnoi Botanike* 1:157–81. (in Russian).

Reig, O. A. 1982. The reality of biological species: A conceptualistic and a systemic approach. *Studies in Logic and the Foundations of Mathematics* 104:479–99.

Rekovets, L. and L. Kuzmenko. 2021. Species as a system within a system. *Novitates Theriologicae* 12:97–104.

Remane, A. 1943. Die Bedeutung der Lebensformtypen für die Orologi. *Biologia Generalis* 17:164–82.

Remes, P. 2008. *Neoplatonism.* Stocksfield: Acumen Publisher.

Reninger, E. 2017. *Tao Te Ching, Verse 42.* https://www.learnreligions.com/tao-te-ching-verse-42-3183165.

Rensch, B. 1929. *Das Prinzip geographischer Rassenkreise und das Problem der Artbildung.* Berlin: Gebrüder Borntraeger.

Rensch, B. 1932. Über die bedeutung des Prinzips geographischer Rassenkreise. *Geographische Zeitschrift* 38:157–66.

Rensch, B. 1934. *Kurze Anweisung für zoologisch-systematische Studien.* Leipzig: Akademische Verlagsgesellschaft.

Resnick, I. M., and K. F. Kitchell. 2008. Introduction. In *Questions concerning Aristotle's On animals, by Albert the Great*, eds I.M. Resnick and K. F. Kitchell, 1–10. Washington, DC: Catholic University of America Press.

Reydon, T. A. C. 2003. Species are individuals – or are they? *Philosophy of Science* 70:49–56.

Reydon, T. A. C. 2004. Why does the species problem still persist? *BioEssays* 26:300–5.

Reydon, T. A. C. 2005a. *Species as units of generalization in biological science: A philosophical analysis.* Doctoral dissertation. Leiden: Leiden University. https://scholarlypublications.universiteitleiden.nl/access/item%3A3075804/view.

Reydon, T. A. C. 2005b. On the nature of the species problem and the four meanings of species. *Studies in History and Philosophy of Biological and Biomedical Sciences* 36:135–58.

Reydon, T. A. C. 2006. Generalizations and kinds in natural science: The case of species. *Studies in History and Philosophy of Biological and Biomedical Sciences* 37:230–55.

Reydon, T. A. C. 2009. Species and kinds: A critique of Rieppel's "one of a kind" account of species. *Cladistics* 25:660–7.

Reydon, T. A. C. 2022. 'Species' as a technical term: Multiple meanings in practice, one idea in theory. In *Species problems and beyond. Contemporary issues in philosophy and practice*, eds J. S. Wilkins, F. E. Zachos, and I. Ya. Pavlinov, 65–90. Boca Raton, FL: CRC Press.

Reydon, T. A. C., and W. Kunz. 2019. Species as natural entities, instrumental units and ranked taxa: New perspectives on the grouping and ranking problems. *Biological Journal of the Linnean Society* 126:623–36.

Rheins, J. G. 2011. Similarity and species concepts. In *Carving nature at its joints: Natural kinds in metaphysics and science*, eds J. K. Campbell, M. O'Rourke, and M. H. Slater, 252–288. Cambridge, MA: MIT Press.

Richards, R. A. 2008. Species and taxonomy. In *The Oxford Handbook of Philosophy of Biology*, ed. M. Ruse. https://www.oxfordhandbooks.com/view/10.1093/oxfordhb/9780195182057.001.0001/oxfordhb-9780195182057-e-008.

Richards, R. A. 2010. *The species problem: A philosophical analysis.* Cambridge: Cambridge University Press.

Richards, R. A. 2013. The species problem: A conceptual problem? In *The species problem: Ongoing issues*, ed. I. Ya. Pavlinov, 41–63. Rijeka: InTech Open Access Publ.

Richards, R. A. 2016. *Biological classification: A philosophical introduction*. Cambridge: Cambridge University Press.

Richards, R. J. 1992. *The meaning of evolution: The morphological construction and ideological reconstruction of Darwin's theory*. Chicago, IL: University Chicago Press.

Riddle, B. R., and D. J. Hafner. 1999. Species as units of analysis in ecology and biogeography: Time to take the blinders off. *Global Ecology and Biogeography* 8:433–41.

Ridley, M. 1989. The cladistic solution to the species problem. *Biology and Philosophy* 4:1–16.

Riegner, M. F. 2013. Ancestor of the new archetypal biology. *Goethe's dynamic typology as a model for contemporary evolutionary developmental biology Studies in History and Philosophy of Biological and Biomedical Sciences* 44: 735–44.

Rieppel, O. 1988. Louis Agassiz (1807–1873) and the reality of natural groups. *Biology and Philosophy* 3:29–47.

Rieppel, O. 2005. Monophyly, paraphyly, and natural kinds. *Biology and Philosophy* 20: 465–87.

Rieppel, O. 2006. On concept formation in systematics. *Cladistics* 22:474–92.

Rieppel, O. 2007. Species: Kinds of individuals or individuals of a kind. *Cladistics* 23:373–84.

Rieppel, O. 2009. Species as a process. *Acta Biotheoretica* 57:33–49.

Rieppel, O. 2010a. Species monophyly. *Journal of Zoological Systematics and Evolutionary Research* 48:1–8.

Rieppel, O. 2010b. Reydon on species, individuals and kinds: A reply. *Cladistics* 26:341–3.

Rieppel, O. 2011a. *Evolutionary theory and creation controversy*. Berlin–Heidelberg: Springer-Verlag.

Rieppel, O. 2011b. Species are individuals—the German tradition. *Cladistics* 27:629–45.

Rieppel, O. 2016. *Phylogenetic systematics: Haeckel to Hennig*. Boca Raton, FL: CRC Press.

Rieppel, O., and M. Kearney. 2002. *Similarity. Biological Journal of the Linnean Society* 75:59–82.

Rieseberg, L. H., and J. M. Burke. 2001. The biological reality of species: Gene flow, selection, and collective evolution. *Taxon* 50:47–67.

Rieseberg, L. H., T. E. Wood, and E. J. Baack. 2006. The nature of plant species. *Nature* 440:524–27.

Rivinus, A. 1696. *Introductio generalis in Rem Herbariam [...]*. Lipsiae: Johannis Heinichii.

Roberge, J. M., and P. Angelstam. 2004. Usefulness of the umbrella species concept as a conservation tool. *Conservation Biology* 18:6–85.

Robson, G. C. 1928. *The species problem: An introduction to the study of evolutionary divergence in natural populations*. Edinburgh: Oliver & Boyd.

Roger, J. and J. L. Fischer (eds). 1987. *Histoire du concept d'espèce dans les Sciences de la Vie*. Paris: Fondation Singer-Polignac.

Rojas, M. 1992. The species problem and conservation: What are we protecting? *Conservation Biology* 6:170–8.

Rollins, R. C. 1965. On the bases of biological classification. *Taxon* 14:1–6.

Romanes, G. J. 1895. *Darwin, and after Darwin: An exposition of the Darwinian theory and a discussion of the post-Darwinian questions*, Vol. 2. [...]. London: Longmans & Green.

Rorem, P. 1993. *Pseudo-Dionysius: A commentary on the texts and an introduction to their influence*. New York, NY: Oxford University Press.

Rosán, L. J. 1949. *The philosophy of Proclus: The final phase of Ancient thought*. New York, NY: Cosmos.

Rosenberg, A. 1985. *The structure of biological science*. New York, NY: Cambridge University Press.

Rosenberg, A. 1987. Why does the nature of species matter? Comments on Ghiselin and Mayr. *Biology and Philosophy* 2:192–7.

Ross, H. A. 2014. The incidence of species-level paraphyly in animals: A re-assessment. *Molecular Phylogenetics and Evolution* 76:10–7.
Rosselló-Mora, R., and R. Amann. 2001. The species concept for prokaryotes. *FEMS Microbiology Reviews* 25:39–67.
Rozanova, M. A. 1946. [*Experimental foundations of plant systematics.*] Moscow–Leningrad: USSR Academy of Sciences. (in Russian).
Rozov, M. A. 2002. [*Philosophy and theory of history. Prolegomen.*] Moscow: Logos. (in Russian).
Rozova, S. S. 1980. [Philosophical reflection on the classification problem.] *Voprosy Filosofii* 8:49–56. (in Russian).
Rozova, S. S. 1986. [*Classification problem in modern science.*] Moscow: Nauka. (in Russian).
Rozova, S. S. 2014. [*Theory of social relay in epistemology and philosophy of science.*] Saarbrucken: Lambert Academic Publ. (in Russian).
Rubtsov, A. S. 1996. [The biological species concept in ornithology: An analysis of the theoretical foundations.] *Zhurnal Obshchei Biologii* 57:747–59. (in Russian).
Rudebusch, G., and F. Muniz. 2020. *Genos and eidos in Plato.* https://www.academia.edu/43147196/Genos_and_Eidos_in_Plato.
Rupke, N. A. 1993. Richard Owen's vertebrate archetype. *Isis* 84:231–51.
Ruse, M. (ed.). 2008. *The Oxford handbook of philosophy of biology.* Oxford: Oxford University Press. 656 p.
Ruse, M. (ed.). 1989. *What the philosophy of biology is: Essays dedicated to David Hull.* Dordrecht: Kluwer Academic Publishers.
Ruse, M. 1981. Species as individuals: Logical, biological, and philosophical problems. *The Behavioral and Brain Sciences* 4:299–300.
Ruse, M. 1987. Biological species: Natural kinds, individuals, or what? *The British Journal for the Philosophy of Science* 38:225–42.
Ruse, M. 1988. *Philosophy of biology today.* Albany, NY: State University of New York Press.
Ruse, M. 1995. The species problem. In *Concepts, theories, and rationality in the biological sciences*, eds G. Wolters and J. G. Lennox, 171–93. Pittsburgh, PA: University of Pittsburgh Press.
Russell, B. 1967. *History of Western philosophy.* New York, NY: Simon & Schuster.
Russell, B. 1992. *Human knowledge: Its scope and limits.* London: Routledge.
Sachs, J. 1906. *History of botany, 1530–1860.* Oxford: Clarendon Press.
Saikia, U., N. Sharma, and A. Das. 2008. What is a species? An endless debate. *Resonance* 13:1049–64.
Saint Augustine. 2000. *The City of God.* New York, NY: The Modern Library.
Saint Augustine. 2011. *The Trinity.* New York, NY: New City Press.
Saint-Andre, P. 2002. Conceptualism in Abelard and Rand. *The Journal of Ayn Rand Studies* 4:123–40.
Samadi, S., and A. Barberousse. 2006. The tree, the network, and the species. *Biological Journal of the Linnean Society* 89:509–21.
Sandmel, S. 1979. *Philo of Alexandria: An introduction.* Oxford: Oxford University Press.
Sancti Thomae de Aquino. 1970. *Opera omnia. Tomus XXII. Quaestiones disputatae de veritate.* Vol. I. Fasc. 2. QQ. 1–7. Roma: Commissio Leonina.
Sandford, S. 2019. From Aristotle to contemporary biological classification: What kind of category is "sex"? *Redescriptions: Political Thought, Conceptual History and Feminist Theory* 22:4–17.
Sankey, H. 1998. Taxonomic incommensurability. *International Studies in the Philosophy of Science* 12:7–16.
Sanson, A. 1868. La notion philosophique de l'espèce. *La Philosophie positive* 2:27–53.
Sanson, A. 1900. *L'espèce et la race en biologie générale.* Paris: C. Reinwald.
Sarkar, S. 2005. *Biodiversity and environmental philosophy: An introduction.* New York, NY: Cambridge University Press.

Sarkar, S., and C. Margules. 2001. Operationalizing biodiversity for conservation planning. *Journal of Biosciences* 27 (Suppl. 2):299–308.

Savolainen, V, R. S. Cowan, A. P. Vogler, et al. 2005. Towards writing the encyclopaedia of life: An introduction to DNA barcoding. *Philosophical Transactions of the Royal Society B* 360:1805–11.

Schafer, A. R. D. 2003. The "kinds" of Genesis 1: What is the meaning of mîn? *Journal of the Adventist Theological Society* 14:86–100.

Schlegel, H. 1844. *Kritische Übersicht der europäischen Vogel*. Leiden: Arnz.

Schulze, E.-D., and H. A. Mooney. 1994. Ecosystem function of biodiversity: A summary. In *Ecosystem function of biodiversity*, eds E.-D. Schulze and H. A. Mooney, 497–510. New York, NY: Springer.

Schwartz, S. S. 1959. [*Some issues in the species problem in terrestrial vertebrates.*] Proceedings of the Institute of Biology of the UV, issue 11. Sverdlovsk: Ural Branch of USSR Academy of Sciences. (in Russian).

Schwartz, S. S. 1980. [*Ecological patterns of evolution.*] Moscow: Nauka. (in Russian).

Sciacca, G. 2020. Why the debate about the metaphysics of biological species should not be deflated. *History of Philosophy and Logical Analysis* 23:474–97.

Scopoli, A. J. 1777. *Introductio ad Historia naturalem systens Genera Lapidu, Plantarum, et Animalium […]*. Pragae: Wolfgangvm Gerle.

Scopoli, J. A. 1783. *Fundamenta botanica preleciones publicis accomodata*. Pavia: Stamperia del Monasterio di San Salvator.

Scudder, G. G. E. 1974. Species concepts and speciation. *Canadian Journal of Zoology* 52:1121–34.

Secord, J. A. 2000. *Victorian sensation. The extraordinary publication, reception and secret authorship of the vestiges of the natural history of creation*. Chicago, IL: University of Chicago Press.

Seely, P. 1997. The meaning of min, 'kind'. *Science and Christian Belief*, 9:47–56.

Seifert, B. 2014. A pragmatic species concept applicable to all eukaryotic organisms independent from their mode of reproduction or evolutionary history. *Soil Organisms* 86:85–93.

Seifert, B. 2020. The gene and gene expression (GAGE) species concept: An universal approach for all eukaryotic organisms. *Systematic Biology* 69:1033–8.

Semenov-Tyan-Shansky, A. A. 1910. [Taxonomic boundaries of species and its subdivisions: An experience of precise categorization of lower taxonomic units.] *Notes of the Imperial Academy of Sciences, Phys.-Math. Dept* 25:1–29. (in Russian).

Shanker, K., S. P. Vijayakumar and. K. N. Ganeshaiah 2017. Unpacking the species conundrum: Philosophy, practice and a way forward. *Journal of Genetics* 96:413–30.

Shaposhnikov, Ch. G. 1975. [Living systems with a low degree of integrity.] *Zhurnal Obshchei Biologii* 3: 323–35. (in Russian).

Sharp, D. 1873. *The object and method of zoological nomenclature*. London: E. W. Janson, Williams & Norgate.

Shatalkin, A. I. 1983a. [To the question of the taxonomic species.] *Zhurnal Obshchei Biologii* 54:172–86. (in Russian).

Shatalkin, A. I. 1983b. [Methodological aspects of the application of mathematical methods in taxonomy.] In [*Theory and methodology of biological classifications*,] 46–55. Moscow: Moscow Society of Naturalists. (in Russian).

Shatalkin, A. I. 1984. [Evolutionary species as a system of individual nature.] *Zhurnal Obshchei Biologii* 66: 291–305. (in Russian).

Shatalkin, A. I. 1988. [*Biological systematics.*] Moscow: Moscow State University Publ. (in Russian).

Shatalkin, A. I. 2009. [*"Zoological Philosophy" by Jean Baptiste Lamarck: A view from the 21st century.*] Moscow: KMK Sic. Press. (in Russian).

Shatalkin, A.I. 1996 [Essentialism and typology.] In [*Contemporary systematcis: Methodological aspects.*] ed. I. Ya. Pavlinov, 23–154. Moscow: Moscow University Press.

Shatalkin, A.I. 2012. [*Taxonomy. Foundations, principles, and rules.*] Moscow: KMK Sci. Press. (in Russian).

Shaw, K. L. 1996. What are 'good' species. *Trends in Ecology and Evolution* 11:174.

Shaw, K. L. 2012. Species and speciation: Overview. *eLS.* 1–7. Chichester: John Wiley & Sons. https://onlinelibrary.wiley.com/doi/abs/10.1002/9780470015902.a0001742.pub2.

Shcherbakov, V. P. 2010. Biological species is the only possible form of existence for higher organisms: The evolutionary meaning of sexual reproduction. *Biology Direct* 5:14.

Shcherbakov, V. P. 2013. Biological species as a form of existence, the higher form. In *The species problem: Ongoing issues*, ed. I. Ya. Pavlinov, 65–92. Rijeka: InTech Open Access Publ.

Shifman, I. S. 2007. [*The Old Testament and its world.*] Saint Petersburg: Saint Petersburg State University Publ. (in Russian).

Shkorbatov, G. 1976. [*Species in nature and in science.*] Kharkiv: Vischaya Shkola. (in Russian).

Shkorbatov, G. 1979. [*Species and speciation.*] Ivanovo: Ivanovo State Pedagogical University. (in Russian).

Shmonin, D. V. 2006. [*In the shadow of the Renaissance: The Second Scholasticism in Spain.*] Saint Petersburg: Saint Petersburg University Publisher. (in Russian).

Shuman, A. N. 2001. [*Philosophical logic: Origins and evolution.*] Minsk: Ekonompress. (in Russian).

Sigwart, J. D. 2018. *What species mean: A user's guide to the units of biodiversity.* Boca Raton, FL: CRC Press.

Sigwart, J. D., and A. Garbett. 2018. Biodiversity assessment, DNA barcoding, and the minority majority. *Integrative and Comparative Biology* 58:1146–56.

Silcox, M. T. 2014. A pragmatic approach to the species problem from a paleontological perspective. *Evolutionary Anthropology* 23:24–6.

Simões, J. A. M. (ed.). 1992. *Energetics of organometallic species.* Dordrecht: Springer Science+Business Media.

Simonetta, A. M. 1982. The myth of objective taxonomy and cladism: Much ado about nothing. *Atti della Società toscana di scienze naturali*, Ser. B 89:175–86.

Simonetta, A. M. 1992. Problems of systematics: Part 1. A critical evaluation of the "species problem" and its significance in evolutionary biology. *Italian Journal of Zoology* 59:447–63.

Simonetta, A. M. 1995. Some remarks on the influence of historical bias in our approach to systematics and the so called "species problem". *Bolletino di Zoologia* 62:37–44.

Simonetta, A. M. 2014. Something to think about (Part 1). Do "species" really exist in Nature or are they just necessary intellectual tools? *Italian Journal of Zoology* 81:162–81.

Simons, P. 2000. *Parts. A study in ontology.* New York, NY: Oxford University Press.

Simons, P. 2013. Vague kinds and biological nominalism. *Metaphysica* 14:275–82.

Simpson, G. G. 1951. The species concept. *Evolution* 5:285–98.

Simpson, G. G. 1953. *The major features of evolution.* New York, NY: Columbia University Press.

Simpson, G. G. 1961. *Principles of animal taxonomy.* New York, NY: Columbia University Press.

Sinskaya, E. N. 1961. [*The doctrine of species and taxa.*] Leningrad: VNIIR. (in Russian).

Sites, J. W., and J. C. Marshall. 2004. Operational criteria for delimiting species. *Annual Review of Ecology, Evolution, and Systematics* 35:199–227.

Skarlato, O. A., and Ya. I. Starobogatov. 1974. [Phylogenetics and principles of natural system construction.] In [*Theoretical issues in systematics and phylogeny of animals,*] 30–46. Leningrad: Nauka. (in Russian).

Skvortsov, A. K. 1967. [The main stages in the development of the species concept.] *Bulleten Moskovskogo obshestva ispytatelej prirody, Biol.* 72:11–27. (in Russian).

Skvortsov, A. K. 1972. [V. L. Komarov and the species problem.] [*Komarov readings*] 24: 48–81. Leningrad: Nauka. (in Russian).

Slater, M. H. 2013. *Are species real? An essay on the metaphysics of species.* Basingstoke: Palgrave Macmillan.

Slater, M. H. 2015. Natural kindness. *The British Journal for the Philosophy of Science* 66:375–411.

Slaughter, M. 1982. *Universal languages and scientific taxonomy in the seventeenth century.* Cambridge: Cambridge University Press.

Sloan, P. 2009. Originating species: Darwin on the species problem. In *The Cambridge companion to the "Origin of Species"*, eds. M. Ruse and R. Richards, 67–86. Cambridge: Cambridge University Press.

Sloan, P. 2013. The species problem and history. *Studies in History and Philosophy of Biological and Biomedical Sciences* 44:237–41.

Sloan, P. 2019. Evolutionary thought before Darwin. In *Stanford encyclopedia of philosophy*, ed. E. N. Zalta. https://plato.stanford.edu/entries/evolution-before-darwin/.

Sloan, P. R. 1970. *The history of the concept of the biological species in the Seventeenth and Eighteenth centuries, and the origin of the species problem.* Doctoral dissertation. San Diego, CA: University of California. https://www.worldcat.org/title/history-of-the-concept-of-the-biological-species-in-the-seventeenth-and-eighteenth-centuries-and-the-origin-of-the-species-problem/oclc/1157490552/editions?referer=di&editionsView=true.

Sloan, P. R. 1972. John Locke, John Ray, and the problem of the natural system. *Journal of the History of Biology* 5:1–53.

Sloan, P. R. 1976. The Buffon–Linnaeus controversy. *Isis* 67:356–75.

Sloan, P. R. 1979. Buffon, German biology, and the historical interpretation of biological species. *British Journal of History of Science* 12:109–53.

Sloan, P. R. 1986. From logical universals to historical individuals: Buffon's idea of biological species. In *Histoire du concept d'espèce dans les sciences de la vie*, eds J. Roger and J. L. Fisher, 101–40. Paris: Fondation Singer-Polignac.

Sloan, P. R. 1995. The gaze of natural history. In *Inventing human science: Eighteenth-century domains*, eds C. Fox, R. Porter, and R. Wokler, 112–151. Berkeley, CA: University of California Press.

Sloan, P. R. 1997. From logical universals to historical individuals: Buffon's idea of biological species. In *Histoire du concept d'espece dans les sciences de la vie*, eds S. Atran, J.-L. Fischer, and J. Roger, 101–40. Paris: Fondation Singer-Polignac.

Sloan, P. R. 2002. Reflections on the species problem: What Marjorie Grene can teach us about a perennial issue. In *The philosophy of Marjorie Grene*, eds R. E. Auxier and L. E. Hahn, 225–55. La Salle: Open Court.

Sloan, P. R. 2006. Kant on the history of nature: The ambiguous heritage of the critical philosophy for natural history. *Studies in History and Philosophy of Science*, Part C 37:627–48.

Sluys, R. 1991. Species concepts, process analysis, and the hierarchy of nature. *Experientia* 47:1162–70.

Sluys, R. and C. J. Hazevoet. 1999. Pluralism in species concepts: Dividing nature at its diverse joints. *Species Diversity* 4:243–56.

Smith, H. M. 1990. The universal species concept. *Herpetologica* 46:122–4.

Smith, J. A. 1822. *Grammar of botany, illustrative of artificial, as well as natural classification. With explanation of Jussieu's System [...].* London: Longman.

Smith, J. E. H. 2009. "The unity of the generative power": Modern taxonomy and the problem of animal generation. *Perspectives on Science* 17:78–104.

Smith, J. E. H. 2011. *Divine machines. Leibniz and the sciences of life*. Princeton, NJ: Princeton University Press.

Sneath, R. H. A., and R. R. Sokal. 1973. *Numercial taxonomy. The principles and methods of numerical classification*. San Francisco, CA: W.H. Freeman & Co.

Sober, E. (ed.). 2006. *Conceptual issues in evolutionary biology*. Cambridge, MA: MIT Press.

Sober, E. 1980. Evolution, population thinking, and essentialism. *Philosophy of Science* 47:350–83.

Sober, E. 1984. Sets, species, and evolution. Comments on Philip Kitcher's 'species'. *Philosophy of Science* 51:334–41.

Sober, E. 2000. *Philosophy of biology*, 2nd ed. Boulder, CO: Westview Press.

Sober, E. 2015. *Ockham's razors: A user's manual*. Cambridge: Cambridge University Press.

Sokal, R. R. 1974. The species problem reconsidered. *Systematic Zoology* 22:360–74.

Sokal, R. R., and R. H. A. Sneath. 1963. *Principles of numercial taxonomy*. San Francisco, CA: W. H. Freeman & Co.

Sokal, R. R., and T. J. Crovello. 1970. The biological species concept: A critical evaluation. *The American Naturalist* 104:127–53.

Sokolov, D. 2008. [*A concept of natural theology in the biological works of John Ray*.] http://nature.web.ru/db/msg.html?mid=1190019&s=121600000. (in Russian).

Sokolov, V. V. 1979. [*Medieval philosophy*.] Moscow: Vysshaya Shkola. (in Russian).

Solbrig, O. T. 1970. *Principles and methods of plant biosystematics*. New York, NY: Macmillan Co.

Solonchak, T., and S. Pesina. 2015. Concept and its structure. *Procedia: Social and Behavioral Sciences* 192:352–8.

Sonneborn, T. M. 1957. Breeding systems, reproductive methods, and species problems in Protozoa. In *The species problem. A symposium presented at the Atlanta Meeting of the American Association for the Advancement of Science*, ed. E. Mayr, 155–324. Washington, DC: American Association for the Advancement of Science.

Sorabji, R. (ed.). 1990. *Aristotle transformed. The ancient commentators and their influence*. Ithaca, NY: Cornell University Press.

Spade, P. V. 1994. *Five texts on the Mediaeval problem of universals: Porphyry, Boethius, Abelard, Duns Scotus, Ockham*. Indianapolis, IN: Hackett Publishing.

Spencer, H. 1864. *The principles of biology*, Vol. 1. London: Williams and Norgate.

Spencer, Q. 2016. Genuine kinds and scientific reality. In *Natural kinds and classification in scientific practice*, ed. C. Kendig, 184–99. Abingdon: Routledge.

Splitter, L. J. 1988. Species and identity. *Philosophy of Science* 55:323–48.

Sprengel, K. 1807. *Curtii Sprengel Historia rei herbariae*, Vol. 1. Amsteldomi: Tabernae librariae et atrium.

Spring, A. F. 1838. *Ueber die naturhistorischen Begriffe von Gattung, Art und Abart, und die Ursachen der Abartungen in den organischen Reichen*. Leipzig: Friedrich Fleischer.

Spruit, L. 1994. *Species intelligibilis. From perception to knowledge*. Vol. 1. *Classical roots and Medieval discussions*. Leiden: E. J. Brill.

Spruit, L. 1995. *Species intelligibilis. From perception to knowledge*. Vol. 2. *Renaissance controversies, later scholasticism, and the elimination […]*. Leiden: E. J. Brill.

Stace, C. A. 1989. *Plant taxonomy and biosystematics*, 2d ed. London: Cambridge University Press.

Stafleu, F. A. 1963. Adanson and the "Familles des plantes". In *Adanson: The Bicentennial of Michel Adanson's "Familles des Plantes"*, pt. 1, ed. G. H. M. Lawrence, 123–263. Pittsburg, PA: Hunt Botanical Library.

Stafleu, F. A. 1966. Introduction. In *Michele Adanson, Familles des plantes, 1763–1764*, v–xv. Lehre: J. Cramer.

Stafleu, F. A. 1969. A historical review of systematic biology. In *Systematic Biology: Proceedings of the International Conference*, 16–44. Washington: Natl. Acad. Sci.

Stafleu, F. A. 1971. *Linnaeus and Linnaeans*. Utrecht: A. Oosthoek.

Staley, J. T. 2006. The bacterial species dilemma and the genomic-phylogenetic species concept. *Philosophical Transactions of the Royal Society of London B* 361:1899–909.

Staley, J. T. 2009a. The phylogenomic species concept for bacteria and archaea. *Microbe* 4:361–5.

Staley, J. T. 2009b. Universal species concept: Pipe dream or a step toward unifying biology? *Journal of Industrial Microbiology and Biotechnology* 36:1331–6.

Staley, J. T. 2013. Transitioning toward a universal species concept for the classification of all organisms. In *The species problem: Ongoing issues*, ed. I. Ya. Pavlinov, 131–40. Rijeka: InTech Open Access Publ.

Stamos, D. N. 1996. Was Darwin really a species nominalist? *Journal of the History of Biology* 29:127–44.

Stamos, D. N. 1998. Buffon, Darwin, and the non-individuality of species—A reply to Jean Gayon. *Biology and Philosophy* 13:443–70.

Stamos, D. N. 2002. Species, languages, and the horizontal/vertical distinction. *Biology and Philosophy* 17:171–98.

Stamos, D. N. 2003. *The species problem. Biological species, ontology, and the metaphysics of biology*. Oxford: Lexington Books.

Stamos, D. N. 2005. Pre-Darwinian taxonomy and essentialism—A reply to Mary Winsor. *Biology and Philosophy* 20:79–96.

Stamos, D. N. 2007. *Darwin and the nature of species*. Albany, NY: State University of New York.

Stamos, D. N. 2013. Darwin's species concept revisited. In *The species problem: Ongoing issues*, ed. I. Ya. Pavlinov, 251–80. Rijeka: InTech Open Access Publ.

Stanford, P. K. 1995. For pluralism and against realism about species. *Philosophy of Science* 62:70–91.

Stanley, S. M. 1978. Chronospecies' longevities, the origin of genera, and the punctuational model of evolution. *Paleobiology* 4:26–40.

Stannard, J. 1980. Albertus Magnus and Medieval herbalism. In *Albertus Magnus and the sciences*, ed. J. A. Weisheipl, 355–77. Toronto: Pontifical Institute of Mediaeval Studies.

Stannard, J., and K. E. Stannard. 1999. *Herbs and herbalism in the Middle Ages and Renaissance*. Aldershot: Ashgate Variorum.

Star, S. L. 2010. This is not a boundary object: Reflections on the origin of a concept. *Science, Technology, & Human Values* 35:601–17.

Starobogatov, Ya. I. 1980. [On the ratio of stationary and evolutionary aspects in the study of the living.] [*Micro- and macroevolution*,] 42–5. Tartu: USSR Academy of Sciences. (in Russian).

Starobogatov, Ya. I. 1985. [The problem of speciation.] *Results of Science and Technology, Ser. General Geology* 20: 5–96. Moscow: VINITI. (in Russian).

Starobogatov, Ya. I. 1996. [Species in theory and in nature.] In *Contemporary systematics: Methodological aspects*, ed. I. Ya. Pavlinov, 165–82. Moscow: Moscow State University Publ. (in Russian).

Starrett, A. 1958. What *is* the subspecies problem? *Systematic Biology* 7:111–5.

Stearn, W. T. 1959. The background of Linnaeus's contributions to the nomenclature and methods of systematic biology. *Systematic Zoology* 8:4–22.

Stearn, W. T. 1986. John Wilkins, John Ray and Carl Linnaeus. *Notes and Records of the Royal Society of London* 40:101–23.

Stebbings, L. 1987. Species concepts: Semantics and actual situations. *Biology and Philosophy* 2:198–203.

Stegny, V. N. 1993. [*Genome architectonics, systemic mutations and evolution*.] Novosibirsk: Novosibirsk State University Press. (in Russian).

Stekolnikov, A. A. 2003. [The problem of truth in biological systematics.] *Zhurnal Obshchei Biologii* 64:357–68.

Stepanyan, L. S. 2011. [The concept of superspecies and the limits of its application.] *Russian Ornithological Journal* 20:1019–27. (in Russian).
Stepin, V. S. 2005. *Theoretical knowledge*. Heidelberg: Springer.
Sterelny, K. 1994. The nature of species. *Philosophical Books* 35:9–20.
Sterelny, K. 1999. Species as ecological mosaics. In *Species, new interdisciplinary essays*, ed. R. A. Wilson, 119–38. Cambridge, MA: MIT Press.
Sterelny, K., and P. E. Griffiths. 1999. *Sex and death: An introduction to philosophy of biology*. Chicago, IL: University of Chicago Press.
Stevens, P. F. 1994. *The development of biological systematics. Antoine-Laurent de Jussieu, Nature, and the Natural System*. New York, NY: Columbia University Press.
Stich, M., C. Briones and S. C. Manrubia 2007. Collective properties of evolving molecular quasispecies. *BMC Evolutionary Biology* 7:110.
Stijn, C. 2018. Values, regulation, and species delimitation. *Zootaxa* 4415:390–2.
Strange S. K. 1989. Plotinus on the articulation of being. *The Society for Ancient Greek Philosophy Newsletter* 155. https://orb.binghamton.edu/sagp/155.
Strange, S. K. 1987. Plotinus, Porphyry, and the Neoplatonic interpretation of the 'Categories.' In *Philosophie, Wissenschaften, Technik*, ed. W. Haase, 955–74. Berlin: De Gruyter.
Strange, S. K. 1992. *Porphyry on Aristotle categories*. London: Bloomsbury.
Stresemann, E. 1936. The Formenkreis theory. *Auk* 53:150–8.
Strickland, H. E. 1841. On the true method of discovering the natural system in zoology and botany. *The Annals and Magazin of Natural History*, Ser. 5, 6:184–94.
Stuessy, T. F. 1989. Comments on specific categories in flowering plant. *Plant Systematics and Evolution* 167:69–74.
Stuessy, T. F. 2008. *Plant taxonomy. The systematic evaluation of comparative data*, 2d ed. New York, NY: Columbia University Press.
Sucker, U. 1978. *Philosophische Probleme der Arttheorie*. Jena: Gustav Fischer.
Sulloway, F. J. 1979. Geographic isolation in Darwin's thinking: The vicissitudes of a crucial idea. *Studies in the History of Biology* 3:23–65.
Summa. 1981. *Summa Theologica by Saint Thomas Aquinas*. Public Domain. http://www.ccel.org/ccel/aquinas/summa.html.
Svenson, H. K. 1953. Linnaeus and the species problem. *Taxon* 2:55–8.
Swoyer, C. 2006. Conceptualism. In *Universals, concepts, and qualities: New essays on the meaning of predicates*, eds E. S. Trawson and A. Chakrabarti, 127–54. Aldershot: Ashgate Publishing.
Sykora, P. 2007. *Are species-kind and species-individual talks equivalent*? https://www.academia.edu/48805513/Are_Species_kind_and_Species_individual_Talks_Equivalent_.
Sylvester-Bradley, P. C. 1952. *The classification and coordination of infraspecific categories*. London: Syst. Assoc.
Szalay, F. S. 1993. Species concepts: The tested, the untestable, and the redundant. In *Species, species concepts and primate evolution*, eds W. H. Kimbel and L. B. Martin, 21–42. New York, NY: Springer Science + Business Media.
Szalay, F. S. and W. Bock. 1991. Evolutionary theory and systematics: Relationships between process and patterns. *Zeitschrift für zoologische Systematik und Evolutionsforschung* 29:1–39.
Takhtajyan, A. L. 1955. [Some questions of the theory of species in systematics of the Recent and fossil plants.] *Botanical Journal* 40: 24–38. (in Russian).
Takhtajyan, A. L. 1970. [Biosystematics: past, present and future.] *Botanical Journal* 55:331–45. (in Russian).
Tallerman, M., and K. R. Gibson. 2012. Introduction to Part IV: Launching language: The development of a linguistic species. In *Language evolution*, eds M. Tallerman and K. R. Gibson, 353–60. New York, NY: Oxford University Press.

Talpsepp, E. 2013. *Species, essentialism and evolutionary theory.* Doctoral dissertation. Bristol: University of Bristol. https://www.academia.edu/11743593/Species_essentialism_and_evolutionary_theory.

Tarán, L. 1981. *Speusippus of Athens. A critical study with a collection of the related texts and commentary.* Leiden: E. J. Brill.

Taylor, A. E. 1928. *The commentary of Proclus on the Timaeus of Plato in five books [...],* Vols. 1, 2. London: A. J. Valpy.

Tazhurizina, Z. A. (ed.). 1979. [*Nikolas of Cuza,*] 2 Vols. Moscow: Mysl. (in Russian).

Templeton, A. R. 1989. The meaning of species and speciation: A genetic perspective. In *Speciation and its consequences*, eds D. Otte and J. A. Endler, 3–27. Sunderland, MA: Sinauer Associates.

Terentyev, P. V. 1968. [In defense of the species as a finite taxon.] *Zoological Journal* 47:887–9. (in Russian).

Theophrastus. 1916. *Enquiry into plants, and minor works on odours and weather signs,* Vol. 1, ed. A. Hort. London: William Heinemann.

Theophrastus. 2010. *On first principles (known as his Metaphysics): Greek text and medieval Arabic translation [...].* Leiden: E. J. Brill.

Thomas Aquinas. 2001. [*Ontology and theory of knowledge: Fragments of essays.*] Moscow: Institute of Philosophy RAS. (in Russian).

Thorburn, W. M. 1918. The myth of Occam's Razor. *Mind* 27:345–53.

Timiryazev, K. A. 1904. An outline of Darwin's theory. In [*Theory of development,*] ed. V. A. Fausek, 49–143. Saint Petersburg: Brockhaus-Efron Publ. (in Russian).

Timonin, A. K. 1998. [Is a nomotetic taxonomy possible?] *Zhurnal Obshchei Biologii* 59:341–61. (in Russian).

Tipton, J. A. 2014. *Philosophical biology in Aristotle's Parts of Animals.* New York, NY: Springer International Publ.

Tkacz, M. W. 2012. Albert the Great on logic, knowledge, and science. In *A companion to Albert the Great. Theology, philosophy, and the sciences*, ed. I. M. Resnick, 507–40. Leiden: E. J. Brill.

Tonoyan, L. 2008. [Development of an operation of division in the treatises of Severin Boethius.] [*Logical-Philosophical Studies,*] 5:67–83. (in Russian).

Toulmin, S. 1972. *Human understanding: The collective use and evolution of concepts.* Princeton, NJ: Princeton University Press.

Tournefort, J. 1694. *Élémens de botanique, ou Methode pour connoître les plantes.* Paris: De l'Imprimerie Royale.

Tournefort, J. 1719. *Institutiones rei herbariae. Editio altera [...],* Tomus Primus. Parisiis: De typo. Regia.

Tremblay, B. 2008. Albertus Magnus on the subject of Aristotle's categories. In *Medieval commentaries on Aristotle's categories*, ed. L. A. Newton, 73–98. Leiden: E. J. Brill.

Trinkaus, C. 1993. Lorenzo Valla's anti-Aristotelian natural philosophy. *I Tatti Studies in the Italian Renaissance* 5:279–325.

Trueman, A. E. 1924. The species concept in palaeontology. *Geological Magazine* 61:355–60.

Tsvelev, N. N. 1995. [Species as one of the taxa.] *Bulleten Moskovskogo obshestva ispytatelej prirody, Biology* 100:62–8. (in Russian).

Turesson, G. 1922. The species and the varieties as ecological units. *Hereditas* 3:100–13.

Turesson, G. 1930. Genecological units and their classificatory value. *Svensk botanisk tidskrift* 24:511–8.

Turner, K. J. 2009. The kind-ness of God: A theological reflection of *mîn*, "kind". In *Genesis Kinds: Creationism and the origin of species*, eds T. C. Wood and P. A. Stock, 31–64. Eugene, OR: Wipf &. Stock.

Turrill, W. B. 1925. Species. *The Journal of Botany British and Foreign* 63:359–66.

Turrill, W. B. 1938. The expansion of taxonomy with special reference to spermatophyta. *Biological Reviews* 13:342–73.

Turrill, W. B. 1940. Experimental and synthetic plant taxonomy. In *The new systematics*, ed. J. Huxley, London: Oxford University Press. P. 47–71.

Tversky, A. 1977. Features of similarity. *Psychological Review* 84:327–52.

Tversky, B. 1989. Parts, partonomies, and taxonomies. *Developmental Psychology* 25:983–95.

Tversky, B. 1990. Where partonomies and taxonomies meet. In *Meaning and prototypes. Studies in linguistic categorization*, ed. S. L. Tsohatzidis, 334–44. London: Routledge.

Tversky, B., and K. Hemenway. 1984. Objects, parts and categories. *Journal of Experimental Psychology General* 113:169–91.

Tweedale, M. 2013. Avicenna Latinus on the ontology of types and tokens. In *Later Medieval metaphysics. Ontology, language, and logic*, eds C. Bolyard and R. Keele, 103–36. New York, NY: Fordham University Press.

Twetten, D., S. Baldner, and S. C. Snyder. 2012. Albert's physics. In *A companion to Albert the Great. Theology, philosophy, and the sciences*, ed. Ir. M. Resnick, 173–220. Leiden: E. J. Brill.

Uranov, A. A. 1979. *Methodological foundations of plant systematics (in their historical development)*. Moscow: Moscow State Pedagogical Institute.

Ure, A., and C. Davidson (eds). 2001. *Chemical speciation in the environment*, 2d ed. London: Blackwell Science.

Urmantsev, Yu. A. 1988. [General system theory: state, application, and prospects of the development.] In [*System. Symmetry. Harmony*,] 38–124. Moscow: Mysl. (in Russian).

Usov, S. A. 1888. [*Essays, Vol. 1. Taxonomic units and groups.*] Moscow: IOLEAO. (in Russian).

Valentine, D. H. 1949. The units of experimental taxonomy. *Acta Biotheoretica* 9:75–88.

Van Fleteren, F. 1992. Augustine (Aurelius Augustinus) (354–430). In *Medieval philosophers*, ed. J. Hackett, 53–67. Detroit, MI: A Bruccoli Clark Layman Book.

Van Fraassen, B. C. 2008. *Scientific representation: Paradoxes of perspective*. New York, NY: Oxford University Press.

Van Regenmortel, M. H. V. 1997. Viral species. In *Species: The units of biodiversity*, eds M. F. Claridge, H. A. Dawah, and M. R. Wilson, 17–24. London: Chapman & Hall.

Van Valen, L. 1976. Ecological species, multispecies, and oaks. *Taxon* 25:233–9.

Van Valen, L. 1988. Species, sets, and the derivative nature of philosophy. *Biology and Philosophy* 3:49–66.

van Winden, J. C. M. 1965. *Calcidius on matter, his doctrine and sources. A chapter in the history of Platonism*. Leiden: E. J. Brill.

Vasilyeva, L. N. 2002. Crisis of the species problem: Causes and consequences. *Evolutionary Biology* 2: 31–50.

Vasilyeva, L. N., and S. L. Stephenson. 2013. An essentialistic view of the species problem. In *The species problem: Ongoing issues*, ed. I. Ya. Pavlinov, 141–69. Rijeka: InTech Open Access Publ.

Vavilov, N. I. 1931. [*Linnaean species as a system.*] Moscow–Leningrad: Gosudarstv. Publ. of Agricultural and Collective-Farm Cooperative Literature. (in Russian).

Vdovina, G. V. 2009. [*The language of the non-obvious. Teachings on signs in scholasticism of the 17th century.*] Moscow: St. Thomas Institute of Philosophy, Theology, and History. (in Russian).

Vdovina, G. V. 2019. [*Intentionality and life. Philosophical psychology of post-medieval scholasticism.*] Moscow: Center for Humanitarian Initiatives. (in Russian).

Velasco, J. D. 2008. Species concepts should not conflict with evolutionary history, but often do. *Studies in History and Philosophy of Biological and Biomedical Sciences* 39:407–14.

Velichkovsky, B. M. 2006. [*Cognitive science. Foundations of psychology of knowing.*] Moscow: Academia Publ. (in Russian).

Verboon, A. 2014. The Medieval tree of Porphyry: An organic structure of logic. In *The tree. symbol, allegory and structural device in Medieval art and thought*, eds A. Worm and P. Salonis, 83–101. Turnhout: Brepols Publ.

Verdier, N. 2005. Hierarchy: a short history of a word in Western thought. In *Hierarchy in natural and social sciences*, ed. D. Pumain, 13–37. Dordrecht: Springer.

Villuendas, M. M. 2019. Una discusión en torno a los límites del concepto especie. *Revista de Humanidades de Valparaíso* 14:241–73.

Vinarskiy, M. V. 2013. [The problem of variability in classical and non-classical systematics.] *Historical and Biological Research* 5:7–27. (in Russian).

Vinarskiy, M. V. 2015. [The fate of the subspecies category in zoological systematics.] 1. History. *Zhurnal Obshchei Biologii* 76:3–14. (in Russian).

Vogt, L. 2008. Learning from Linnaeus: towards developing the foundation for a general structure concept for morphology. *Zootaxa* 1950:123–52.

Volkova, E. V., and A. I. Filyukov. 1966. [*Philosophical issues in the theory of species.*] Minsk: Science and Technology Press. 211 p. (in Russian).

von Vaupel, K. J. C. 1987. Phylogenetic analysis and its foundations. In *Systematics and evolution: A matter of diversity*, ed. P. Hovenkamp, 159–72. Utrecht: Utrecht University Press.

von Vaupel, K. J. C. 1994. Punctuated equilibria and phyletic gradualism: Even partners can be good friends. *Acta Biotheoretica* 42:15–48.

von Wettstein, R. 1898. *Grundzüge der geographisch-morphologischen Methode der Pflanzensystematik*. Jena: G. Fischer.

Vorontsov, N. N. 1980. [Synthetic theory of evolution: Its sources, basic postulates, and unsolved problems.] *Journal Vesesoyusn. Khimicheskigo Obshestva* 25:293–312. (in Russian).

Vorontsov, N. N. 2004. [*The development of evolutionary ideas in biology.*] Moscow: URSS. (in Russian).

Vos, A. 2006. *The philosophy of John Duns Scotus*. Edinburgh: Edinburgh University Press.

Voultsiadou, E., V. Gerovasileiou, L. Vandepitte, et al. 2017. Aristotle's scientific contributions to the classification, nomenclature and distribution of marine organisms. *Mediterranean Marine Science* 18:468–78.

Voyshvillo, E. K. 1989. [*Concept as a form of thinking: Logical and epistemological analysis.*] Moscow: Moscow State University Publ. (in Russian).

Vrba, E. S. 1995. Species as habitat specific, complex systems. In *Speciation and the recognition concept*, eds D. Lambert and H. G. Spencer, 3–44. Baltimore, MD: Johns Hopkins University Press.

Wagner, G. P. and Tomlinson, G. 2020. *Extending the explanatory scope of evolutionary theory: The origination of historical kinds in biology and culture*. Preprints, 2020040025. https://www.preprints.org/manuscript/202004.0025/v2.

Wallace, A. R. 1858. On the tendency of varieties to depart indefinitely from the original type. *Proceedings of the Linnean Society of London* 3:53–62.

Wallace, W. A. 1999. Introduction. In *Albertus Magnus. On animals. A Medieval Summa Zoologica*, Vol. I, 1–42. Baltimore, MD: Johns Hopkins Universily Press.

Wallis, R. T. 1972. *Neoplatonism*. London: Duckworth.

Walton, C. 1970. Ramus and the art of judgment. *Philosophy and Rhetoric* 3:152–64.

Wang, X. 2002. Taxonomy, truth-value gaps and incommensurability: a reconstruction of Kuhn's taxonomic interpretation of incommensurability. *Studies in the History and Philosophy of Sciences* 33:465–85.

Wang, X., Z. He, S. Shi, et al. 2020. Genes and speciation: Is it time to abandon the biological species concept? *National Science Review* 7:1387–97.

Ward, D. F., and M. C. Stanley. 2004. The value of RTUs and parataxonomy versus taxonomic species. *New Zealand Entomology* 27:3–9.

Wartofsky, M. W. 1979. *Models: Representation and scientific understanding*. Boston, MA: Springer Science + Business Media.

Watson, H. C. 1843. Remarks on the distinction of species in nature and in books; preliminary to the notice of some variations and transitions of character, observed in the native plants of Britain. *The London Journal of Botany* 2:613–22.

Weatherby, C. A. 1942. Subspecies. *Rhodora* 44:157–67.

Weinberg, S. 1992. *Dreams of a final theory: The scientist's search for the ultimate laws of nature*. New York, NY: Pantheon.

Weisheipl, J. (ed.). 1980. *Albertus Magnus and the sciences: Commemorative essays*. Toronto: Pontifical Institute of Medieval Studies.

Wells, T., T. Carruthers, P. Muñoz-Rodríguez, et al. 2021. Species as a heuristic: Reconciling theory and practice. *Systematic Biology* DOI: https://doi.org/10.1093/sysbio/syab087.

Wheeler, Q. D. 1999. Why the phylogenetic species concept? Elementary. *Journal of Nematology* 31:134–41.

Wheeler, Q. D., and N. I. Platnick. 2000. The phylogenetic species concept. In *Species concepts and phylogenetic theory: A debate*, eds Q. D. Wheeler and R. Meier, 55–69. New York, NY: Columbia University Press.

Wheeler, Q. D., and R. Meier (eds). 2000. *Species concepts and phylogenetic theory: A debate*. New York, NY: Columbia University Press.

Whewell, W. 1859. *The history of the inductive sciences: From earliest to the present time*, 3d ed., Vol. 2. London: D. Appleton.

Whittaker, R. H. 1972. Evolution and measurement of species diversity. *Taxon*, 21 (2/3): 213–251.

Wierzbicka, A. 1996. *Semantics: Primes and universals*. New York, NY: Oxford University Press.

Wildberg, C. 2016. Neoplatonism. In *Stanford encyclopedia of philosophy*, ed. E. N. Zalta. https://plato.stanford.edu/entries/neoplatonism/.

Wiley, E. O. 1978. The evolutionary species concept reconsidered. *Systematic zoology* 27:17–26.

Wiley, E. O. 1980. Is the evolutionary species fiction? A consideration of classes, individuals and historical entities. *Systematic Zoology* 29:76–80.

Wiley, E. O. 1981. *Phylogenetics: the theory and practice of phylogenetic systematics*. New York, NY: John Wiley & Sons.

Wiley, E. O. and R. L. Mayden. 2000. The evolutionary species concept. In *Species concepts and phylogenetic theory: A debate*, eds Q. D. Wheeler and R. Meier, 70–89. New York, NY: Columbia University Press.

Wilkins, J. 1668. *An essay towards a real character and a philosophical language*. London: Royal Society.

Wilkins, J. S. 2002. *Summary of 26 species concepts*. https://researchdata.museum.vic.gov.au/forum/wilkins_species_table.pdf.

Wilkins, J. S. 2003a. *The origins of species concepts. History, characters, modes, and synapomorphies*. Doctoral dissertation. https://philarchive.org/archive/WILTOO-32.

Wilkins, J. S. 2003b. How to be a chaste species pluralist-realist: the origins of species modes and the synapomorphic species concept. *Biology and Philosophy* 18:621–38.

Wilkins, J. S. 2006a. The concept and causes of microbial species. *History and Philosophy of Life Science* 28:389–408.

Wilkins, J. S. 2006b. Species, kinds, and evolution. *Reports of the National Center for Science Education* 26:36–45.

Wilkins, J. S. 2007. The dimensions, modes and definitions of species and speciation. *Biology and Philosophy* 22:247–66.

Wilkins, J. S. 2009a. *Species: A history of the idea*. Berkeley, CA: University California Press. (Wilkins, J. S. 2018. *Species The evolution of the idea*. 2nd ed.: Boca Raton, FL: CRC Press)

Wilkins, J. S. 2009b. *Defining species. A sourcebook from Antiquity to today*. New York, NY: Peter Lang.

Wilkins, J. S. 2010. What is a species? Essences and generation. *Theory in Biosciences* 129:141–8.

Wilkins, J. S. 2011. Philosophically speaking, how many species concepts are there? *Zootaxa* 2765:58–60.

Wilkins, J. S. 2013a. Essentialism in biology. In *The philosophy of biology. A companion for educators*, ed. K. Kampourakis, 395–19. Dordrecht: Springer.

Wilkins, J. S. 2013b. Biological essentialism and the tidal change of natural kinds. *Science & Education* 22:221–40.

Wilkins, J. S. 2018. The reality of species: Real phenomena not theoretical objects. In *Routledge Handbook of evolution and philosophy*, ed. R. Joyce, 167–81. Abingdon: Routledge.

Wilkins, J. S. 2022. The good species. In *Species problems and beyond. Contemporary issues in philosophy and practice*, eds J. S. Wilkins, F. E. Zachos, and I. Ya. Pavlinov, 105–24. Boca Raton, FL: CRC Press.

Wilkins, J. S., F. E. Zachos, and I. Ya. Pavlinov (eds). 2022. *Species problems and beyond. Contemporary issues in philosophy and practice.* Boca Raton, FL: CRC Press.

Wilks, I. 2008. Peter Abelard and his contemporaries. In *Handbook of the history of logic*, Vol. 2, eds D. M. Gabbay and J. Woods, 83–156. Oxford: Elsevier.

Willdenow, C. L. 1792. *Grundriss der Kräuterkunde: zu Vorlesungen entworfen.* Berlin: Haude u. Spener.

Williams, D. M. 2007. Otto Kleinschmidt (1870–1954), biogeography and the 'origin' of species: From Formenkreis to progression rule. *Biogeografía* 1:3–9.

Williams, M. B. 1985. Species are individuals: Theoretical foundations for the claim. *Philosophy of Science* 52:578–90.

Williams, M. B. 1989. Evolvers are individuals: Extension of the species as individuals claim. In *What the philosophy of biology is: Essays dedicated to David Hull*, ed. M. Ruse, 301–8. Dordrecht: Kluwer.

Williams, M. B. 1992. Species: current usages. In *Keywords in evolutionary biology*, eds E. F. Keller and E. A. Lloyd, 318–23. Cambridge, MA: Harvard University Press.

Williams, W. T., and M. B. Dale 1965. Fundamental problems in numerical taxonomy. *Advances in Botanical Research* 2:35–68.

Williamson, T. 2000. *Knowledge and its limits.* Oxford: Oxford University Press.

Wilson, B. E. 1995. A (not-so-radical) solution to the species problem. *Biology and Philosophy* 10:339–56.

Wilson, B. E. 1996. Changing conceptions of species. *Biology and Philosophy* 11:405–20.

Wilson, D. S. 1990. Species of thought: A comment on evolutionary epictemology. *Biology and Philosophy* 5:37–62.

Wilson, E. O., 1992. *The diversity of life.* Cambridge, MA: Harvard University Press.

Wilson, E. O., and W. L. Brown 1953. The subspecies concept and its taxonomic implication. *Systematic Zoology* 2:97–111.

Wilson, L. G. 1971. Sir Charles Lyell and the species question [...]. *American Scientist* 59:43–55.

Wilson, R. A. (ed.). 1999a. *Species: New interdisciplinary essays.* Cambridge, MA: MIT Press.

Wilson, R. A. 1996. Promiscuous realism. *British Journal for the Philosophy of Science* 47:303–16.

Wilson, R. A. 1999b. Realism, essence, and kind: Resuscitating species essentialism? In *Species: New interdisciplinary essays*, ed. R. A. Wilson, 187–207. Cambridge, MA: MIT Press.

Wilson, R. A. 2007. The biological notion of individual. In *Stanford encyclopedia of philosophy*, ed. E. N. Zalta. https://plato.stanford.edu/archives/fall2007/entries/biology-individual/#EvoBioInd.

Wilson, R. A., M. J. Barker, and I. Brigandt. 2007. When traditional essentialism fails: Biological natural kinds. *Philosophical Topics* 35:189–215.

Windelband, W. 1892. *Lehrbuch der Geschichte der Philosophie*. Freiburg: J. B. C. Mohr.

Winker, K., D. A. Rocque, T. M. Braile, et al. 2007. Vainly beating the air: Species-concept debates need not impede progress in science or conservation. *Ornithological Monographs* 30–44.

Winsor, M. P. 1979. Louis Agassiz and the species question. *Studies in the History of Biology* 3:89–117.

Winsor, M. P. 2001. Cain on Linnaeus: The scientist-historian as unanalysed entity. *Studies in History and Philosophy of Biological and Biomedical Sciences* 32:239–54.

Winsor, M. P. 2003. Non-essencialist methods in pre-Darwinian taxonomy. *Biology and Philosophy* 18:387–400.

Winsor, M. P. 2006a. Creation of the essentialism story: An exercise in metahistory. *History and Philosophy of the Life Sciences* 28:149–74.

Winsor, M. P. 2006b. Linnaeus' biology was not essentialist. *Annals of the Missouri Botanical Garden* 93:2–7.

Winston, J. E. 1999. *Describing species. Practical taxonomic procedure for biologists*. New York, NY: Columbia University Press.

Wippel, J. F. 2000. *The metaphysical thought of Thomas Aquinas*. Washington, DC: Catholic University of America Press.

Wise, K. P. 1990. Baraminology: A young-earth creation biosystematic method. In *Proceedings of the Second International Conference on Creationism*, eds R. E. Walsh and C. L. Brooks, 345–58. Pittsburgh, PA: Creation Science Fellowship.

Witt, C. 2021. Aristotle's biological metaphysics. In *The Cambridge companion to Aristotle's biology*, ed. S. Connell, 114–26. Cambridge: Cambridge University Press.

Wolters, G., and J. G. Lennox (ed.). 1995. *Concepts, theories, and rationality in the biological sciences*. Pittsburgh: University of Pittsburgh Press.

Wood, S. W. 1994. A hierarchical theory of systematics. *Evolutionary Theory* 10: 273–7.

Wood, T. C. 2002. A baraminology tutorial with examples from the grasses (Poaceae). *Journal of Creation* 16:15–25.

Wood, T. C. 2008. Animal and plant baramins. *CORE Issues in Creation* 3:1–258.

Wood, T. C. 2016. A list and bibliography of identified baramins. *Journal of Creation Theology and Science, Series B: Life Sciences* 6:91–101.

Wood, T. C., Wise K.P., Sanders R., et al. 2003. A refined baramin concept. *Occasional Papers of the Baraminology Study Group* 3:1–14.

Woodger, J. H. 1937. *The axiomatic method in biology*. Cambridge: Cambridge University Press.

Woodger, J. H. 1952. From biology to mathematics. *British Journal for the Philosophy of Science* 3:1–21.

Woods, M. 1993. Form, species, and predication in Aristotle. *Synthese* 96:399–415.

Wu, C.-I. 2001. The genic view of the process of speciation. *Journal of Evolutionary Biology* 14: 851–65.

Yang, Y., and Z.-K. Zhou. 2010. New insights into the species problem. *Science China: Life Sciences* 53:964–72.

Yudin, E. G. 1997. [*Methodology of science. Systemity. Activity.*] Moscow: Editorial URSS. (in Russian).

Yuzepchuk, S. V. 1939. [The problem of the species in the light of Darwin's teachings. [*Sovetskaya Botanica* 6–7:12–34. (in Russian).

Yuzepchuk, S. V. 1957. [Linnaeus and the species problem.] *Bulletin of the USSR Academy of Sciences* 5:44–50. (in Russian).

Yuzepchuk, S. V. 1958. [Komarov's concept of species, its historical development and reflection in the "Flora of the USSR",] in [*The problem of species in botany*,] ed. P. A. Baranov, 130–204. Moscow–Leningrad: USSR Academy of Sciences. (in Russian).

Zachos, F. E. 2016. *Species concepts in biology. Historical development, theoretical foundations and practical relevance*. Basel: Springer.
Zachos, F. E. 2018. (New) species concepts, species delimitation and the inherent limitations of taxonomy. *Journal of Genetics* 97:811–5.
Zadeh L. A. 1992. Knowledge representation in fuzzy logic. In *An introduction to fuzzy logic applications in intelligent systems*, eds R. R. Ygaer and L. A. Zadeh, 1–25. New York, NY: Springer Science + Business Media.
Zarenkov, N. A. 1988. [*Theoretical biology*.] Moscow: Moscow State University Publ. (in Russian).
Zavadsky, K. M. 1961. [*The doctrine of species*.] Leningrad: Nauka. (in Russian).
Zavadsky, K. M. 1968. [*Species and speciation*.] Leningrad: Nauka. (in Russian).
Zink, R. M., and M. C. McKitrick. 1995. The debate over species concepts and its implications for ornithology. *Auk* 112:701–19.
Zirkle, C. 1959. Species before Darwin. *Proceedings of the American Philosophical Society* 103:636–44.
Zuev, V. V. 2002. [*The problem of reality in biological taxonomy*.] Novosibirsk: Novosibirsk State University Publ. (in Russian).
Zuev, V. V. 2015. [*Introduction to the theory of biological taxonomy*.] Moscow: Infra-M. (in Russian).

Index

The Index includes only those persons' names who appear in the main text, and not in the references.

D

E

I

J

K

L

M

Q

R

S

T

U

V

W

X

Z